Advances in String Theory

The First Sowers Workshop in Theoretical Physics

AMS/IP

Studies in Advanced Mathematics

Volume 44

Advances in String Theory

The First Sowers Workshop in Theoretical Physics

Eric Sharpe
Arthur Greenspoon
Editors

American Mathematical Society • International Press

Shing-tung Yau, General Editor

2000 *Mathematics Subject Classification.* Primary 81T30, 83E30.

Library of Congress Cataloging-in-Publication Data

Sowers Workshop in Theoretical Physics (1st : 2007 : Virginia Polytechnic Institute and State University)

Advances in string theory : the First Sowers Workshop in Theoretical Physics / Eric Sharpe and Arthur Greenspoon, editors.

p. cm. — (AMS/IP studies in advanced mathematics ; v. 44)

Includes bibliographical references.

ISBN 978-0-8218-4764-0 (alk. paper)

1. String models—Congresses. I. Sharpe, Eric R., 1970– II. Greenspoon, Arthur, 1943– III. Title.

QC794.6.S85S645 2007
539.7′258—dc22 2008039170

∞ The paper used in this book is acid-free and falls within the guidelines
established to ensure permanence and durability.
Visit the AMS home page at `http://www.ams.org/`
Visit the International Press home page at URL: `http://www.intlpress.com/`

10 9 8 7 6 5 4 3 2 1 13 12 11 10 09 08

Contents

Preface

Over the last decade string theory has, despite its purely theoretical content, started to make a strong impact on many areas of physics: high energy and hadronic physics, gravitation and cosmology, mathematical physics and even condensed matter physics. The impact has been through many major conceptual and methodological developments in quantum field theory in the past fifteen years. The ideas of duality, holography, extra dimensions, conformal field theory, gauge theory/gravity correspondence, *etc.* have excited the imagination of many theoretical and experimental physicists in these diverse fields of physics. In addition, string theory has exerted a dramatic influence on developments in contemporary mathematics, ranging from mirror symmetry and enumerative geometry in algebraic geometry to Seiberg-Witten theory in four-manifolds.

Nevertheless, despite these advances the fundamentals of string theory are still largely unknown. Thus it seemed appropriate to have a gathering of younger leading practitioners of various aspects of the field around the common theme: "What is string theory?" This gathering took place at Virginia Tech on May 14–18, 2007. This unique event was made possible by a generous donation from a friend and benefactor of the physics department at Virginia Tech, Mr. Mark Sowers, in whose honor the workshop was named, "The first Sowers workshop in theoretical physics."

As can be seen from the proceedings the range of topics was very wide, cutting through many aspects of string theory. We thank the contributors for making this volume possible in a timely manner. We also thank the colleagues at the Department of Physics and the College of Science at Virginia Tech for support and help. Most of all, we sincerely thank our donor and our friend Mr. Mark Sowers for making this meeting, and thus this proceedings possible.

We warmly dedicate this volume to Mark Sowers.

The organizers,
D. Minic, E. Sharpe, T. Takeuchi, and A. Yelnikov

Acknowledgements

This workshop was made possible through a meeting of two worlds: the world of business and the world of science. Mr. Mark Sowers, a highly successful business man, is fascinated by the frontiers of theoretical physics. The organizers, Djordje Minic, Eric Sharpe, Tatsu Takeuchi, and Alexandr Yelnikov, had a vision: To bring together some of the brightest young minds in string theory to discuss their most recent findings and generate new ideas, in a relaxed and supportive setting. Thanks to Mark Sowers's generous support, the vision became reality, and the first Sowers Theoretical Physics Workshop "What is String Theory?" was held at Virginia Tech in May 2007. The College of Science and the Department of Physics gratefully acknowledge both worlds: We thank Mark Sowers most warmly for his generosity, and the organizers for running a successful and enjoyable workshop. We hope that these proceedings will help continue the discussions.

Lay Nam Chang, Dean, College of Science
Beate Schmittmann, Chair, Department of Physics

AMS/IP Studies in Advanced Mathematics
Volume 44, 2008

Puff Field Theory

Ori J. Ganor

ABSTRACT. Puff Field Theory is a string-theoretic construction of a nonlocal QFT where fundamental particles puff-up to occupy a three-dimensional volume. This conjectured theory is not Lorentz invariant, but Lorentz invariance is restored in the IR limit. A proposal for the supergravity dual also exists. It has peculiar properties near the boundary, such as infinite redshift in frequency combined with infinite blueshift in wavelength.

1. Introduction

On large scales FRW cosmology breaks the microscopic Lorentz symmetry, preserving only spatial rotations, and leaving a preferred time direction. It is therefore interesting to search for quantum field theories with a similar symmetry breaking $SO(3,1) \to SO(3)$. Denoting by Λ the typical scale at which Lorentz violating interactions might become important (if they exist), we will assume that $\Lambda \ll M_{\text{Planck}}$. The question of UV completeness of the QFT is then pertinent. We would like to only consider UV complete theories, and we will also require that in the IR limit they approach $N = 4$ Super Yang-Mills. This generally requires the introduction of nonlocality. Below, we will present a string-theoretic construction that suggests the existence

2000 *Mathematics Subject Classification.* Primary 81T60; Secondary 81T30, 81R40, 83E50.

Key words and phrases. String theory.

I would like to thank Mark Sowers for his generosity and to thank the organizers of the Sowers workshop, Djordje Minic, Eric Sharpe, Tatsu Takeuchi, and Oleksandr Yelnykov for the wonderful hospitality. These notes are based on [**1**]–[**2**], which include work in collaboration with Akikazu Hashimoto (Madison, WI), Sharon Jue (UC Berkeley), Bom Soo Kim (UC Berkeley), and Anthony Ndirango (UC Berkeley). This work was supported in part by the Center of Theoretical Physics at UC Berkeley, and in part by the Director, Office of Science, Office of High Energy and Nuclear Physics, of the U.S. Department of Energy under Contract DE-AC03-76SF00098, and in part by the NSF under grant PHY-0098840.

of such a theory. We will also see that such a theory allows superluminal velocities.

2. Lorentz violation

Our theory will come with a dimensionful parameter

$$\zeta \sim \Lambda^{-\Delta},$$

where Λ has dimensions of mass and sets the scale of Lorentz violations, and we will consider only the case $\Delta > 0$, which corresponds to a UV-relevant parameter.

In the real world, Lorentz-violating interactions can affect the dispersion relation of the photon. There are, however, excellent experimental bounds on this dispersion relation. One technique, for example, uses the observations of gamma ray bursts (GRBs). There have been observations of GRBs at distances of billions of light-years away, which last on the order of milliseconds, during which photons in the energy range of typically 100 KeV–100 MeV are detected [**3**]. This implies that photons with different energies must be traveling at almost exactly the same speed. The bounds from these observations on the variation in the speed of the photon are [**3**],

$$\frac{\Delta c}{c} < \left|\frac{E}{10^{16}\mathrm{GeV}}\right|, \qquad E < 200\,\mathrm{KeV}.$$

These bounds have also been extended to higher energies [**4**],

$$\frac{\Delta c}{c} < \left|\frac{E}{10^{17}\,\mathrm{GeV}}\right|, \qquad 1\mathrm{MeV} < E < 17\,\mathrm{MeV}.$$

So, if we wish our theories to have potentially realistic applications in phenomenology, we had better preserve the Lorentz invariant photon dispersion relation, at least to a very good accuracy.

There is a nice generic way to construct theories with Lorentz violation using an extra dimension and a brane [**5**]. Letting x_4 denote the extra (fifth) dimension, one postulates a 3-brane at $x_4 = 0$ with g_{00} a function of x_4 such that at $x_4 = 0$, $g_{00} = -1$ (by a choice of coordinates). If $-g_{00} > 1$ somewhere in the bulk ($x_4 \neq 0$) then particles may appear to travel at superluminal velocities, from the point of view of an observer on the brane. The new bound on velocities is $|v| < \sqrt{-g_{00}}$. Our construction is related to this idea, but we will recast it in a dual setting where we will have a better control of the background.

3. Nonlocality

There are several examples of self-consistent nonlocal QFTs that can be constructed within the framework of string theory. $U(N)$ $\mathcal{N} = 4$ Super-Yang-Mills theory on a noncommutative $\mathbf{R}^4$ (NCSYM) [**6**]–[**8**] is perhaps the most famous one. In this theory fundamental particles can become extended objects – electric dipoles whose length L in one particular direction is proportional to their momentum $p_\perp$ in another [**9**], so that $L \sim \theta p_\perp$, where θ is the fundamental noncommutativity parameter. Noncommutative geometry can be constructed directly in field theory, by replacing products of $U(N)$ matrices with the "star-product" [**6**]–[**8**],

$$\Phi_1 \star \Phi_2 = \left(\exp \frac{i}{2}\theta^{ij}\partial_i^{(1)}\partial_j^{(2)}\right)[\Phi_1\Phi_2]$$

where Φ_1 and Φ_2 are $N \times N$ matrix fields and θ^{ij} is an antisymmetric spacetime tensor.

Another related example of a nonlocal theory that can be constructed from string theory is *dipole-theory* [**10**]–[**11**]. There, fundamental particles are dipoles of length proportional not to momentum, but to R-symmetry charge. It can be constructed from $U(N)$ Yang-Mills theory by replacing covariant derivatives of adjoint fields Φ with

$$D_\mu\Phi(x) \longrightarrow \partial_\mu\Phi(x) - iA_\mu\left(x - \frac{1}{2}QL\right)\Phi(x) + i\Phi(x)A_\mu\left(x + \frac{1}{2}QL\right)$$

where Q is (a certain component of) the R-charge of Φ and L is a constant 4-vector, which is the fundamental parameter of the theory.

In both these examples nonlocality is *linear* – fundamental particles expand into nonlocal objects which are segments with a linear dimension. *Puff Field Theory* (PFT), the theory that we will construct in these notes, is different in that fundamental particles expand into objects with volume, and the volume is proportional to (a component of) the R-charge of the particle, as we shall see.

At low energies, all the examples above of nonlocal QFTs, including the conjectured PFT, are deformations of $\mathcal{N} = 4$ SYM with gauge group $U(N)$. Their Lagrangians are schematically of the form

$$L = L_{\mathcal{N}=4} + \zeta\mathcal{O}^{(\Delta+4)} + \cdots \tag{3.1}$$

where ζ is a parameter and $\mathcal{O}^{(\Delta+4)}$ is an operator of dimension $\Delta + 4$. The various dimensions as well as the schematic structure of the operators $\mathcal{O}$ are listed below (see the references above for more details):

Theory	Deformation $\zeta\mathcal{O}$	Δ	
NCYM	$\frac{1}{2}g_{\text{YM}}^2\theta^{\mu\nu}\ \text{tr}\{F_{\mu\sigma}F_{\nu\tau}F^{\sigma\tau} + F_{\sigma\tau}F^{\mu\nu}F_{\mu\nu} + \cdots\}$	2	$\zeta \to \theta$
Dipole	$\frac{1}{2}g_{\text{YM}}^2 L_{IJ}^\mu\ \text{tr}\{F_{\mu\nu}(\Phi^I D^\nu \Phi^J + \cdots)\}$	1	$\zeta \to L$
PFT	$\zeta\mathcal{O}^{(7)}$	3	

Here g_{YM} is the coupling constant, $F_{\mu\nu}$ are the field strength components $(\mu, \nu = 0, \ldots, 3)$, Φ^I are the scalars $(I, J = 1, \ldots, 6)$, fermions as well as other less illuminating interaction terms have been suppressed, and we will discuss $\mathcal{O}^{(7)}$ in more detail below.

4. Constructing PFT through string theory

Our construction can be regarded as a variant of the Douglas-Hull construction [**7**] of NCSYM, so we will begin by reviewing the latter. One starts with type-IIA string theory compactified on T^2, which for simplicity we take to be of the form $S^1 \times S^1$. We denote the compactification radii by R_1', R_2'. Now add N coincident D0-branes, and take the limit $R_1', R_2' \to 0$. T-duality can be applied if there are no other fluxes, and then one gets N D2-branes compactified on a large T^2 with compactification radii α'/R_i' $(i = 1, 2)$. The low-energy limit of this construction then has a decoupled sector of 2+1D $U(N)$ $\mathcal{N} = 4$ SYM. But Douglas and Hull added an obstruction to this small/large area T-duality, in the form of a constant NSNS B-flux. Then, T-duality does not yield a simpler description, but instead Douglas and Hull showed that the low-energy description of the system in the limit $R_1', R_2' \to 0$ can be described by the nonlocal NCSYM.

We now turn to the construction of PFT. We start with type-IIA T^3 in the form of $S^1 \times S^1 \times S^1$ with compactification radii R_i' $(i = 1, 2, 3)$ and string coupling constant g_{st}'. We add N Kaluza-Klein particles in the 1st direction and seek the low-energy description of this configuration in the limit

$$\alpha'^{-1/2}R_1' \longrightarrow 0, \quad \alpha'^{-1/2}R_2' \to \text{finite}, \quad \alpha'^{-1/2}R_3' \to \text{finite}, \quad g_{\text{st}}' \to \text{finite}. \tag{4.1}$$

By 'low-energy' here we generally mean energies low compared to $\alpha'^{-1/2}$, but in our case we actually need a somewhat stronger condition – energies small compared to $(\alpha'^{-1/2}R_1')^{1/2}\alpha'^{-1/2}$. Using U-duality, it is not hard to see that the sought-after description is simply 3+1D $U(N)$ $\mathcal{N} = 4$ SYM. In fact, T-duality in the 1st direction, followed by S-duality, followed by T-dualities in the 2nd and 3rd directions, converts the system to N D3-branes in type-IIB compactified on another T^3 with compactification radii

$$R_1 = \alpha'^{\frac{3}{4}} g_{\text{st}}'^{-\frac{1}{2}} R_1'^{-\frac{1}{2}}, \qquad R_k = \alpha'^{\frac{5}{4}} g_{\text{st}}'^{\frac{1}{2}} R_1'^{-\frac{1}{2}} R_k'^{-1} \qquad (k = 2, 3),$$

and string coupling constant $g_{\text{st}} = \alpha' R_3'^{-1} R_2'^{-1}$. The limit (4.1) was chosen so that the ratios R_2/R_1 and R_3/R_1 and the coupling constant g_{st} remain finite, while $\alpha'^{-1/2}R_k \longrightarrow \infty$ $(k = 1, 2, 3)$.

Following Douglas and Hull, we now add an obstruction to U-duality. The obstruction that seems to lead to interesting results in our case is not a flux but a geometrical twist. Let x_1, x_2, x_3 be the compact coordinates on T^3 with periodicities $2\pi R_1, 2\pi R_2, 2\pi R_3$ and let $\vec{y}$ denote the vector of coordinates in the six transverse directions. We then replace the original periodicity conditions of T^3 on the type-IIA side with

$$(x_1, x_2, x_3, \vec{y}) \sim (x_1 + 2\pi R_1, x_2, x_3, \Omega\vec{y})$$

where Ω is some element of the rotation group SO(6). Strictly speaking, we should let Ω be an element of Spin(6), but since we will eventually only need an infinitesimal Ω, the distinction will be immaterial. We have to decide, though, on how to scale Ω in the limit (4.1). We want Ω to be infinitesimally close to the identity element such that

$$\Omega = \exp\left(\frac{2\pi}{R_1 R_2 R_3}\zeta\right) \longrightarrow I, \qquad \zeta \to \text{finite}.$$

Here,

$$\zeta \equiv \begin{pmatrix} & \beta_1 & & & & \\ -\beta_1 & & & & & \\ & & & \beta_2 & & \\ & & -\beta_2 & & & \\ & & & & & \beta_3 \\ & & & & -\beta_3 & \end{pmatrix} \in so(6). \tag{4.2}$$

We note in passing that if ζ is in an appropriate $su(3)$ ($su(2)$) subgroup of $so(6)$ then $\mathcal{N} = 1$ ($\mathcal{N} = 2$) SUSY is preserved, respectively. For the $\mathcal{N} = 2$ case, we set $\beta \equiv \beta_1 = \beta_2$ and $\beta_3 = 0$.

With the introduction of ζ, U-duality is less useful in the limit (4.1). However, we will argue below that the low-energy limit still describes a decoupled QFT on $\mathbf{R}^{3,1}$, but a nonlocal one.

Heuristically, the construction is designed so that fundamental particles that carry R-charge acquire a fundamental volume proportional to that R-charge and to ζ. To see this, consider the original type-IIA geometry with N Kaluza-Klein particles. For simplicity, let's focus on two out of the six transverse directions, and combine them into a complex variable $z \equiv y_1 + iy_2$. The twist Ω reduces to an element of $SO(2)$ which we write as $\omega \equiv \exp(\pi i\beta/R_1R_2R_3)$. Now consider a field ϕ in this geometry. Suppressing the coordinates x_2, x_3, we get the boundary conditions $\phi(x_1 + 2\pi R_1', \omega z) = \phi(x_1, z)$. Expanding in a Fourier series we get

$$\phi(x_1, z) = \sum_{p,\ell} C_{n,\ell}(|z|) z^{-\ell} e^{ipx_1/R_1'}, \qquad p \in \mathbf{Z} + \frac{2\beta\ell}{R_1R_2R_3}.$$

Thus, formally the Kaluza-Klein momentum is not an integer anymore, and its fractional part is proportional to both the angular momentum ℓ and the parameter β, which is a component of ζ. If we formally perform the U-duality transformation that we used in the $\zeta = 0$ case and if we interpret our result above literally, we get a non-integer number of D3-branes that occupy a volume of $(2\pi)^3(R_1R_2R_3N + 2\beta\ell)$. In the next section we will make this statement more concrete.

5. Electric and magnetic fluxes

So, we treat PFT as a formal QFT. If we replace the ∞ in (4.1) with "very large" we get a compactified (on T^3) version of PFT. As above, we denote the compactification radii by R_1, R_2, R_3. We also denote $V \equiv R_1 R_2 R_3$. In the $\mathcal{N} = 2$ supersymmetric PFT we can calculate the energies of BPS states that include electric and magnetic fluxes, as well as momentum. Suppose we have ℓ units of R-charge, k_i units of momentum, e_i units of electric flux, and m_i units of magnetic flux in the ith direction, for $i = 1, 2, 3$. (All of these are, of course, integers.) With the notation

$$\mathbf{P} \equiv \sum_{i=1}^{3} \frac{k_i}{R_i} \hat{\mathbf{n}}_i, \qquad \mathbf{E} \equiv \sum_{i=1}^{3} \frac{e_i R_i}{2\pi V} \hat{\mathbf{n}}_i, \quad \mathbf{B} \equiv \sum_{i=1}^{3} \frac{m_i R_i}{2\pi V} \hat{\mathbf{n}}_i,$$

we get the BPS energy

$$\begin{aligned} E = {} & \frac{2\ell\beta}{g_{\rm st}\alpha'^2} + \frac{2\pi^2 V^2}{|NV + 2\ell\beta|} \left(\frac{g_{\rm YM}^2}{2\pi} \mathbf{E}^2 + \frac{2\pi}{g_{\rm YM}^2} \mathbf{B}^2 \right) \\ & + \left| \mathbf{P} - \frac{4\pi^2 V^2}{|NV + 2\ell\beta|} \mathbf{E} \times \mathbf{B} \right| . \end{aligned}$$

Analogous formulas for NCSYM have been derived, for instance, in [**12**]–[**14**]. The first term is a manifestation of the Volume/R-charge relation discussed above, and the appearance of ℓ in the denominator of the remaining terms clearly shows that the theory is nonlocal, since ℓ is an integral of the R-charge density over the entire volume. We also see from the last term that the dispersion relation of massless Kaluza-Klein particles remains relativistic.

6. Supergravity dual

The supergravity dual was found in [**2**] using techniques similar to those applied in the NCSYM case [**15**]–[**16**] and the dipole-theory case [**11**]. The result is type-IIB supergravity on a space with metric

$$\begin{aligned} ds^2 = {} & \frac{R^2}{r^2} K^{-\frac{1}{2}} \left[dx^2 + dy^2 + dz^2 - \left(dt - \frac{4\pi N}{r^2} \vec{n}^T \zeta d\vec{n} \right)^2 \right] \\ & + \frac{R^2}{r^2} K^{\frac{1}{2}} dr^2 + R^2 K^{\frac{1}{2}} d\Omega_5^2, \\ C_4' = {} & \frac{\pi N}{r^4} K^{-1} dt \wedge dx \wedge dy \wedge dz - \frac{\pi N}{g_{\rm st}\alpha'^2 r^6} K^{-1} \vec{n}^T \zeta d\vec{n} \wedge dx \wedge dy \wedge dz, \end{aligned} \tag{6.1}$$

where C_4' is the RR flux, and

$$K \equiv 1 + \frac{16\pi^2 N^2}{r^6} \vec{n}^T \zeta^T \zeta \vec{n}, \quad \vec{n} \in S^5, \quad d\Omega_5^2 = \sum_{I=1}^{6} dn_I^2, \quad R^4 \equiv 4\pi g_{\rm st} N \alpha'^2.$$

We can now explore in more detail the IR and UV limits of PFT. In the IR limit $r \to \infty$ we expand the background as

$$ds^2 = \frac{R^2}{r^2}[dr^2 + dx^2 + dy^2 + dz^2 - dt^2] + R^2 d\Omega_5^2 + \frac{8\pi N R^2}{r^4}\vec{n}^T\zeta d\vec{n}\, dt + \cdots$$
$$C_4' = \frac{\pi N}{r^4} dt \wedge dx \wedge dy \wedge dz - \frac{\pi N}{g_{\text{st}}\alpha'^2 r^6}\vec{n}^T\zeta d\vec{n} \wedge dx \wedge dy \wedge dz + \cdots$$

The last term on each line corresponds to a deformation of $\mathcal{N} = 4$ by an irrelevant operator. This is the operator $\mathcal{O}^{(7)}$ from (3.1) which can be identified using the general AdS/CFT rules [**17**]–[**18**]. It has dimension 7, it transforms as a vector under Lorentz transformations, and it is in the adjoint representation of the R-symmetry group $SU(4)$. Further analysis using supersymmetry [**2**] shows that in the case of a $U(1)$ gauge symmetry

$$\zeta\mathcal{O} \to \zeta^{\mu}_{AB} T_{\mu\nu} J^{\nu AB} + \zeta^{\mu}_{AB}\epsilon^{ABCDEF}\epsilon_{\mu\nu\sigma\tau}\Phi^C\partial_\nu\Phi^D\partial_\sigma\Phi^E\partial_\tau\Phi^F + \cdots, \tag{6.2}$$

where fermions have been suppressed, Φ^A $(A, \dots, F = 1, \dots, 6)$ are the scalars of the $\mathcal{N} = 4$ vector multiplet, $J^{\nu AB} = -J^{\nu BA}$ is the R-charge, and $T^{\mu\nu}$ is the energy-momentum tensor. In the case of a generic $U(N)$ gauge group, the expression is similar, but requires a symmetrized trace and additional commutator terms, and has not been calculated.

7. UV limit

The supergravity dual allows us to make a few plausible statements with regard to the UV behavior of the (large N limit of) PFT on $\mathbf{R}^{3,1}$. Whereas in the IR the metric (6.1) approaches $AdS_5 \times S^5$, in the UV limit $r \to 0$ the metric differs markedly from $AdS_5 \times S^5$, and in fact becomes singular. For simplicity, it is convenient to restrict the form of the parameter ζ and set $\beta_1 = \beta_2 = \beta_3 \equiv \beta$ in (4.2). This form does not preserve any supersymmetry, but it is convenient because it preserves a $U(3)$ subgroup of the R-symmetry group $SU(4)$. This subgroup has a simple geometrical interpretation in terms of the S^5 component of the supergravity dual.

S^5 can be realized as an S^1 fibration over $\mathbf{CP}^2$ (the *Hopf* fibration). The R-symmetry $SU(4)$ is a double cover of the isometry group $SO(6)$ of S^5, and $U(3) \subset SU(4)$ is the subgroup of isometries that preserve the Hopf fibration structure. It is the isometry group of the base $\mathbf{CP}^2$.

In $AdS_5 \times S^5$ the radius of S^5 is a constant R. When ζ is turned on, the coordinates $\vec{n}$ in (6.1) still describe a manifold that is diffeomorphic to S^5, but the metric is different. If, in addition, ζ preserves $U(3) \subset SU(4)$ as above, the metric on the deformed S^5 is determined by specifying both the size of the fiber S^1 and the size of the base $\mathbf{CP}^2$. From (6.1) we find

$$R_{\text{fiber}} = R\left(1 + \frac{16\pi^2 N^2 \beta^2}{r^6}\right)^{-1/4}, \qquad R_{\text{base}} = R\left(1 + \frac{16\pi^2 N^2\beta^2}{r^6}\right)^{1/4}.$$

In the UV limit we find $R_{\text{fiber}} \sim r^{3/2}$ and $R_{\text{base}} \sim r^{-3/2}$. (Note that, in contrast, in the IR limit both R_{fiber} and R_{base} approach the constant R.) Thus, in the UV limit the base $\mathbf{CP}^2$ expands to infinite size, and the fiber S^1 shrinks to zero. The type-IIB supergravity description is therefore inadequate. However, we can contemplate a local duality transformation along the fiber that would convert type-IIB on a small S^1 to M-theory on T^2. A similar transformation in the context of $AdS_5 \times S^5$ was discussed in [**20**]. The final result is an M-theory background with

$$
\begin{aligned}
\ell_P{}^{-2} ds_M^2 &= (4\pi N)^{-1/3} g_{\text{st}}^{-1} \rho^2 \big[\Delta^{-2/3}(dx_1^2 + dx_2^2 + dx_3^2) - \Delta^{1/3} dx_0^2\big] \\
&\quad + (4\pi N)^{2/3} \Delta^{1/3} \rho^{-2} d\rho^2 + (4\pi N)^{2/3} \Delta^{1/3} ds_B^2 \\
&\quad + (4\pi N)^{-1/3} \Delta^{1/3} (g_{\text{st}}^{-1} d\xi^2 + g_{\text{st}} d\eta^2), \\
\ell_P{}^{-3} G_4 &= 2\pi N \omega \wedge \omega + 2\omega \wedge d\eta \wedge d\xi \\
&\quad + \frac{2}{3} (4\pi N)^{-2} g_{\text{st}}^{-3} \beta d \left(\frac{\rho^6}{\Delta}\right) \wedge dx_1 \wedge dx_2 \wedge dx_3,
\end{aligned} \tag{7.1}
$$

where

$$
\Delta \equiv 1 + (4\pi N)^{-1} g_{\text{st}}^{-3} \beta^2 \rho^6,
$$

ℓ_P is the Planck length, x_0, x_1, x_2, x_3 are coordinates on $\mathbf{R}^{3,1}$, η and ξ are periodic coordinates with period 2π (parameterizing T^2), ds_B^2 is the metric on $\mathbf{CP}^2$, ω is a harmonic 2-form whose cohomology class generates $H^2(\mathbf{CP}^2, \mathbf{Z})$, and G_4 is the 4-form flux. We also replaced the coordinate r with $\rho \propto 1/r$, so that $\rho \to \infty$ is the UV limit. The background (7.1) becomes weakly coupled as $\rho \to \infty$. It is not completely clear that local duality is reliable in our case, especially since the background is not supersymmetric, but we take the fact that the final result is weakly coupled as encouraging.

Let us study the UV limit in more detail. As $\rho \to \infty$ we can approximate the background (7.1) as

$$
\begin{aligned}
\ell_P{}^{-2} ds_M^2 &\approx -(4\pi N)^{-2/3} g_{\text{st}}^{-2} \beta^{2/3} \rho^4 dx_0^2 \\
&\quad + (4\pi N)^{1/3} g_{\text{st}} \beta^{-4/3} \rho^{-2} (dx_1^2 + dx_2^2 + dx_3^2) \\
&\quad + (4\pi N)^{1/3} g_{\text{st}}^{-1} \beta^{2/3} (d\rho^2 + \rho^2 ds_B^2) \\
&\quad + (4\pi N)^{-2/3} \beta^{2/3} \rho^2 (g_{\text{st}}^{-2} d\xi^2 + d\eta^2), \\
\ell_P{}^{-3} G_4 &\approx 2\pi N \omega \wedge \omega + 2\omega \wedge d\eta \wedge d\xi - 4 g_{\text{st}}^3 \beta^{-3} \rho^{-7} d\rho \wedge dx_1 \wedge dx_2 \wedge dx_3.
\end{aligned} \tag{7.2}
$$

This background exhibits several interesting properties as $\rho \to \infty$:

- It is geodesically complete;
- The curvature tends to zero;
- There is an infinite redshift in frequency $g_{00} \to \infty$;

FIGURE 1. The Penrose diagram of the conjectured supergravity dual of PFT. The IR region is described by (6.1), while the UV region is described by (7.1). In the intermediate region neither description is weakly coupled.

- There is an infinite blueshift in wavelength $g_{11}, g_{22}, g_{33} \to 0$.

The Penrose diagram of this background is depicted in Figure 1, and is similar to that of AdS space.

8. Degrees of freedom

We argued in the previous section that the UV regime of PFT is holographically dual to a weakly-coupled M-theory background. This raises the question of whether the spectrum of PFT is really discrete. Since the ρ direction of (7.1) is infinite in extent, it looks as if there is an "infinite amount" of weakly coupled M-theory in the holographic dual, which would suggest a continuous spectrum. (Similar issues arise in Little String Theory [**21**].) A more careful analysis, however, suggests that the spectrum of PFT is indeed discrete. First, let's be more precise about what we mean by "discrete." The question is whether when compactified on T^3 with generic boundary conditions (that should include R-symmetry twists in order to remove zero modes of scalar fields) the spectrum of the theory is discrete. And the answer might very well be "yes," due to the following remarkable property of (7.1).

Consider particle trajectories with fixed energy E and spatial momentum $\vec{p}$ (defined with respect to the Killing vectors $\partial/\partial x_0$ and $\partial/\partial x_1, \dots \partial/\partial x_3$, respectively). Then the following inequality can easily be derived from the condition that the trajectory be timelike or lightlike:

$$\rho \leq \rho_{\max} \equiv (4\pi N)^{1/6} g_{\mathrm{st}}^{1/2} \beta^{-1/3} \left(\frac{E}{|\vec{p}|} \right)^{1/3} . \tag{8.1}$$

Returning to the question of the spectrum, note that if PFT is compactified on a fixed T^3 and all possible zero modes of bosonic fields are removed, then $|\vec{p}|$ is bounded from below (by a value of the order of the inverse of the longest side of the T^3). If we now also put an upper bound on E, we see from (8.1) that ρ is bounded from above, and only a finite portion of the background (7.1) is accessible. This suggests a discrete spectrum.

9. Superluminal velocities

Without Lorentz invariance there is no a priori bound on velocities. For example, it was demonstrated in NCYM [**22–24**] that under various circumstances certain particles can have a nonrelativistic dispersion relation, and can therefore travel faster than the speed of light.

Superluminal velocities are also possible in PFT. To see this, consider the IR expansion (3.1) with the operator given in (6.2). We will concentrate on the first term of (6.2) – the TJ coupling. In a background with a nonvanishing R-charge density $\langle\zeta J^0\rangle \equiv \langle\zeta_{AB}J^{AB\,0}\rangle$ we have

$$L = L_{(N=4)} + \langle\zeta J_0\rangle T^{00} + \cdots$$

Thus, in R-charged matter the effect of PFT, to lowest order in ζ, can be mimicked by a shift of the time-time component of the background metric,

$$g_{00} \longrightarrow g_{00} - \zeta\langle J^0\rangle + \cdots,$$

and the upper bound on velocities is therefore

$$v_{\max} \approx 1 + \frac{1}{2}\langle\zeta J^0\rangle.$$

Thus, superluminal velocities are possible inside R-charged matter, at least if $\langle\zeta J^0\rangle$ is positive.

Using the extreme UV supergravity dual (7.2), we can estimate the energy required to travel at a velocity $v \gg 1$. Consider a supergravity particle with mass $m > 0$ and energy E in the metric (7.2). We have

$$-{\ell_P}^2 m^2 \geq -(4\pi N)^{2/3} g_{\mathrm{st}}^2 \beta^{-2/3}\frac{E^2}{\rho^4},$$

(derived by requiring a timelike worldline). The upper bound on velocity is given by

$$|v| < v_{\max} = \sqrt{-\frac{g_{00}}{g_{11}}} = (4\pi N)^{-1/2} g_{\mathrm{st}}^{-3/2}\beta\rho^3.$$

Combining the last two inequalities we get

$$E \geq \ell_P m|v|^{2/3}\beta^{-1/3}.$$

Thus, in this context, higher velocity (surpassing the speed of light) requires higher energy. To be sure, this discussion does not address the massless case $m = 0$.

10. D3-branes in strong RR flux

Formally, we can set $N = 0$ in the construction of section §4, perform the U-duality transformation on the background with Ω, and place N D3-branes in the resulting (strongly coupled) type-IIB background, instead of N Kaluza-Klein particles in the original type-IIA background.

The resulting type-IIB background without D3-branes is given, formally, by

$$ds^2 = f^{-\frac{1}{2}} \sum_{i=1}^{3} dx_i^2 + f^{\frac{1}{2}}(dr^2 + r^2 d\Omega_5^2 - dx_0^2) - \frac{f^{-\frac{1}{2}} r^4}{g_{\text{st}}^2 \alpha'^4} (\vec{n}^T \zeta d\vec{n})^2,$$

$$C_4' = \frac{1}{4 g_{\text{st}}^2 \alpha'^2} f^{-1} r^2 \vec{n}^T \zeta d\vec{n} \wedge dx_1 \wedge dx_2 \wedge dx_3,$$

where C_4' is a 4-form such that the 5-form field strength is the self-dual part of dC_4', and

$$f \equiv 1 + \frac{1}{g_{\text{st}}^2 \alpha'^4} r^2 \vec{n}^T \zeta^T \zeta \vec{n}.$$

Backgrounds of this kind have been studied in [**25**]. It is a *flux-tube* whose characteristic feature is a nonvanishing RR 5-form flux with three components along x_1, x_2, x_3 and two components perpendicular to it. PFT can formally be described as the theory on D3-brane probes of this flux-tube. The nonlocal effects are then associated with the strong RR flux.

11. Summary

We have argued for the existence of a possibly novel type of quantum field theory, which is: (i) Lorentz violating; (ii) $SO(3)$-invariant; (iii) decoupled from gravity; (iv) nonlocal; and (v) exhibits a characteristic TJ coupling (energy-momentum times R-current) at the leading order in low-energy.

Many open issues remain. First, the fundamental formulation of this theory is unknown. It would appear to be qualitatively different from NCSYM or dipole theories. Those can be described in terms of matrix fields, because essentially those are theories of dipoles with two endpoints. But in PFT pointlike objects formally expand to a finite volume, as opposed to finite dipole moment.

Constructing PFT-like deformations of theories other than $\mathcal{N} = 4$ SYM could also be of interest. It would also perhaps be interesting to understand if there is any relation to Little String Theory, or any way to realize PFT starting from a 5 + 1D theory. The duals of (p, q) 5-branes described by Witten [**26**] contain the same geometric twist ingredient as in the construction of PFT, and there might be a stronger link.

Disclaimer

information, apparatus, product, or process disclosed, or represents that its use would not infringe privately owned rights. Reference herein to any specific commercial product, process, or service by its trade name, trademark, manufacturer, or otherwise, does not necessarily constitute or imply its endorsement, recommendation, or favoring by the United States Government or any agency thereof, or The Regents of the University of California. The views and opinions of authors expressed herein do not necessarily state or reflect those of the United States Government or any agency thereof or The Regents of the University of California.

References

[1] O. J. Ganor, "A new Lorentz violating nonlocal field theory from string-theory," Phys. Rev. D **75**, 025002 (2007) [arXiv:hep-th/0609107].

[2] O. J. Ganor, A. Hashimoto, S. Jue, B. S. Kim and A. Ndirango, "Aspects of puff field theory," JHEP **0708**, 035 (2007) [arXiv:hep-th/0702030].

[3] G. Amelino-Camelia, J. R. Ellis, N. E. Mavromatos, D. V. Nanopoulos and S. Sarkar, "Potential Sensitivity of Gamma-Ray Burster Observations to Wave Dispersionin Vacuo," Nature **393**, 763 (1998) [arXiv:astro-ph/9712103].

[4] S. E. Boggs, C. B. Wunderer, K. Hurley and W. Coburn, "Testing Lorentz Non-Invariance with GRB021206," Astrophys. J. **611**, L77 (2004) [arXiv:astro-ph/0310307].

[5] C. Csaki, J. Erlich and C. Grojean, "Gravitational Lorentz violations and adjustment of the cosmological constant in asymmetrically warped spacetimes," Nucl. Phys. B **604**, 312 (2001) [arXiv:hep-th/0012143].

[6] A. Connes, M. R. Douglas and A. Schwarz, "Noncommutative geometry and matrix theory: Compactification on tori," JHEP **9802**, 003 (1998) [arXiv:hep-th/9711162].

[7] M. R. Douglas and C. M. Hull, "D-branes and the noncommutative torus," JHEP **9802**, 008 (1998) [arXiv:hep-th/9711165].

[8] N. Seiberg and E. Witten, "String theory and noncommutative geometry," JHEP **9909**, 032 (1999) [arXiv:hep-th/9908142].

[9] D. Bigatti and L. Susskind, "Magnetic fields, branes and noncommutative geometry," Phys. Rev. D **62**, 066004 (2000) [arXiv:hep-th/9908056].

[10] A. Bergman and O. J. Ganor, "Dipoles, twists and noncommutative gauge theory," JHEP **0010**, 018 (2000) [arXiv:hep-th/0008030].

[11] A. Bergman, K. Dasgupta, O. J. Ganor, J. L. Karczmarek and G. Rajesh, "Nonlocal field theories and their gravity duals," Phys. Rev. D **65**, 066005 (2002) [arXiv:hep-th/0103090].

[12] B. Morariu and B. Zumino, "Super Yang-Mills on the noncommutative torus," arXiv:hep-th/9807198.

[13] A. Konechny and A. S. Schwarz, "1/4-BPS states on noncommutative tori," JHEP **9909**, 030 (1999) [arXiv:hep-th/9907008].

[14] B. Pioline and A. S. Schwarz, "Morita equivalence and T-duality (or B versus Theta)," JHEP **9908**, 021 (1999) [arXiv:hep-th/9908019].

[15] A. Hashimoto and N. Itzhaki, "Non-commutative Yang-Mills and the AdS/CFT correspondence," Phys. Lett. B **465**, 142 (1999) [arXiv:hep-th/9907166].

[16] J. M. Maldacena and J. G. Russo, "Large N limit of non-commutative gauge theories," JHEP **9909**, 025 (1999) [arXiv:hep-th/9908134].

[17] J. M. Maldacena, "The large N limit of superconformal field theories and supergravity," Adv. Theor. Math. Phys. **2**, 231 (1998) [Int. J. Theor. Phys. **38**, 1113 (1999)] [arXiv:hep-th/9711200].

[18] S. S. Gubser, I. R. Klebanov and A. M. Polyakov, "Gauge theory correlators from non-critical string theory," Phys. Lett. B **428**, 105 (1998) [arXiv:hep-th/9802109].

[19] E. Witten, "Anti-de Sitter space and holography," Adv. Theor. Math. Phys. **2**, 253 (1998) [arXiv:hep-th/9802150].

[20] M. J. Duff, H. Lu and C. N. Pope, "AdS(5) x S(5) untwisted," Nucl. Phys. B **532**, 181 (1998) [arXiv:hep-th/9803061].

[21] O. Aharony, M. Berkooz, D. Kutasov and N. Seiberg, "Linear dilatons, NS5-branes and holography," JHEP **9810**, 004 (1998) [arXiv:hep-th/9808149].

[22] K. Landsteiner, E. Lopez and M. H. G. Tytgat, "Excitations in hot non-commutative theories," JHEP **0009**, 027 (2000) [arXiv:hep-th/0006210].

[23] D. Bak, K. M. Lee and J. H. Park, "Noncommutative vortex solitons," Phys. Rev. D **63**, 125010 (2001) [arXiv:hep-th/0011099].

[24] A. Hashimoto and N. Itzhaki, "Traveling faster than the speed of light in non-commutative geometry," Phys. Rev. D **63**, 126004 (2001) [arXiv:hep-th/0012093].

[25] J. G. Russo and A. A. Tseytlin, "Magnetic flux tube models in superstring theory," Nucl. Phys. B **461**, 131 (1996) [arXiv:hep-th/9508068].

[26] E. Witten, "New 'gauge' theories in six dimensions," JHEP **9801**, 001 (1998) [Adv. Theor. Math. Phys. **2**, 61 (1998)] [arXiv:hep-th/9710065].

DEPARTMENT OF PHYSICS, UNIVERSITY OF CALIFORNIA, BERKELEY, CA 94720, AND LAWRENCE BERKELEY NATIONAL LABORATORY, BERKELEY, CA 94720

E-mail address: `origa@socrates.berkeley.edu`

AMS/IP Studies in Advanced Mathematics
Volume 44, 2008

Mottness and Strong Coupling

Robert G. Leigh, Ting-Pong Choy, and Philip Phillips

ABSTRACT. There are interesting condensed matter systems that display strong coupling phenomena. Understanding the dynamics of these systems is just as challenging as the strongly coupled problems more familiar to particle physicists. Examples of such a system are the materials (cuprates) that have a superconducting phase with a relatively high transition temperature. The standard Bardeen-Cooper-Schrieffer (BCS) theory that is so successful in explaining more familiar superconductors is incapable of explaining the physics of the cuprates. One of the key features of these materials is that the normal state is not a metal (described by free electron (Fermi liquid) theory), but a *Mott insulator*. This is an inherently strongly coupled system, and in fact, there is feedback from high to low energy scales, that is, UV-IR mixing. In this talk, we show that a new degree of freedom, that cannot be thought of as a composite of electron degrees of freedom, emerges in a proper low energy description of this system, and this degree of freedom is responsible for the exotic phenomena associated with these materials.

1. Introduction

The intrigue of strong coupling dynamics is one of the driving forces in research in formal high energy particle physics and string theory and has been for some time. Over the years, we have developed a vast array of techniques and insight into such problems, even if we have not made the final steps to an understanding of the strongly coupled particle physics systems of direct interest, such as quantum chromodynamics. It is perhaps refreshing to discover that there are physical systems that are experimentally accessible that display similar strong coupling dynamics. This similarity is not always fully appreciated, and one even finds attempts at explanations of such systems by ad hoc use of what amounts to free field theory. Interesting physics is rarely that easy.

An example of such a system is contained in models of materials (the so-called cuprates) that display a superconducting phase with a high

2000 *Mathematics Subject Classification.* Primary 82D55.

transition temperature. A useful model of these materials is a simple lattice model (the Hubbard model) that we will describe fully later. This is a *two-dimensional* model, because electrons are much more mobile in (the copper-oxide) planes within the crystalline structure. These materials may be modified by *doping*, effectively adding or subtracting to the number of electrons on the lattice. The resulting phase diagram, plotting temperature versus (hole) doping is shown in Figure 1. At zero doping (corresponding to one electron per lattice site, also referred to as half-filling), we find antiferromagnetic order at low temperatures. At high hole doping, the system becomes a more or less ordinary metal, and can be described by Fermi liquid theory. The bulk of the phase diagram shown in Figure 1, though, is an enigma. We see the superconducting dome, but there are large regions with strange properties – the pseudogap and strange metal regimes. These regions of the phase diagram, which we will refer to as the *normal state*, are significantly different from a Fermi liquid.

We do not understand the nature of the superconducting phase. In developing an understanding though, it is certainly important to understand first the normal state. The normal state, though, has an array of strange properties. First, over a wide range of temperatures, the resistivity is a *linear* function of temperature. In fact, this is one of the hardest features to explain; we will comment on this in more detail later on in the talk. Second, in the pseudogap regime, if one plots the density of states, one finds a precipitous drop at one point. This is not a true gap, and an explanation of this phenomenon has been difficult. As well, there are other more detailed features of the normal state that we will give details on later. In any case, it should

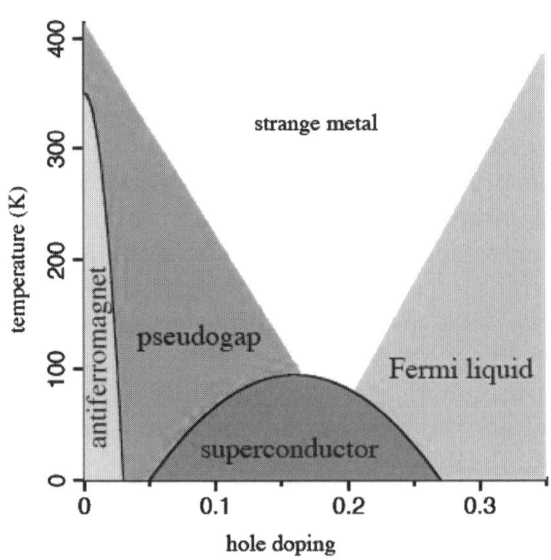

FIGURE 1. The phase diagram of a typical cuprate material, plotted versus doping.

be obvious that a proper theory of these materials should be able to explain each of these phenomena in a natural way.

2. The Hubbard Model

The model that we will consider here is the (one-band) Hubbard model in two spatial dimensions, that has Hamiltonian

$$H_{\text{Hubb}} = -t \sum_{i,j,\sigma} g_{ij} c^\dagger_{i,\sigma} c_{j,\sigma} + U \sum_{i,\sigma} c^\dagger_{i,\uparrow} c^\dagger_{i,\downarrow} c_{i,\downarrow} c_{i,\uparrow} \tag{2.1}$$

Here, sites of the lattice are labelled by $i, j, \ldots$ and $c^\dagger_{i,\sigma}$ creates an electron at site i with z-component of spin $\sigma \in \{\uparrow, \downarrow\}$. We will denote occupation numbers by $n_{i,\sigma} \equiv c^\dagger_{i,\sigma} c_{i,\sigma}$. That this is an appropriate model is exemplified by the fact that the materials of interest are *Mott insulators* at half-filling. At half-filling, there is one electron per site (and we suppose that the system is unpolarized). The Pauli principle allows for up to two electrons per site, so if the electrons were free, they could easily move from site to site giving a non-zero charge conductivity, and we would have a metal. However, there are materials that are insulators at half-filling. Mott explained this phenomenon by introducing a large energy cost for doubly occupying sites; in this way, there is a gap, and thus vanishing conductivity. This is what the Hamiltonian (2.1) models: the term proportional to t is a hopping term (the matrix g_{ij} determines the extent of this hopping; we will take this to correspond to nearest-neighbour hopping for simplicity), while the term proportional to U contributes only on sites which have two electrons. Thus the Hubbard model describes a Mott insulator in the regime $U \gg t$. What is unusual about the materials of interest is that not only are they insulators at half-filling, but this behaviour persists for non-zero doping.

Now, we note that in a naive continuum limit, the regime $U \gg t$ corresponds to strong coupling – the hopping term is kinetic energy, and the U term is an interaction. So, to understand this model, we must employ some sort of strong coupling techniques.

What has often been done (and indeed is still championed by many) is to replace the Hubbard model by a simpler spin model, the tJ-model. This is motivated by considering that as $U \to \infty$, states which involve doubly-occupied sites become more and more unlikely, and it would seem a good idea to simply project these states out. More precisely, one can do a similarity transformation on (2.1) to bring it into block-diagonal form, in which the energies of a given block are approximately (in the limit $U \gg t$) some number times U. Since the Hamiltonian is block diagonal it makes sense to project to the lowest energy block. The resulting Hamiltonian is, at leading order, that of the tJ-model.

This procedure though is problematic, for several reasons. First, although one can do this block diagonalization, the resulting eigenstates are complicated linear combinations of the original electronic states. Although t/U is small (of order 1/10), it is big enough to have important observable consequences, some of which we will describe later. So systematically taking t/U effects into account (and doing so correctly) is certainly important. A second problem with this approach is that the $U \to \infty$ limit is tricky: there is an order of limits problem [**1, 2**]. Taking N (the size of the system) to infinity does not commute with taking U to infinity. For these reasons, we choose to study the Hubbard model directly; the payoff, as we will find, is that there is emergent low-energy dynamics that is relatively easy to quantify. Although this could be understood from treating projective models carefully, it is much more difficult, and it masks the true character of the low energy physics [**3, 4**].

3. It's not just electrons!

About fifteen years ago, Polchinski [**5**] (see also [**6**]) considered the high T_c materials from the point of view of the renormalization group. One of the basic features of effective field theories that we all understand very well is that, for such a theory to be natural (technically), there should be no relevant perturbations. In a strongly interacting theory, we do not have the luxury of calculability. But one thing that we could do is make an assumption about the low energy physics, and then study whether or not the resulting theory is natural in this sense. Now, in condensed matter physics, an often made assumption is that if we start with a system of electrons, the effect of renormalization is simply to dress the electrons into weakly interacting 'quasiparticles' (this gives the 'Fermi liquid' theory). From the point of view of a particle physicist, this is a bizarre suggestion, but indeed it works very well for metals. In fact, as Polchinski showed, this is very easily understood from the point of view of the renormalization group.

In condensed matter, the notion of the renormalization group is somewhat more complicated than in particle physics – we must renormalize towards the Fermi surface. One can organize operators in an effective action via their scaling properties, but because of the presence of the Fermi surface, there can be subtleties. One of these is responsible in fact for the (natural) existence of the Cooper pairing mechanism in the BCS theory of ordinary superconductors.

Polchinski noted [**5**] quite clearly that if one makes the assumption that the high T_c materials are described by Fermi liquid theory, then there can be no explanation of linear-T resistivity. Over the years, there have been a variety of attempts made to effectively circumvent this statement, but it seems to me that what it really points to is simply that the low energy theory is not just a Fermi liquid – there are other emergent degrees of freedom that have nothing to do with electrons. The theory that we will describe here fits this bill, and as we will argue, explains a lot more.

In fact, it is fairly simple to demonstrate that new low energy physics is required. In a non-interacting system, the number of low-energy addition states per electron per spin is equal to one. Should the number of low-energy addition states per electron per spin exceed unity, Fermi liquid theory fails and new electronic states emerge at low energy that cannot be constructed from the non-interacting system. To show that this happens in a doped Mott insulator, we compare the number of electrons per site (n_h) that can be added to the holes created by the dopants with the number of single-particle electron addition states per site at low energy,

$$L = \int_{\mu}^{\Lambda} N(\omega) d\omega, \tag{3.1}$$

defined as the integral of the single-particle density of states ($N(\omega)$) from the chemical potential, μ, to a cutoff energy scale, Λ, demarcating the IR and UV scales. In a Fermi liquid or non-interacting system, $L = n_h = 2 - n$ (see Figure 2), where n is the electron filling. Consequently, the number of low-energy states per electron per spin is identically unity.

For a doped Mott insulator in the 'atomic limit,' i.e., one electron per site with infinite on-site repulsion U, it is still true that $L/n_h = 1$, because creating a hole leaves behind an empty site which can be occupied by either a spin-up or a spin-down electron. Hence, when x electrons (see Figure 2) are removed, $L = n_h = 2x = 2 - 2n$ in the atomic limit [**9**]. (Here, we define $x = 1 - n$; the hole doping is measured relative to half-filling.)

However, real Mott systems do *not* have $U \to \infty$. Finite hopping with matrix element t creates double occupancy, and as a result, empty sites with weight t/U. Such empty sites with fractional weight contribute to L. Consequently, when $0 < t/U \ll 1$, L is strictly larger than $2x$. The actual behaviour [**7**], shown by the red curve in Figure 2, has been confirmed both experimentally [**8**] and theoretically [**9, 10**]. The number of *electrons* that can be added remains twice the number of dopants, $n_h = 2x$, even when

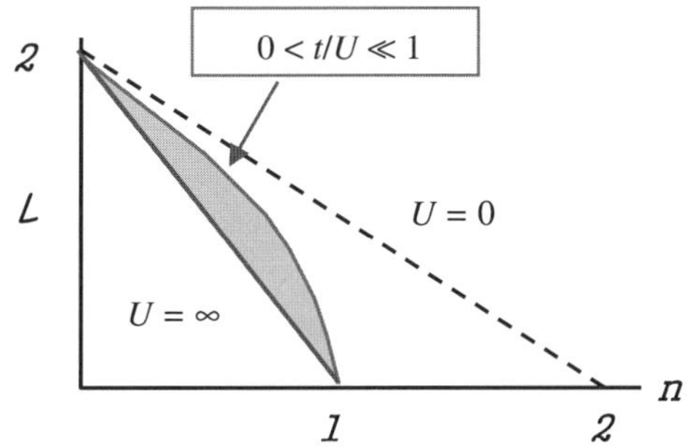

FIGURE 2. Integrated low-energy spectral weight, L, defined in Eq. (3.1), as a function of the electron filling, n: 1) the dashed line is the non-interacting limit, vanishing on-site interaction ($U = 0$), 2) atomic limit (blue line) of a doped Mott insulator, $U = \infty$, and 3) a real Mott insulator in which $0 < t/U \ll 1$, red curve.

the hopping is finite. As a result, for a doped Mott insulator, $L/n_h > 1$. Consequently, in contrast to a Fermi liquid, simply counting the number of electrons that can be added does *not* exhaust the available phase space to add an electron at low energy. Thus, additional degrees of freedom at low energy, not made out of the elemental excitations, must exist. They arise through exchanges with the doubly occupied sites; because of this we might suppose that this physics could be associated with a doubly charged collective excitation. In the theory that we will describe next, there is in fact a charge $2e$ bosonic field present in the low energy description. We believe that this boson binds to holes giving rise to a new charge e excitation. It is the physics of this new charge e state that mediates the non-Fermi liquid behaviour in a doped Mott insulator, and we will show a series of experimental evidences for this picture.

4. Our theory

A determination of a low energy theory for the Hubbard model is difficult. We would like to integrate out the high energy degrees of freedom, but these are not easily isolated. As we mentioned, it is possible to do a similarity transformation on the Hubbard Hamiltonian which block-diagonalizes it. The characteristic energy scales of each block are separated by multiples of U. In this way, we could obtain a low energy description by projecting to the lowest block. We will refer to such schemes as projective models. The problem with such a scheme is that one must appreciate that the projection is performed on the transformed Hilbert space, and the transformed fermions are not the original electrons – a transformed fermion has admixtures (weighted by t/U) of multiple electron operators. This significantly complicates the low energy physics, and although one could construct perturbation theory in t/U, the formalism masks the true character of the low energy physics. In particular, the transformed fermions are not free, and do not lead to a Fermi liquid. There are a variety of experimental results that cannot be easily explained in projective models. So we seek an alternative to projective models.

The basic idea of our construction [**3, 4**] is to rewrite the Hubbard model in such a way as to isolate the high energy degrees of freedom so that they can be simply integrated out. It turns out that a similar procedure has been used in the past, to elucidate collective plasma oscillations in electron gases [**12**]. Here, we will need to involve some new theoretical structures. We do this by first introducing a new oscillator that represents the degrees of freedom at high energy and including a constraint which ensures that the extended theory is *equivalent* to the Hubbard model. If we simply solve this constraint we return to the description (2.1) of the Hubbard model, while if instead, we integrate out the high energy degrees of freedom, we will obtain the low energy effective theory. In fact, it is possible to formulate the extension in such a way that this latter integration can be done *exactly* – that is, we will obtain a low energy effective theory that is precisely equivalent to the Hubbard model. The central idea that we are employing here is familiar in non-linear σ-models, which are of course constrained systems, as will be our extended theory. In the σ-model, the field which is introduced as a Lagrange multiplier for the constraint ends up determining the phase structure of the theory – the field σ in this sense dominates the infrared, even though it cannot be thought of as a composite of the original fields. It is a collective excitation. In our extended theory, we will similarly find a field that in the UV is a Lagrange multiplier field. In the low energy effective theory, it plays a very important role, as we will see, although we believe that its nature is quite different from what is seen in σ-models.

To this end, we extend the Hilbert space $\otimes_i (\mathcal{F}_\uparrow \otimes \mathcal{F}_\downarrow \otimes \mathcal{F}_D)$, where $\mathcal{F}$ denotes a fermionic Fock space. We associate $D^\dagger$ with the creation of double-occupation. Since a site cannot be doubly occupied more than once, we will take D to be a fermionic oscillator (it would also be possible to take D to be bosonic, but a constraint would have to be added in some way to limit the occupation number; this would unnecessarily complicate the theory). We will formulate a Lagrangian for this extended theory in which the high energy scale U is isolated in terms involving D; thus, integrating over the high-energy scale will be accomplished by integrating over D. The Hamiltonian for the extended theory will be such that if we were to solve the constraint, precisely the Hubbard Hamiltonian (1) would be recovered.

So the task is to understand how to conveniently formulate such a theory. In the left frame of Figure 3, we have indicated the states of the extended Hilbert space for a single site, as well as the action of the standard electron creation operator, $c_{i\sigma}^\dagger$ and the new fermionic operator, $D^\dagger$. The new extended states are indicated in red. In the Hamiltonian, we will include terms that correspond to hoppings between nearest-neighbour sites. In the right frame of Figure 3, we have indicated the hoppings that we will include: the notation (x, y) denotes the occupation on two nearest neighbour sites, with σ denoting a spin-σ electron and D an occupied D-state. There are of course other states than those indicated in the right frame in the Hilbert space, but we need not include them in describing the hoppings.

There is a complication in writing a Lagrangian, because we have taken D to be a fermionic oscillator. We wish to write a constraint that identifies D with double occupation. However, the operator that creates double occupation, $c_{i,\uparrow}^\dagger c_{i,\downarrow}^\dagger$, at a site i is bosonic. To account for this in an elegant way, we introduce a complex Grassmann parameter θ, such that the constraint is of the form

$$D_j \sim \theta c_{j,\uparrow} c_{j,\downarrow}. \tag{4.1}$$

The Lagrangian will be written as an integral over these Grassmann parameters, in a way that bears some resemblance to superspace. However, we emphasize that this is being introduced merely as a formality – at the end

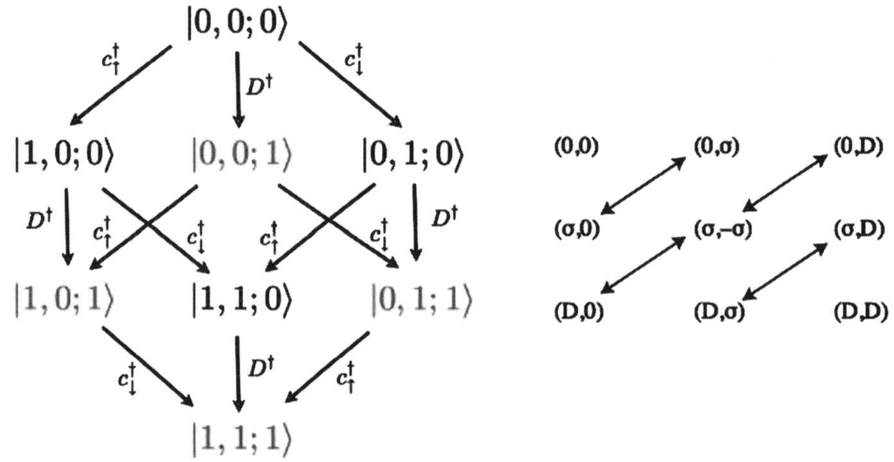

FIGURE 3. Extended Hilbert space (left) which allows an explicit integration of the high energy scale. Hopping processes (right) included in the Lagrangian.

of the construction, we will obtain an effective Lagrangian in which θ does not appear.

We will take the Lagrangian to be

$$
\begin{aligned}
L = \int d^2\theta \Bigg[& \bar{\theta}\theta \sum_{i\sigma} (1 - n_{i\bar{\sigma}}) c_{i\sigma}^\dagger \dot{c}_{i\sigma} + \sum_i D_i^\dagger \dot{D}_i + U \sum_j D_j^\dagger D_j \\
& - t \sum_{i,j,\sigma} g_{ij} \Big[C_{ij\sigma} c_{i,\sigma}^\dagger c_{j,\sigma} + D_i^\dagger c_{j,\sigma}^\dagger c_{i,\sigma} D_j \\
& + (D_j^\dagger \theta c_{i,\sigma} V_\sigma c_{j,\bar{\sigma}} + h.c.) \Big] + H_{\rm con} \Bigg].
\end{aligned} \tag{4.2}
$$

The parameter V_σ has values $V_\uparrow = 1$, $V_\downarrow = -1$, and simply ensures that D couples to the spin singlet. The operator $C_{ij\sigma}$ is of the form $C_{ij\sigma} \equiv \bar{\theta}\theta\alpha_{ij\sigma} \equiv \bar{\theta}\theta(1 - n_{i,\bar{\sigma}})(1 - n_{j,\bar{\sigma}})$. Note that the dynamical terms that appear in the Lagrangian are non-traditional because the dynamics with the $c_{i\sigma}$ operators must exclude those sites which contain the occupancy $c_{i\downarrow}^\dagger c_{i\uparrow}^\dagger|0\rangle$. Finally, the constraint $H_{\rm con}$ is taken to be

$$
H_{\rm con} = s\bar{\theta} \sum_j \varphi_j^\dagger (D_j - \theta c_{j,\uparrow} c_{j,\downarrow}) + h.c. \tag{4.3}
$$

where φ_j is a charge $2e$ bosonic field. Although we will not describe the details here, it turns out the energy scale s should be identified with (order) t.

To see how this constraint removes unphysical states that arise from the extended Hilbert space, we compute the partition function,

$$Z = \int [\mathcal{D}c\, \mathcal{D}c^\dagger\, \mathcal{D}D\, \mathcal{D}D^\dagger\, \mathcal{D}\varphi\, \mathcal{D}\varphi^\dagger] \exp\left(- \int_0^\tau L dt \right), \tag{4.4}$$

in Euclidean signature. If we first integrate[1] over φ_i, we find a series of δ-functions which makes the integral over D trivial. Performing the D integrations then results in a Lagrangian given by $\int d^2\theta\, \bar{\theta}\theta L_{\text{Hubb}} = \sum_{i\sigma} c^\dagger_{i\sigma}\dot{c}_{i\sigma} + H_{\text{Hubb}}$, identical to that of the Hubbard model. This constitutes the **ultraviolet (UV) limit** of our theory. As is evident, in this limit the extended Hilbert space contracts, unphysical states such as $|1,0,1\rangle$, $|0,1,1\rangle$, $|1,1,1\rangle$ are set to zero, and we identify $|1,1,0\rangle$ with $|0,0,1\rangle$. Note there is no contradiction between treating D as fermionic and the constraint in Eq. (4.3). The constraint never governs the commutation relation for D. The value of D is determined by Eq. (4.3) only when φ is integrated over. This is followed immediately by an integration over D, at which point D is eliminated from the theory.

So the theory that we have formulated is indeed equivalent to the Hubbard model. Its utility however is that it permits us to coarse-grain the system cleanly for $U \gg t$. The energy scale associated with D is the large on-site energy U – that is, U appears in the Lagrangian as a 'mass' for D. Hence, it makes sense, instead of solving the constraint, to integrate out D. This will result in the **low-energy (IR) theory**. Such an integration may be done exactly, as the theory is Gaussian in D! Since this integral is Gaussian, it suffices to complete the square in the D-field. To accomplish this, we define the matrix

$$\mathcal{M}_{ij} = \left(\delta_{ij} - \frac{t}{(\omega + U)} g_{ij} \sum_\sigma c^\dagger_{j,\sigma} c_{i,\sigma} \right) \tag{4.5}$$

where ω is the frequency. It will also be convenient to introduce the composite $b_i = \sum_j b_{ij} = \sum_{j\sigma} g_{ij} c_{j,\sigma} V_\sigma c_{i,\bar{\sigma}}$. At zero frequency the Hamiltonian is

$$H_h^{IR} = -t \sum_{i,j,\sigma} g_{ij} \alpha_{ij\sigma} c^\dagger_{i,\sigma} c_{j,\sigma} + H_{\text{int}} - \frac{1}{\beta} \text{Tr} \ln \mathcal{M}$$

where

$$\begin{aligned} H_{\text{int}} &= -\frac{t^2}{U} \sum_{j,k} b_j^\dagger (\mathcal{M}^{-1})_{jk} b_k - \frac{s^2}{U} \sum_{i,j} \varphi_i^\dagger (\mathcal{M}^{-1})_{ij} \varphi_j \\ &\quad - s \sum_j \varphi_j^\dagger c_{j,\uparrow} c_{j,\downarrow} + \frac{st}{U} \sum_{i,j} \varphi_i^\dagger (\mathcal{M}^{-1})_{ij} b_j + \text{h.c.} \end{aligned} \tag{4.6}$$

[1]Care must be taken in properly defining the contour of integration.

The theory presented here corresponds to hole doping, but one can do a similar analysis for electron doping as well by suitable modifications [**3, 4**].

5. Experimental Consequences

The bosonic field φ is the charge $2e$ field that we spoke of in an earlier section. In the low energy theory, the operator which creates a single electron represents a composite excitation. To determine its form, we add to the starting Lagrangian a source term that generates the canonical electron operator when the constraint is solved. The appropriate transformation that yields the canonical electron operator in the UV, [**11**] is

$$L \to L + \sum_{i,\sigma} J_{i,\sigma}\left[\bar{\theta}\theta(1-n_{i,-\sigma})c^\dagger_{i,\sigma} + V_\sigma D^\dagger_i c_{i,-\sigma}\theta\right] + \text{h.c.}$$

However, in the IR in which we only integrate over the heavy degree of freedom D_i, the electron creation operator

$$c^\dagger_{i,\sigma} \to (1-n_{i,-\sigma})c^\dagger_{i,\sigma} + V_\sigma \frac{t}{U} b^\dagger_i \mathcal{M}^{-1}_{ij} c_{j,-\sigma} - V_\sigma \frac{s}{U}\varphi^\dagger_i \mathcal{M}^{-1}_{ij} c_{j,-\sigma}. \tag{5.1}$$

The electron operator contains the standard term for motion in the lower Hubbard band (LHB), with a renormalization from spin fluctuations (second term) and a new charge e term, $c_{i,-\sigma}\mathcal{M}^{-1}_{ij}\varphi^\dagger_j$.

Note that our low energy theory does not contain dynamical terms for φ. Thus, at least at leading order, it is a localized quantity. In the non-linear σ-model, the σ-field attains dynamics through interactions and can condense. One might suppose that the same occurs here. However, there is another possibility, which is that φ becomes bound to holes, in such a way that the extra term in the electron operator, $c_{i,-\sigma}\mathcal{M}^{-1}_{ij}\varphi^\dagger_j$, becomes identified with a propagating degree of freedom. The 'confining' mechanism has yet to be firmly established, but as we will see, this interpretation has strong merits. Consequently, we predict that an electron at low energies is in a superposition of the standard lower Hubbard band state (modified with spin fluctuations) and a new composite charge e state described by $c_{i,-\sigma}\mathcal{M}^{-1}_{ij}\varphi^\dagger_j$. This is a strong claim. In any case, it is the presence of these two distinct excitations that gives rise to the spectral weight transfer and the non-Fermi liquid behaviour. We will review here a number of additional experimental consequences.

Consider the single-particle electron Green function

$$\begin{aligned} G(k,\omega) &= -iFT\langle Tc_i(t)c^\dagger_j(0)\rangle \\ &= -iFT\int[D\varphi^*_i][D\varphi_i]\int[Dc^*_i][Dc_i]c_i(t)c^*_j(0)\exp\left(-\int L[c,\varphi]dt\right). \end{aligned} \tag{5.2}$$

A detailed analysis is difficult but suppose as a simple approximation, we treat φ_i to be spatially independent with no dynamics of its own. We will organize the calculation of $G(k,\omega)$ by first integrating out the fermions (holding φ fixed)

(5.3)
$$G(k,\omega)=\int[D\varphi^*][D\varphi]FT\left(\int[Dc_i^*][Dc_i]c_i(t)c_j^*(0)\exp\left(-\int L[c,\varphi]dt\right)\right)$$

where now

$$L=\sum_{i\sigma}(1-n_{i\bar\sigma})c^*_{k\sigma}\dot c_{k\sigma}-\left(2\mu+\frac{s^2}{U}\right)\varphi^*\varphi-\sum_{k\sigma}(g_t t\alpha_k+\mu)c^*_{k\sigma}c_{k\sigma}$$
$$+s\varphi^*\sum_k(1-\frac{2t}{U})c_{-k\downarrow}c_{k\uparrow}+c.c. \tag{5.4}$$

This effective Lagrangian can be diagonalized and written as

$$L=\sum_{k\sigma}(1-n_{i\bar\sigma})\gamma^*_{k\sigma}\dot\gamma_{k\sigma}+\sum_k(E_0+E_k-\lambda_k-\frac{2}{\beta}\ln(1+e^{-\beta\lambda_k}))$$
$$+\sum_{k\sigma}\lambda_k\gamma^*_{k\sigma}\gamma_{k\sigma}, \tag{5.5}$$

in terms of a set of Bogoliubov quasiparticles,

$$\gamma^*_{k\uparrow}=+\left(\cos^2\theta_k\right)c^*_{k\uparrow}+\left(\sin^2\theta_k\right)c_{-k\downarrow} \tag{5.6}$$
$$\gamma_{k\downarrow}=-\left(\sin^2\theta_k\right)c^*_{k\uparrow}+\left(\cos^2\theta_k\right)c_{-k\downarrow} \tag{5.7}$$

where $\cos^2\theta_k=\frac{1}{2}(1+\frac{E_k}{\lambda_k})$. Here,

$$\begin{aligned}\alpha_k&=2(\cos k_x+\cos k_y),\\ \eta_k&=2(\cos k_x-\cos k_y),\\ E_0&=-\left(2\mu+\frac{s^2}{U}\right),\\ E_k&=-g_t t\alpha_k-\mu,\\ \lambda_k&=\sqrt{E_k^2+\Delta_k^2},\\ \Delta_k&=s\varphi^*\left(1-\frac{2t}{U}\alpha_k\right),\end{aligned}$$

and $g_t=\frac{2\delta}{1+\delta}$ when

$$\delta=1-n\rightarrow 1-Q+2\varphi^*\varphi$$

is a renormalized factor which originates from the correlated hopping term $(1-n_{i\bar\sigma})c^\dagger_{i\sigma}c_{j\sigma}(1-n_{j\bar\sigma})$. Starting from Eq. (5.5), we integrate over the fermions in Eq. (5.3) to obtain,

$$G(k,\omega) = \frac{1}{Z}\int [D\varphi^*][D\varphi]G(k,\omega,\varphi) \cdot \exp\left(-\sum_k (E_0 + E_k - \lambda_k - \frac{2}{\beta}\ln(1+e^{-\beta\lambda_k}))\right) \tag{5.8}$$

where

$$G(k,\omega,\varphi) = \frac{\sin^2\theta_k[\varphi]}{\omega+\lambda_k[\varphi]} + \frac{\cos^2\theta_k[\varphi]}{\omega-\lambda_k[\varphi]} \tag{5.9}$$

is the exact Green function corresponding to the Lagrangian, Eq. (5.5). The two-pole structure of $G(k,\omega,\varphi)$ will figure prominently in the structure of the electron spectral function. To calculate $G(k,\omega)$, we numerically evaluated the remaining φ integral in Eq. (5.8). Since Eq. (5.8) is averaged over all values of φ, we are not just doing mean-field or a saddle-point analysis. Physically, Eq. (5.8) serves to mix (through the integration over φ) all subspaces with varying number of double occupancies into the low-energy theory. Hence, it should retain the full physics inherent in the bosonic degree of freedom.

The spectral function for $U = 10t$ evaluated from Eq. (5.9) and displayed in Figure 4 exhibits four key features. First, regardless of the doping, there is a low-energy kink in the electron dispersion. In fact, two kinks exist. The low-energy kink occurs at roughly $0.2t \approx 100$ meV. By treating the mass term for the boson as a variable parameter, we verified that the low-energy kink is determined by the bare mass. In the effective low-energy theory, the bare mass is t^2/U. This mass is independent of doping. Experimentally, the low-energy kink [**14**] is impervious to doping. Consequently, the boson offers a natural explanation of this effect. A second high-energy kink appears at roughly $0.5t \approx 250$ meV which closely resembles the experimental kink at 300 meV [**15**]. At sufficiently high doping (see Figures (4a) and (4b)), the high-energy kink disappears.

Third, experimentally, the high-energy kink is accompanied by a splitting of the electron dispersion into two branches [**15**]. This bifurcation is evident in Figure 4. The energy difference between the two branches achieves a maximum at $(0,0)$ as is seen experimentally. A computation of the spectral function at $U = 20t$ and $n = 0.9$ reveals that the dispersion as well as the bifurcation still persist. Further, the magnitude of the splitting does not change, indicating that the energy scale for the bifurcation and the maximum energy splitting are set by t and not U. The two branches below the chemical potential correspond to the standard band in the LHB on which φ vanishes and a branch on which $\varphi \neq 0$. The two branches indicate that there are two local maxima in the integrand in Eq. (5.8). One of the maxima, $\varphi = 0$, arises from the extremum of $G(k,\omega,\varphi)$ whereas the other, the effective free energy (exponent in Eq. (5.8)) is minimized ($\varphi \neq 0$). Above

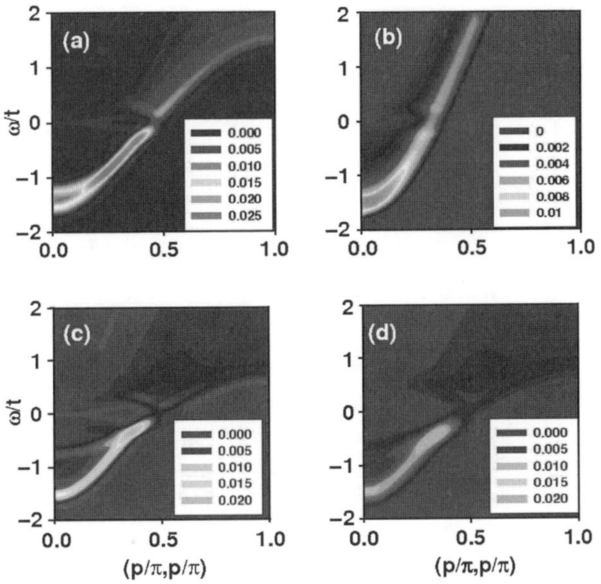

FIGURE 4. Spectral function for two different fillings (a) $n = 0.8$ and (b) $n = 0.4$ along the nodal direction. The absence of a splitting in the electron dispersion at $n = 0.4$ indicates the bifurcation ceases beyond a critical doping. The spectral functions for two different values of the on-site repulsion, (c)$U = 10t$ and (d)$U = 20t$ for $n = 0.9$ reveals that the high-energy kink and the splitting of the electron dispersion have at best a weak dependence on U. This indicates that this physics is set by the energy scale t rather than U.

the chemical potential only one branch survives. The split electron dispersion below the chemical potential is consistent with the composite nature of the electron operator dictated by Eq. (5.1). At low energies, the electron is a linear superposition of two states, one the standard band in the LHB described by excitations of the form, $c^\dagger_{i\sigma}(1 - n_{i\bar{\sigma}})$ and the other a composite excitation consisting of a bound hole and the charge $2e$ boson, $\sim c_{i\bar{\sigma}}\varphi^\dagger_i$.

The formation of the composite excitation, $c_{i\bar{\sigma}}\varphi^\dagger$, leads to a pseudogap at the chemical potential primarily because the charge $2e$ boson is a local non-propagating degree of freedom. The spectral functions for $n = 0.9$ and $n = 0.8$ both show an absence of spectral weight at the chemical potential. Non-zero spectral weight resides at the chemical potential in the heavily overdoped regime, $n = 0.4$, consistent with the vanishing of the pseudogap beyond a critical doping away from half-filling. There are a number of other experimental signatures of our low energy theory that we will not describe in detail here. One such signature is the appearance of a peak in the mid-infrared in the optical conductivity.

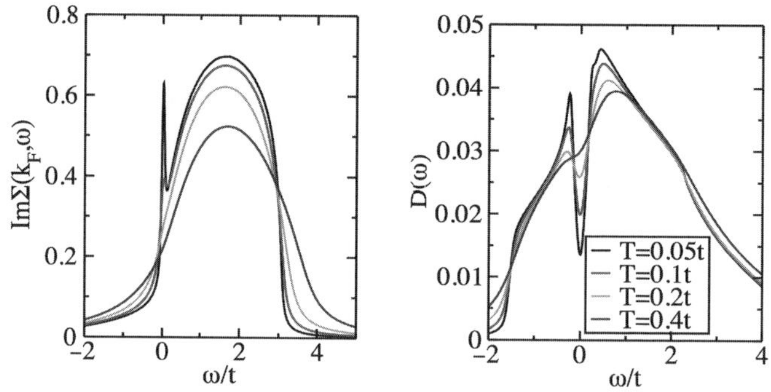

FIGURE 5. The imaginary part of the electron self-energy as a function of temperature for $n = 0.7$. A peak develops at $\omega = 0$ at low temperature which is the signature of the opening of the pseudogap. The density of states explicitly showing the pseudogap is shown in the adjacent figure.

As we mentioned earlier, a resistivity that is linear in temperature cannot be obtained in Fermi liquid theory. It has been suggested that linear-T resistivity should be attributed to 'quantum criticality.' However, it has been shown [**13**] that one-parameter scaling gives (assuming that the critical degrees of freedom carry the current, and charge is conserved) a DC conductivity of the form $\sigma(\omega = 0) \sim T^{1/z}$ where the dynamical exponent z appears in a dispersion relation $E \sim |p|^z$. Thus for quantum criticality to explain linear-T resistivity (given these mild assumptions), we must have $z < 0$. This is not completely impossible (it can occur for example in non-commutative field theories), but perhaps a safer conclusion to draw is that one or more of the assumptions made here is violated. In fact, in the low energy theory that we have described, we have multiple charge carriers, and the physics is indeed quite different than the assumptions referred to.

In ordinary metals, the temperature dependence of the resistivity is set by phonons. Below the Debye temperature, electron-phonon scattering gives a T^5 dependence, but above the Debye temperature, we do find a linear dependence. In the present case, phonons cannot be the answer, because of the scales involved. However, we do have the charge $2e$ bosonic field, and we can identify the appearance of linear-T resistivity with electron-boson scattering.

6. Conclusions

We have described here our determination of the low energy effective description of the Hubbard model, which describes the strongly interacting physics of cuprate materials. We have shown that this theory explains several important experimental observations of the normal state. In fact, the theory makes several interesting predictions as well [**16**]. Work towards understanding the superconducting phase and strong coupling dynamics continues.

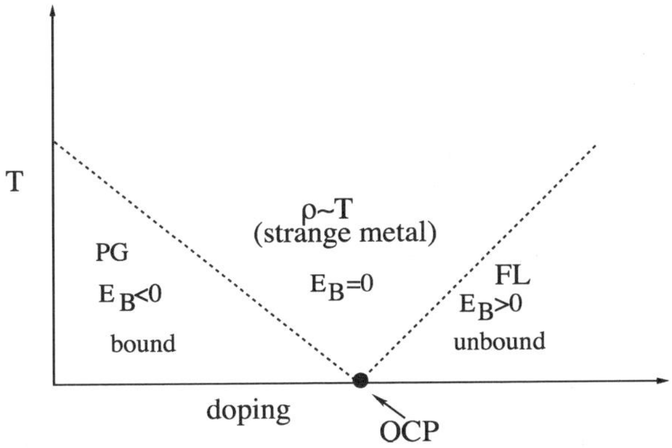

FIGURE 6. Proposed phase diagram for the binding of the holes and bosons that result in the formation of the pseudogap phase. Once the binding energy vanishes, the energy to excite a boson vanishes. In the critical regime, the dominant scattering mechanism is still due to the interaction with the boson. T-linear resistivity results anytime $T > \omega_b$, where ω_b is the energy to excite a boson. To the right of the quantum critical regime (QCP), the boson is irrelevant and scattering is dominated by electron-electron interactions indicative of a Fermi liquid. The QCP signifies the end of the binding of fermionic and bosonic degrees of freedom that result in the pseudogap phase.

Acknowledgments

This is a contribution by RGL to the proceedings of the first Sowers Theoretical Physics Workshop on String Theory, held at Virginia Tech in May 2007. We wish to thank the organizers of this workshop for hospitality and their kind latitude in subject matter.

References

[1] P. W. Anderson, The Theory of Superconductivity in the High-Cuprates (Princeton University Press, Princeton NJ, 1997).
[2] T.-P. Choy and P. Phillips, Phys. Rev. Lett. 95, 196405 (2005).
[3] R. G. Leigh, P. Phillips and T.-P. Choy, Phys. Rev. Lett. 99 (2007) 046404.
[4] T.-P. Choy, R. G. Leigh, and P. Phillips, Phys. Rev. **B77** (2008) 14512.
[5] J. Polchinski, hep-th/9210046.
[6] G. Benfatto and G. Gallavotti, J. Stat. Phys. **59** (1990) 541; Phys. Rev. **B42** (1990) 9967; R. Shankar, Physica **A177** (1991) 530.
[7] P. Phillips, T. -P. Choy, R. G. Leigh, arXiv/0802.3405.
[8] C. T. Chen, et al. Phys. Rev. Lett. **66**, 104 (1991).
[9] M. B. J. Meinders, H. Eskes, and G. A. Sawatzky, Phys. Rev. B **48**, 3916-3926 (1993).
[10] A. B. Harris and R. V. Lange, Phys. Rev. **157**, 295 (1967).
[11] T.-P. Choy, R. G. Leigh, and P. Phillips, Phys Rev. **B77** (2008) 104524.
[12] D. Bohm and D. Pines, Phys. Rev. **92**, 609 (1953).

[13] P. Phillips and C. Chamon, Phys. Rev. Lett. 95, 107002 (2005).
[14] A. Lanzara, et al. Nature **412**, 510 (2001).
[15] J. Graf, et al., Phys. Rev. Lett. **98**, 67004 (2007).
[16] T.-P. Choy, P. Phillips and R.G. Leigh, in preparation.

Department of Physics, University of Illinois at Urbana-Champaign, 1110 West Green Street, Urbana, IL 61801-3080
E-mail address: `rgleigh@uiuc.edu`

Department of Physics, University of Illinois at Urbana-Champaign, 1110 West Green Street, Urbana, IL 61801-3080
E-mail address: `tingchoy@uiuc.edu`

Department of Physics, University of Illinois at Urbana-Champaign, 1110 West Green Street, Urbana, IL 61801-3080
E-mail address: `dimer@uiuc.edu`

AMS/IP Studies in Advanced Mathematics
Volume 44, 2008

Holographic Aspects of Generalized Electric-Magnetic Dualities

Anastasios C. Petkou

ABSTRACT. We review recent work on various holographic aspects of electric-magnetic dualities in models that involve conformally coupled scalars and abelian gauge fields in AdS_4 spaces. Such models are relevant for the holographic description of M-theory. We also briefly comment on some new results on the holographic properties of generalized electric-magnetic duality in gravity.

1. Introduction and summary

The standard AdS_5/CFT_4 paradigm may be viewed as the holographic image of the decoupling of massive degrees of freedom in four-dimensional YM theories. In YM the decoupling takes place moving from the asymptotically free UV limit to the strongly coupled IR regime. In string theory the decoupling takes place running down to the low-energy supergravity limit. The field and string theory decoupling limits are holographically identified, hence supergravity corresponds to strongly coupled field theories. The presence of D-branes, singularities and black holes in string theory is naturally associated to phenomena such as symmetry breaking, confinement and finite-temperature in field theory.

Recent work on AdS_4/CFT_3 brought to light a possible new kind of holographic correpondence. Bulk theories that exhibit a generalized form of electric-magnetic duality correspond to boundary theories whose correlation functions have special transformation properties [**1, 2**]. The latter transformations are induced by certain "double-trace" deformations [**2**]. This implies the existence of "duality-related fixed points" in three dimensions, which would be the holographic images of the electric-magnetic dual descriptions

---

2000 *Mathematics Subject Classification.* Primary 81T30; Secondary 83E30.

A. C. P. was supported in part by the grant PYTHAGORAS II by the Greek Ministry of Education.

of the bulk theories [**4**]. More recently, some intriguing transport properties of three-dimensional theories have also been attributed to bulk electric-magnetic duality [**5, 6, 7**]. Such studies, combined with the observation that three-dimensional field theories have a quite different quantum structure from their four- and two-dimensional counterparts,[1] imply that the bulk theory is rather exotic. As electric-magnetic duality appears to be relevant for M-theory compactifications it is natural to conjecture that the quantum structure of three-dimensional field theories provides crucial information on the non-linear dynamics of M-theory.

In this note we review recent work on various holographic aspects of generalized electric-magnetic duality. In Section 2 we present models with conformally coupled scalars and in Section 3 the case of a $U(1)$ gauge field in AdS_4. These models can be embedded into M-theory, so our results provide a hint for the role of generalized electric-magnetic duality in the holographic description of the latter. The $U(1)$ gauge field is the first step towards the discussion of YM, gravity and higher-spin gauge fields in AdS_4. We summarize and put our findings into perspective in Section 4.

2. Conformally coupled scalars

2.1. A toy model. A conformally coupled scalar with quartic self-interaction in fixed Euclidean AdS_4 simplifies, after a suitable conformal rescaling of the field and metric, to the model of a massless scalar $\Phi(r,\bar{x})$ with quartic self-interaction $(\lambda/4!)\Phi^4$ on the upper half of $\mathbb{R}^4$ with $r \geq 0$ [**3**]. The boundary is the hyperplane $r=0$. Φ behaves near the boundary as

$$\Phi(r,\vec{x}) = \alpha(\vec{x}) + r\,\beta(\vec{x}) + \cdots \tag{2.1}$$

with α and β arbitrary functions. Requiring regularity of (2.1) for $r \to \infty$ gives a non-local and generally invertible relationship $\beta = \beta[\alpha]$. It was pointed out in [**3**] that the on-shell bulk action $W_{\text{ren}}[\alpha]$ of the model evaluated as a functional of α gives (minus) the *effective action* of a dual boundary theory that contains an operator O_1 with dimension $\Delta = 1$. The vacuum structure of the dual boundary theory is determined by

$$\langle O_1(\vec{x})\rangle_{\alpha=\alpha_0} = -\alpha_0(\vec{x}), \qquad \left.\frac{\delta W[\alpha]}{\delta\alpha(\vec{x})}\right|_{\alpha_0} = 0. \tag{2.2}$$

For $\lambda < 0$ our toy model possesses a non-trivial classical instanton solution

$$\Phi(r,\vec{x}) = \sqrt{\frac{48}{-\lambda}}\frac{b}{b^2+r^2+\vec{x}^2} \implies \alpha_0(\bar{x}) = \sqrt{\frac{48}{-\lambda}}\frac{b}{b^2+\vec{x}^2}, \tag{2.3}$$

with b an arbitrary parameter with dimensions of length. This implies that the dual boundary theory has a non-trivial vacuum structure where the

[1] For a review of three-dimensional theories see [**8**].

operator O_1 has non-zero expectation value $-\alpha_0$. A three-dimensional model that may reproduce the bulk results is a massless scalar with self-interaction $(g/6!)\varphi^6$. For $g < 0$ this model has the instanton solution

$$\varphi_0(\vec{x}) = \left(\frac{360}{-g}\right)^{1/4} \left(\frac{c}{c^2 + \vec{x}^2}\right)^{1/2} . \tag{2.4}$$

One can show that the parameters of the bulk and boundary models are related as

$$\frac{1}{g} = -\frac{32}{45}\frac{1}{\lambda^2}, \quad c = \kappa b, \quad \varphi_0^2(\kappa\vec{x}) = -\langle O_1(\vec{x})\rangle \Longrightarrow \kappa^2 = \frac{16\pi^2}{3\lambda} . \tag{2.5}$$

It appears that the bulk and boundary theories in this model are both accessible by standard field theoretic methods. This is connected to the fact that *both* α and β can be interpreted as expectation values of boundary operators. In a Hamiltonian analysis of the bulk scalar field theory α and β play the roles of "coordinate" and "momenta", hence their mutual interchange resembles the standard canonical transformation $p \to q$, $q \to -p$ of Hamiltonian mechanics. This may be termed *harmonic oscillator duality,* to be compared and contrasted with the usual *electric-magnetic duality.*

To strengthen the above observation we have radially quantized the bulk and boundary theories [**3**]. The corresponding mode expansions of the free field configurations are

$$\hat{\Phi}(R,\theta,\psi,\omega) = \sum_{jlm} \left(\frac{a^+_{jlm}}{\sqrt{j+1}} R^j\, Y^*_{jlm}(\Omega_3) + \frac{a^-_{jlm}}{\sqrt{j+1}}\frac{1}{R^{j+2}} Y_{jlm}(\Omega_3)\right), \tag{2.6}$$

$$\hat{\varphi}(R,\Omega_2) = \sum_{\ell m} \frac{1}{\sqrt{2\ell+1}} \left(b^\dagger_{\ell m}\, R^\ell\, Y^*_{\ell m}(\Omega_2) + b_{\ell m}\frac{1}{R^{\ell+1}}\, Y_{\ell m}(\Omega_2)\right) \tag{2.7}$$

with Y_{jlm} the hyperspherical harmonics of S^3. The boundary is at $\theta = \pi/2$. Our results above implies that the bulk elementary operator is identified with a properly normal ordered boundary composite operator as

$$\hat{\Phi}\Big(R, \frac{\pi}{2}, \psi, \omega\Big) \equiv -O_1(\vec{x}) =: \hat{\varphi}^2(x):. \tag{2.8}$$

The bulk and boundary creation and annihilation operators are related as

$$a_{jlm} = \sum_{m_1 m_2} c_{lm}^{\ell\ell m_1 m_2}\, b_{\ell m_1} b_{\ell m_2}, \quad a^\dagger_{jlm} \sum_{m_1 m_2} c_{lm}^{\ell\ell m_1 m_2}\, b^\dagger_{\ell m_1} b^\dagger_{\ell m_2}, \quad j = 2\ell \tag{2.9}$$

with constant coefficients $c_{lm}^{\ell\ell m_1 m_2}$. Bulk one-particle states correspond to boundary two-particle states. We have made an effort to extend the above remarkable bulk/boundary quantum correspondence to fluctuations around the non-trivial instanton solutions. We obtained a highly non-trivial classical correspondence between the bulk and the square of the boundary fluctuations around the instantons[2] but we have not yet fully developed the quantum correspondence.

2.2. A model embedded in M-theory. In [**10**] it was pointed out that conformally coupled scalars also appear in M-theory compactifications. Specifically, we considered the model with action

$$S=\frac{1}{2}\int \mathrm{d}^4x\sqrt{g}\left(\frac{-R+2\Lambda}{\kappa^2}+(\partial_\mu\phi)^2+\frac{1}{6}R\phi^2+\lambda\phi^4\right) \tag{2.10}$$

where $\kappa^2=8\pi G_4$, λ is a dimensionless coupling and the cosmological constant is $\Lambda=-3/l^2$. For the special value of the quartic coupling $\lambda=\kappa^2/6l^2>0$ the action (2.10) is obtained by a consistent truncation of 11-dimensional supergravity.

The asymptotic behavior of the scalar ϕ is as in (2.1). Motivated by the existence of the instanton solution (2.3) we can impose the "self-dual" boundary conditions $\beta=-la\alpha^2$ by adding the boundary term

$$S_{\text{bdy}}=-\frac{l^3a}{3}\int \mathrm{d}^3x\,\alpha^3(\vec{x}). \tag{2.11}$$

We consider solutions with vanishing energy-momentum tensor, in which case the background is still AdS, since all extrema of the action (2.10) have constant Ricci scalar $R=-12/l^2$ [**11**]. It is quite remarkable that in this case we are able to calculate *exactly* the effective potential (constant α) of the dual boundary theory as

$$V_{\lambda,a}(\alpha)=\frac{1}{3\lambda}\left[\left(\frac{R}{6}+\lambda\alpha^2\right)^{3/2}-a\lambda\alpha^3-\left(\frac{R}{6}\right)^{3/2}\right], \tag{2.12}$$

as well as its effective action in the double-scaling limit $\lambda-a^2=\mu\to 0$

(2.13)
$$\Gamma_{\text{eff}}[\alpha]=\frac{1}{3a}\int d^3x\sqrt{g_{(0)}}\left(\frac{1}{2}\partial_i\varphi\partial^i\varphi+\frac{1}{16}R[g_{(0)}]\varphi^2+\frac{1}{8}\mu\varphi^6\right)+(a^{-2}),$$

where $\varphi^2=\alpha$ and $g_{(0)ij}$ is the boundary metric. Remarkably, (2.13) coincides with the conformal three-dimensional models used in [**3**].

For $\lambda>0$ the action (2.10) has the instanton solution

$$\phi=\frac{2}{l\sqrt{|\lambda|}}\left(\frac{Br}{-\text{sgn}(\lambda)B^2+(r+A)^2+(\vec{x}-\vec{x}_0)^2}\right), \tag{2.14}$$

[2]This result was obtained by T. Koornwinder and appears in the appendix of [**3**].

where A, B, x_0^i, $i = 1, 2, 3$, are arbitrary constants. This is non-singular provided $A > B \geq 0$ and satisfies the "self-dual" boundary condition. The existence of this solution is rather surprising and implies the instability (*à la* Coleman-de Luccia) of pure AdS_4 towards the spontaneous dressing by a scalar field. It also implies the possible instability of a stack of M2-branes. The decay rate of the vacuum is

$$(2.15) \qquad P \propto \exp(-\Gamma_{\text{eff}}\big|_{\text{inst}}), \quad \Gamma_{\text{eff}}\big|_{\text{inst}} = \frac{4\pi^2 l^2}{\kappa^2}\left(\frac{1}{\sqrt{1-\kappa^2/6l^2a^2}} - 1\right).$$

with $a = \sqrt{|\lambda|}A/B$. Note that the deformation parameter a drives the theory from the regime of marginal stability at $a = \kappa/\sqrt{6}l$ to total instability at $a \to \infty$.

3. Abelian gauge fields

In this case, we encounter the first instance of standard electric-magnetic duality in the bulk. Not only do we have the freedom to choose the leading or subleading term in the gauge field as the source of the operator (*i.e.* "generalized electric-magnetic duality"), but we also can choose between "electric" and "magnetic" boundary conditions. This choice corresponds to adding an abelian AB-type Chern-Simons term [**1, 16**] on the boundary. This term is marginal.

We will first discuss regular instanton bulk solutions. These are special because they provide exact solutions of the coupled gravity-matter field equations whilst allowing for non-trivial dynamics of the gauge field at the boundary [**12**]. Instantons are self-dual under S-duality; therefore we will consider electric-magnetic duality on generic solutions.

Instantons are regular self-dual solutions $F = *F$ of the field equations. The regularity condition for Euclidean solutions relates the boundary value of the electric field to the boundary value of the transverse part of the gauge field:

$$(3.1) \qquad E_i(p) = -\frac{1}{g^2}\,|p|\,A_i(p).$$

The on-shell effective action now gives the generating functional of the boundary theory:

$$(3.2) \qquad W[A] = -\frac{1}{2}\int d^3x\, A_i(x)E_i(x) = \frac{1}{2g^2}\int d^3p\,|p|\,A_i(p)A_i(-p).$$

The partition function is purely a functional of A_i, which is interpreted as a source in the CFT. The above is consistent with a Dirichlet choice of boundary conditions where $\delta A_i = 0$ at the boundary. However, gauge fields in AdS_4 admit a more general choice of boundary conditions, for instance those corresponding to bulk instantons. To get a variational problem that generates instanton boundary conditions we need to modify the action by a boundary Chern–Simons term: $S_{\rm CS} = \frac{\theta}{8\pi^2}\int \mathrm{d}^3x\, \epsilon^{ijk} A_i \partial_j A_k$. The boundary conditions are now modified to

$$\Box^{1/2} A_i + \frac{\theta g^2}{4\pi^2}\, \epsilon_{ijk}\partial_j A_k = 0. \tag{3.3}$$

For $\theta = \pm 4\pi^2/g^2$, this gives regular bulk instanton solutions. As noticed in [**12**], this is the equation of motion for a topologically massive spin-1 particle in three dimensions [**13, 14**], with mass $m \sim \Box^{1/2}$.

To discuss S-duality, we go back to generic solutions of the Dirichlet boundary problem. Holographically, A_i is a source that couples to a conserved current of dimension 2. The two-point function of this current is computed to give the familiar result, plus a parity-breaking term:

$$\langle J_i(p) J_j(-p)\rangle = \frac{1}{g^2}\, |p|\, \Pi_{ij} + \frac{i\theta}{4\pi^2}\, \epsilon_{ijk} p_k. \tag{3.4}$$

Π_{ij} is the projector onto transverse vectors. For simplicity we set $g = 1$ and $\theta = 0$, the general case being discussed in [**2, 12**].

The conserved current J_i is equal to the boundary value of the electric field. The bulk equations have electric-magnetic invariance which interchanges the boundary values of the electric and magnetic fields: $B_i \leftrightarrow \pm E_i$. However, the boundary terms are not invariant; the Dirichlet and Neumann boundary problems get interchanged [**1**].

Having seen the Dirichlet problem, let us now discuss the dual Neumann quantization scheme [**12**]. Now the electric field is held fixed at the boundary whereas the magnetic field is included as part of the Hilbert space. Hence, the path integral formulation will now include an integral over boundary configurations of the gauge field $A_i(x)$ up to gauge transformations. The holographic interpretation of this is that the electric field corresponds to a dual source, and A_i to an operator. There is a problem though. A_i itself cannot be the dual operator because it has dimension 1 and is below the unitarity bound. However, we can construct from it a dual current $\tilde{J} = *\mathrm{d}A$ with dimension 2 [**2, 12**]. This current corresponds to the bulk magnetic field. It is conserved and lies on the marginal line of the unitarity bound, hence its correlators correspond to a unitary theory. Further, since the electric field is conserved, it can be written in terms of a 1-form $J = *\mathrm{d}\tilde{A}$. This 1-form is the one that is now fixed, and is the dual background field [**1**].

In summary, bulk electric-magnetic duality interchanges $A \leftrightarrow \pm\tilde{A}$ on the boundary. The Dirichlet problem corresponds to fixing the gauge field on the boundary, with the electric field corresponding to a dimension-2 current. In the Neumann problem, the magnetic field corresponds to the dual current $\tilde{J} = B$ whereas the dual gauge field is the source. The dual generating functional is $\tilde{W}[\tilde{A}] = \frac{1}{2}\int \mathrm{d}^3x\, \tilde{A}_i B_i$. The two theories are related by a Legendre transform of an AB-type Chern–Simons term $-\int A \wedge \mathrm{d}\tilde{A}$ [**1, 12**].

The embedding of this model in M-theory was discussed in [**5, 12**].

4. Gravity and M-theory holography

Gravity is known to exhibit electric-magnetic properties similar to the ones discussed here [**9**][3]. Whereas we will discuss the implications for duality in the CFT elsewhere [**17**], we will give some new results here. Despite some profound differences with the lower spin cases, there are good reasons to believe that duality in the CFT works in a similar way for gravity. Namely, not only do we have the possibility to choose between leading and subleading terms in the graviton expansion, but also we have the option to choose between "electric" and "magnetic" boundary conditions. The latter corresponds to adding a gravitational Chern-Simons term in the boundary.

We observe a pattern in the M-theory compactifications. Firstly, the three-dimensional boundary theories do not exhibit the decoupling of massive modes that we observe in four-dimensional YM theories. Namely, the theories related by "double-trace" deformations correspond to fixed-points which appear to have very similar operator content, in contrast with YM theories where the weak and strong coupling limits have very different content. Secondly, the Chern–Simons terms in the boundary should correspond to parity anomalies of three-dimensional theories. One may conjecture that the structure of three-dimensional theories is a hologram of the structure of the full M-theory. Namely, we expect that full blown M-theory is duality invariant, however only after its full spectrum is taken into account. This duality, which is a generalization of electric-magnetic duality, is broken at the level of 11-dimensional supergavity, but we see a sign of it in the boundary conditions.

To end we will make some remarks on the generalization of the results in the previous section to the gravitational field. We will consider instanton configurations where the Weyl tensor is self-dual. It can be shown [**17**] that the stress-energy tensor of such configurations is given by the Cotton tensor,

$$\langle T_{ij}\rangle = \frac{\ell^2}{8\pi G_N}\, C_{ij}, \tag{4.1}$$

computed with respect to the boundary metric $g_{ij}(r=0,x)$ in the Graham–Fefferman coordinate system. Recall that the Cotton tensor is symmetric, traceless, and conserved. According to [**15**] this is a boundary condition for the third derivative of the metric $g_{ij}(r,x)$, $g_{(3)}$. Expanding around an AdS background, $g_{ij}(r,x) = \delta_{ij} + h_{ij}(r,x)$, we will look for regular instantons. Using the results in [**18**], one can show that regularity imposes

$$h_{(3)ij} = \frac{1}{3}\, |\Box|^{3/2} h_{(0)ij}, \tag{4.2}$$

and we have projected onto the transverse, traceless part of the graviton. Combining the two conditions above, we get that regular bulk instantons

[3]See also references in [**12**].

satisfy

$$\Box h_{(0)ij} = \alpha\, \epsilon_{ikl} \Box^{1/2} \partial_k h_{(0)jl} + (i \leftrightarrow j). \tag{4.3}$$

This is a non-trivial differential equation for the boundary metric that generalizes the *self-dual* boundary conditions of the scalar and the $U(1)$ case (3.3).

References

[1] E. Witten, "SL(2, Z) action on three-dimensional conformal field theories with Abelian symmetry," pp 1173–1200 in *From fields to strings: circumnavigating theoretical physics*, volume 2, World. Sci. Publ., Singapore, 2005 [arXiv:hep-th/0307041].
[2] R. G. Leigh and A. C. Petkou, "SL(2, Z) action on three-dimensional CFTs and holography," JHEP **0312**, 020 (2003) [arXiv:hep-th/0309177].
[3] S. de Haro and A. C. Petkou, JHEP **0612** (2006) 076 [arXiv:hep-th/0606276].
[4] A. C. Petkou, arXiv:hep-th/0410116.
[5] C. P. Herzog, P. Kovtun, S. Sachdev and D. T. Son, Phys. Rev. D **75** (2007) 085020 [arXiv:hep-th/0701036].
[6] S. A. Hartnoll and P. Kovtun, Phys. Rev. D **76** (2007) 066001 [arXiv:0704.1160 [hep-th]].
[7] S. A. Hartnoll and C. P. Herzog, arXiv:0706.3228 [hep-th].
[8] M. Moshe and J. Zinn-Justin, Phys. Rept. **385** (2003) 69 [arXiv:hep-th/0306133].
[9] R. G. Leigh and A. C. Petkou, arXiv:0704.0531 [hep-th].
[10] S. de Haro, I. Papadimitriou and A. C. Petkou, Phys. Rev. Lett. **98** (2007) 231601 [arXiv:hep-th/0611315].
[11] I. Papadimitriou, JHEP **0705** (2007) 075 [arXiv:hep-th/0703152].
[12] S. de Haro and P. Gao, "Electric-magnetic duality and deformations of three-dimensional CFT's," arXiv:hep-th/0701144. To appear in Phys. Rev. **D**.
[13] P. K. Townsend, K. Pilch and P. van Nieuwenhuizen, "Selfduality In Odd Dimensions," Phys. Lett. **136B** (1984) 38 [Addendum-ibid. **137B** (1984) 443].
[14] S. Deser and R. Jackiw, "'Selfduality' Of Topologically Massive Gauge Theories," Phys. Lett. B **139** (1984) 371.
[15] S. de Haro, S. N. Solodukhin and K. Skenderis, "Holographic reconstruction of space-time and renormalization in the AdS/CFT correspondence," Commun. Math. Phys. **217** (2001) 595 [arXiv:hep-th/0002230].
[16] H. U. Yee, "A note on AdS/CFT dual of SL(2,Z) action on 3D conformal field theories with U(1) symmetry," Phys. Lett. B **598**, 139 (2004) [arXiv:hep-th/0402115].
[17] S. de Haro and A. Petkou, "Gravitational Instantons and Duality in AdS_4", to appear.
[18] G. E. Arutyunov and S. A. Frolov, "On the origin of supergravity boundary terms in the AdS/CFT correspondence," Nucl. Phys. B **544** (1999) 576 [arXiv:hep-th/9806216].

Department of Physics, University of Crete, Heraklion 71003, Greece
E-mail address: petkou@physics.uoc.gr

AMS/IP Studies in Advanced Mathematics
Volume 44, 2008

Null and Spacelike Singularities and Gauge-Gravity Duality

Sumit R. Das

ABSTRACT. Certain null and spacelike singularities in ten dimensional supergravity have natural holographic duals in terms of (3 + 1)-dimensional gauge theory with sources. In many situations the holographic duals appear to be well defined in regions where the supergravity develops singularities. We describe some recent progress in this area.

1. Introduction

Spacelike and null singularities pose a peculiar puzzle. At these singularities, "time" begins or ends – and it is not clear what is the meaning of this. Classic examples of such singularities are those which appear in the interior of neutral black holes and those which appear in cosmology.

It has always been suspected that near singularities usual notions of space and time break down and a consistent quantization of gravity would provide a more abstract structure which replaces spacetime. However we do not know as yet what this abstract structure could be in general. In some situations, string theory has provided concrete ideas about the nature of this structure. These are situations where gravitational physics has a tractable holographic description [**3**] in terms of a non-gravitational theory in a lower number of space-time dimensions. In view of the spectacular success of the holographic principle in black hole physics, it is natural to explore whether this can be used to understand conceptual issues posed by singularities.

In string theory, holography is a special case of a more general duality between open and closed strings. This duality implies that the dynamics of open strings contains the dynamics of closed strings. Since closed strings contain gravity, space-time questions can be posed in the open string theory which does not contain gravity and is therefore conceptually easier. Under special circumstances, the open string theory can be truncated to its low

2000 *Mathematics Subject Classification.* Primary 83E30; Secondary 81T30, 81T60, 83E50.

energy limit – which is a gauge theory on a *fixed* background. In these situations, open-closed duality becomes particularly useful. The simplest example is non-critical closed string theory in two space-time dimensions. Here the holographic theory is gauged matrix quantum mechanics [**4**]. The second class of examples involve string theory or M-theory defined on spacetimes with a compact null direction. Then a sector of the theory with some specified momentum in this null direction is dual to a $(d+1)$-dimensional gauge theory, where d depends on the number of additional (spacelike) compact directions. Using standard terminology we will call them matrix theories [**5**]. Finally, the celebrated AdS/CFT correspondence [**6**] relates closed string theory in asymptotically anti-de Sitter spacetimes to gauge theories living on their boundaries. In all these cases, the dynamical "bulk" spacetime (on which the closed string theory lives) is an approximation which holds in a specific regime of the gauge theory. In this regime, the closed string theory reduces to supergravity. Generically, there is no spacetime interpretation, though the gauge theory may make perfect sense. This fact opens up the possibility that in regions where the bulk gravity description is singular, one may have a well defined gauge theory description and one has an answer to the question : *What replaces spacetime?*

Treating time-dependent backgrounds in string theory, particularly those with singularities, has been notoriously difficult. However, some modest progress has been made recently in both worldsheet formulations as well as holographic formulations of all the three types mentioned above. The key idea in these various types of holography are similar. One looks for toy models where the spacetime background on which the closed string theory is defined is singular, but the holographic gauge theory description does not appear to be problematic. Thus, the gauge theory provides the correct description of the region which would appear singular if the gravity interpretation is extrapolated beyond its regime of validity.

The question of cosmological singularities has been addressed in all these three kinds of holography. In the following, we will discuss recent attempts to understand cosmological singularities using the AdS/CFT correspondence. For a review of the treatment of singularities in matrix model formulations of two-dimensional string theory with references to the original literature, see [**7**]. Some recent progress appears in [**8**]. For a review (and references to the original literature) of approaches based on matrix theory, see [**9**].

2. Null singularities in the AdS/CFT correspondence

In many respects the AdS/CFT correspondence is a more controlled example of the holographic principle. In its simplest setting, the correspondence implies that IIB string theory on $AdS_5 \times S^5$ with a constant 5-form flux is dual to $(3+1)$-dimensional $N = 4$ supersymmetric $SU(N)$ Yang-Mills theory which lives on the boundary of AdS_5. If R_{AdS} denotes the radius of the S^5 as well as the curvature length scale of AdS_5 and g_s

denotes the string coupling, the coupling constant g_{YM} and the rank of the gauge group N of the Yang Mills theory are related by

$$\frac{R_{AdS}^4}{l_s^4} = 4\pi g_{YM}^2 N, \qquad g_s = g_{YM}^2 \tag{2.1}$$

This immediately implies that the gauge theory describes *classical* string theory in the 't Hooft limit

$$N \to \infty, \qquad g_{YM} \to 0, \qquad g_{YM}^2 N = \text{finite} \tag{2.2}$$

The low energy limit of the closed string theory – supergravity – is a good approximation only in the strong coupling regime $g_{YM}^2 N \gg 1$. For small $g_{YM}^2 N$ supergravity and hence conventional spacetime is not a good description of the gauge theory dynamics. Finite N corrections correspond to string loop effects.

There have been several approaches to cosmological singularities by finding appropriate modifications of the AdS solutions which correspond to deformations of the Yang-Mills theory or to states in the theory [**10**]. We will discuss one approach developed in [**2, 1, 11, 12, 13, 14**][1]. The idea is to find bulk solutions which have cosmological singularities where the usual notions of spacetime break down, while the gauge theory description remains tractable.

In the following we will recount the main points in [**2, 1, 14**].

2.1. The Supergravity Background and the Conjecture. The usual $AdS_5 \times S^5$ solution is given by the Einstein frame metric in Poincaré coordinates

$$ds^2 = \left(\frac{r^2}{R_{AdS}^2}\right)\eta_{\mu\nu}dx^\mu dx^\nu + \left(\frac{R_{AdS}^2}{r^2}\right)dr^2 + R_{AdS}^2 d\Omega_5^2 \tag{2.3}$$

and a 5-form field strength and dilaton Φ

$$F_{(5)} = R_{AdS}^4(\omega_5 + *_{10}\omega_5) \quad \Phi = \text{constant} \tag{2.4}$$

This has maximal supersymmetry.

We consider supergravity solutions which are *non-normalizable* deformations of this,

$$\begin{aligned} ds^2 &= \left(\frac{r^2}{R_{AdS}^2}\right)\tilde{g}_{\mu\nu}(x)dx^\mu dx^\nu + \left(\frac{R_{AdS}^2}{r^2}\right)dr^2 + R_{AdS}^2 d\Omega_5^2\ , \\ \Phi &= \Phi(x) \\ F_{(5)} &= R_{AdS}^4(\omega_5 + *_{10}\omega_5) \end{aligned} \tag{2.5}$$

[1]See [**15**] for an interesting approach to find signatures of spacelike singularities inside AdS black holes in the CFT.

The equations of motion then imply that the Ricci tensor $\tilde{R}_{\mu\nu}$ constructed from the metric $\tilde{g}_{\mu\nu}(x)$ must obey the equation

$$\tilde{R}_{\mu\nu} = \frac{1}{2}\partial_\mu \phi \partial_\nu \Phi, \tag{2.6}$$

while the dilaton must satisfy

$$\partial_\mu(\sqrt{-\det(\tilde{g})}\ \tilde{g}^{\mu\nu}\partial_\nu \Phi) = 0. \tag{2.7}$$

It turns out that this solution is the near-horizon limit of the geometry produced by three-branes whose worldvolume metric is given by $\tilde{g}_{\mu\nu}(x)$.

For generic $\tilde{g}_{\mu\nu}(x)$ this solution will have curvature singularities at the Poincaré horizon at $r = 0$. This does not happen when $\tilde{g}_{\mu\nu}(x)$ and $\Phi(x)$ are functions of a null coordinate x^+. We will therefore restrict our attention to such solutions. Such solutions retain half of the supersymmetries with parameters ϵ satisfying $\Gamma^+\epsilon = 0$. Furthermore, for reasons which will soon become clear, we will consider brane metrics which are conformal to flat space

$$\tilde{g}_{\mu\nu}(x)dx^\mu dx^\nu = e^{f(x^+)}\left[-2dx^+dx^- + dx_1^2 + dx_2^2\right] \tag{2.8}$$

The equations of motion (2.6) then require that the dilaton is also a function of x^+ alone. The dilaton equation (2.7) is automatically satisfied, while (2.6) simplifies to

$$\frac{1}{2}(f')^2 - f'' = \frac{1}{2}(\partial_+\Phi)^2. \tag{2.9}$$

where prime denotes derivative with respect to x^+.

The conjecture is that string theory in this null background is dual to $(3+1)$-dimensional $N = 4$ Yang-Mills theory which lives on a background spacetime given by $\tilde{g}_{\mu\nu}(x)$ and an x^+-dependent coupling

$$g_{YM}(x^+) = e^{\Phi(x^+)/2}\sqrt{g_s} \tag{2.10}$$

The bosonic part of the action is

$$S = \int d^4x\ \mathrm{Tr}\left\{\frac{1}{4}e^{-\Phi}F_{\mu\nu}F^{\mu\nu} + \frac{1}{2}(D_\mu\chi^I)(D^\mu\chi^I) + \frac{1}{4}[\chi^I,\chi^J]^2\right\} \tag{2.11}$$

where $\chi^I, I = 1,\ldots,6$ are the adjoint scalars.

There are several pieces of evidence for the validity of this conjecture. First, when $f \ll 1$, we also have $\Phi \ll 1$. In that case, the solution represents small non-normalizable metric and dilaton deformations of standard $AdS_5 \times S^5$. The AdS/CFT correspondence then implies that the dual gauge theory is deformed by operators which are dual to these modes, viz. the energy-momentum tensor $T_{\mu\nu}$ and $\mathrm{Tr}\,F^2$ respectively. This is evident from the action (2.11). Secondly, we may consider the action of a single probe D3-brane in the background and examine the way $f(x^+)$ and $\Phi(x^+)$ appear – it is easy to check that this is consistent with (2.11). Finally, as noted above, our solution is the near-horizon limit of the asymptotically flat geometry of a stack of D3-branes with a curved brane worldvolume[2].

We will consider solutions which are $AdS_5 \times S^5$ in asymptotic null time $x^+ \to \pm\infty$, but develop null singularities for some value of x^+ which may be chosen to be at $x^+ = 0$. However, we will require that $g_s \ll 1$ *and the effective string coupling* $e^\Phi g_s$ *remains weak for all* x^+. This latter feature distinguishes our solution from some others in the literature[3]. A nice example of such a solution is

$$e^{f(x^+)} = \tanh^2 x^+, \qquad e^\Phi = g_s \left|\tanh \frac{x^+}{2}\right|^{\sqrt{8}}. \tag{2.12}$$

At $x^+ = 0$ all local curvature invariants are bounded. However this point may be reached in a finite physical time. For example, the affine parameter λ along a geodesic $x^+(\lambda)$ with all other coordinates constant is given by

$$\lambda = x^+ - \tanh\ x^+ \tag{2.13}$$

Thus $x^+ = 0$ can be reached in a finite value of the affine parameter. Furthermore, it turns out that tidal forces between neighboring geodesics diverge at this point. Therefore $x^+ = 0$ is a genuine null singularity.

Consider the solution as a time evolution in light-cone time x^+. At $x^+ \to -\infty$ the Yang-Mills coupling approaches $\sqrt{g_s}$ exponentially. In the dual Yang-Mills theory, we will always work in the 't Hooft limit $g_s \to 0, N \to \infty$ with $g_s N$ finite and large. Therefore, according to the usual AdS/CFT correspondence, the ground state of the theory is dual to *supergravity* in $AdS_5 \times S^5$ as stated above. This vacuum evolves in time according to the Yang-Mills Hamiltonian whose effective coupling *decreases.* The dual description of this time evolution is the supergravity solution described above. Supergravity, however, makes sense only when the Yang-Mills coupling is large. Thus, as we approach $x^+ \to 0$ the coupling approaches zero and the supergravity description becomes meaningless. The singularity therefore appears at a place where we expect a spacetime interpretation of the gauge theory to break down.

2.2. The gauge theory and Weyl transformations. As emphasized above, one of the salient features of our toy model is that the gauge theory is weakly coupled at the "singularity," pretty much like the Matrix

---

[2]The full supergravity solution is given in [**2**].

[3]These include orbifold models, backgrounds with time-dependent warping, models based on tachyon condensation. References to the original literature can be found in [**2**]. Some of these topics are reviewed in [**16**].

Theory examples given above. This is reassuring, since one would hope that weakly coupled gauge theory makes sense and provides the alternative structure which replaces dynamical bulk space time in this region. However, it is precisely at this point that the nondynamical spacetime of the gauge theory shrinks to zero size!

Normally this would be a disaster since a gauge theory on a zero size space-time would be singular even if it is weakly coupled. What saves the day is the fact that this particular gauge theory is Weyl invariant. This is evident at the classical level – the factor $e^{f(x^+)}$ does not appear in the classical action. If the coupling was constant the theory would have been conformally invariant (in the sense of invariance under conformal diffeomorphisms) as well. Here the x^+ dependence of the coupling breaks these conformal symmetries but retains Weyl invariance at least classically. This means that at the classical level our gauge theory simply does not see the shrinking Weyl factor.

Usually Weyl invariance of quantum field theories is broken at the quantum level by anomalies. Our gauge theory is a special case of $N = 4$ Yang-Mills theory coupled to nondynamical conformal supergravity, where only the metric and the dilaton fields of the background supergravity are turned on. The Weyl anomaly of this theory has been worked out a while ago with the result **[17, 18, 19]**

$$\langle T^\mu_\mu \rangle = -\frac{N^2}{64\pi^2}\left\{2\left(R_{\mu\nu}R^{\mu\nu} - \frac{1}{3}R^2\right) + 4\left[-2\left(R^{\mu\nu} + \frac{1}{3}Rg^{\mu\nu}\right)\partial_\mu\Phi\partial_\nu\Phi \right.\right.$$
$$\left.\left. + (\nabla^2\Phi)^2 + \frac{4}{3}(g^{\mu\nu}\partial_\mu\Phi\partial_\nu\Phi)^2\right]\right\} \tag{2.14}$$

In fact, it turns out that the *operator* T^μ_μ involves only scalars made out of conformal supergravity fields[4]. In our case, the only nonvanishing components of the Riemann tensor are R_{+i+i} with $i = 1, 2$ and the only nonvanishing component of $\partial_\mu\Phi$ is $\partial_+\Phi$. Since there are no nonvanishing components with a contravariant $+$ index, we cannot form a scalar by contracting these tensors. Therefore the Weyl anomaly vanishes for our null background. This implies that correlation functions of dressed conformal operators are equal to those in a *flat* metric with the same x^+ dependent coupling,

$$\left\langle \prod_a e^{f(x_a^+)\Delta_a}\, O_a(x^a) \right\rangle_{e^f\eta_{\mu\nu},\Phi(x^+)} = \left\langle \prod_a O_a(x^a) \right\rangle_{\eta_{\mu\nu},\Phi(x^+)} \tag{2.15}$$

where Δ_a is the conformal dimension of the operator O_a. In other words, the shrinking conformal factor is invisible to these observables at the quantum level.

---

[4]We are grateful to A. Tseytlin for a correspondence about this point.

2.3. PBH transformations and a flat boundary. The fact that the conformal factor can be removed by a Weyl transformation should be reflected in the bulk theory. Indeed, it is well known that Weyl transformations of the boundary are equivalent to a set of coordinate transformations in the bulk [**21, 22**]. Any asymptotically AdS space-time may be written in a standard coordinate system of the Fefferman-Graham form

$$ds^2 = \frac{1}{\bar{\rho}^2} d\bar{\rho}^2 + \frac{1}{\bar{\rho}^2} \tilde{g}_{\mu\nu}(x, \bar{\rho}) dx^\mu dx^\nu \tag{2.16}$$

Now consider the coordinate transformations [**23, 24, 25**]

$$\bar{\rho} \to \bar{\rho}\, e^{-\sigma(x,\bar{\rho})}, \qquad x^\mu \to x^\mu + a^\mu(x, \bar{\rho}) \tag{2.17}$$

which keeps this form of metric invariant. For infinitesimal transformations this is ensured by requiring σ to be a function of x alone, and

$$\frac{1}{\bar{\rho}} \partial_{\bar{\rho}} a^\mu = -\tilde{g}^{\mu\nu} \partial_\nu \sigma \tag{2.18}$$

The transformation of the metric $\tilde{g}_{\mu\nu}$ is given by

$$\delta\tilde{g}_{\mu\nu}(x, \bar{\rho}) = 2\sigma(x, \bar{\rho}) \left(1 - \frac{1}{2}\bar{\rho}\partial_{\bar{\rho}}\right) \tilde{g}_{\mu\nu}(x, \bar{\rho}) + \nabla_{(\mu} a_{\nu)}(x, \bar{\rho}) \tag{2.19}$$

The expression (2.19) explicitly shows that this transformation includes a Weyl transformation of the metric $\tilde{g}_{\mu\nu}$.

Consider now a metric of the form

$$ds^2 = \frac{1}{z^2} \left[dz^2 + e^{f(x)} \eta_{\mu\nu} dx^\mu dx^\nu\right] \tag{2.20}$$

Our aim is to perform a PBH transformation to remove the conformal factor $e^{f(x)}$ in the boundary metric. However, we need to do this for finite PBH transformations.

When the conformal factor $f(x)$ is a function of a single null coordinate x^+, i.e. when the original metric is of the form

$$ds^2 = \frac{1}{z^2} \left[dz^2 + e^{f(x^+)}(-2dx^+dx^- + d\vec{x}^2)\right] \tag{2.21}$$

it turns out that it is easy to figure out the correct finite PBH transformations. These are given by the following:

$$\begin{aligned} z &= w e^{f(y^+)/2} \\ x^- &= y^- - \frac{1}{4} w^2 (\partial_+ f) \\ x^+ &= y^+ && (2.22) \\ \vec{x} &= \vec{y} && (2.23) \end{aligned}$$

These coordinate transformations have been independently found in [**20**].

In these coordinates the metric becomes

$$\begin{aligned} ds^2 &= \frac{1}{w^2}\left[dw^2 - 2dy^+dy^- + d\vec{y}^2 + \frac{1}{4}w^2[(f')^2 - 2f''](dy^+)^2\right] \\ &= \frac{1}{w^2}\left[dw^2 - 2dy^+dy^- + d\vec{y}^2 + \frac{1}{4}w^2(\Phi')^2(dy^+)^2\right] \end{aligned} \tag{2.24}$$

where in the second line we have used (2.9).

The new coordinates provide a new foliation of the space-time. The boundary $w = 0$ is naively the same as the original boundary $z = 0$. However, it is well known that AdS/CFT requires an infrared cutoff in the bulk which corresponds to an ultraviolet cutoff in the dual gauge theory. For any such finite cutoff ϵ, the boundary $w = \epsilon$ is not the same as $z = \epsilon$, and becomes flat in the $\epsilon \to 0$ limit. Consequently the dual Yang-Mills theory lives on a flat space with an x^+-dependent coupling.

Notice that, in these coordinates, there is only one function $\Phi(x^+)$ which we are free to choose. In particular, e^Φ can be chosen to be bounded and vanishing at $x^+ = 0$ in an *analytic* fashion. The proposed dual will then have a coupling which is bounded and vanishes at $x^+ = 0$ in a smooth fashion.

For such solutions, $\partial_+\Phi$ will, however, diverge at $x^+ = 0$. This means that the bulk spacetime will be as usual singular. This may be seen by looking at the behavior of geodesics as in [**1, 2**]. In the new coordinates, these geodesics are

$$\begin{aligned} w &= z_0 F(y^+) \\ y^- &= y_0^- - \frac{1}{4}z_0^2\frac{d}{dy^+}(F(y^+))^2 \end{aligned} \tag{2.25}$$

where $F(y^+) = e^{-f(y^+)/2}$. The affine parameter along such geodesics is given by

$$\lambda = \int^{y^+} \frac{1}{F(y^+)^2}\, dy^+ \tag{2.26}$$

It is easy to find functions $\Phi(x^+)$ so that the singularity at $y^+ = 0$ is reached in finite affine parameter. For such solutions $F(y^+)$ must diverge at $y^+ = 0$. The magnitude of the tidal acceleration between two such geodesics separated along a transverse direction is given by (see equation (2.10) of [**2**])

$$|a| = (F(y^+))^3 F''(y^+) \tag{2.27}$$

and would diverge as well. The form of the metric (2.24), however, shows that this singularity weakens as we approach the boundary $w = 0$, leaving a flat boundary metric.

2.4. Some properties of the Gauge theory. We are therefore left with a gauge theory on flat space with an x^+-dependent coupling. The coupling, however, appears as an overall factor in the gauge field term. Generally, this would imply that the propagator of canonically normalized fields would be unconventional. This could be a danger since the derivatives of $\Phi(x^+)$ diverge at $x^+ = 0$. Luckily this does not happen either. To see this, fix a light cone gauge [**26**]

$$A_- = 0. \tag{2.28}$$

The fields A_+ are then determined in terms of the transverse components by a constraint equation which turns out to be identical to that for the standard $N = 4$ theory *by virtue of the fact that the coupling depends only on* x^+,

$$\frac{1}{2}\partial_- A_+ = \partial_i A_i + \frac{i}{\partial_-}[A_i, \partial_- A_i] \tag{2.29}$$

Let us now define new fields $\bar{A}_i, \bar{A}_+$ as follows:

$$\bar{A}_i(x) = e^{-\Phi(x^+)/2} A_i(x), \qquad \bar{A}_+(x) = e^{-\Phi(x^+)/2} A_+(x) \tag{2.30}$$

Since Φ is a function of x^+ alone, equation (2.29) continues to hold with the replacement $A_i \to \bar{A}_i$, $A_+ \to A_+$. In terms of these new fields it may be easily checked that up to terms which are quadratic in the fields,

$$e^{-\Phi(x^+)}\mathrm{Tr}\, F^2 = \mathrm{Tr}\, \bar{F}^2 - \frac{1}{2}\partial_- \left[(\partial_+\Phi)\bar{A}^i \bar{A}_i\right] \tag{2.31}$$

where $\bar{F}$ is the field strength constructed out of $\bar{A}$. Since the additional term is a total derivative, it does not contribute to the action. *This means that the quadratic terms in the action are identical to those in the light-cone gauge action for standard* $N = 4$ *theory.* The factors of e^Φ and its derivatives appear only in the interaction terms of the $\bar{A}$ fields. Since the coupling $e^{\Phi/2}$ approaches zero at the singularity and is small and bounded everywhere else, one might expect that the correlation functions of the fields $\bar{A}_\mu$ are well behaved.

Generically, time-dependent backgrounds lead to particle production. An initial vacuum state typically evolves into a squeezed state of particle–antiparticle pairs. In our null background, however, such processes do not occur. The argument relies on the fact that in light-front quantization the states are labelled by k_-, k_1, k_2 where $0 \leq k_- \leq \infty$ and $-\infty \leq k_1, k_2 \leq \infty$. Since the background depends only on x^+, the momentum along x^-, k_-, is conserved. The Fock vacuum of the theory has $k_- = k_i = 0$. It is then clear that this state cannot evolve into a state containing particles with nonzero k_- since there are no states with negative k_-. The $k_- = 0$ sector, however, may cause problems with this argument.

2.5. The worldsheet theory. Since the effective 't Hooft coupling of the Yang-Mills theory becomes small at the singularity, it is natural to expect that stringy effects are large. At the same time, in the large N limit string loop effects should be small as well. It is of interest to investigate whether worldsheet string theory could make sense in this background. Unfortunately, because of the presence of RR flux, we do not have a tractable worldsheet formulation of the full worldsheet theory. However, a look at the bosonic part of the action in a physical gauge makes it clear that stringy effects are important near the singularity. For this purpose, let us write the ten-dimensional metric in slightly different coordinates,

$$ds^2 = \frac{1}{\vec{W}^2}\left[-2dy^+dy^- + d\vec{y}^2 + d\vec{W}^2 + \frac{1}{4}\vec{W}^2(dy^+)^2\right], \tag{2.32}$$

where the six-dimensional vector $\vec{W} = (w, \theta_i)$ and θ_i are the five angles on the S^5. Fixing the light cone gauge $y^+ = \tau$ following [**27**] the bosonic part of the action becomes

$$\begin{aligned} S = \frac{1}{2}\int d\sigma d\tau \Bigg[&(\partial_\tau \vec{y})^2 + (\partial_\tau \vec{W})^2 - \frac{1}{y^4}e^{\Phi(\tau)}[(\partial_\sigma \vec{y})^2 + (\partial_\sigma \vec{W})^2] \\ &- \frac{1}{4}\vec{W}^2(\partial_+\Phi)^2\Bigg], \end{aligned} \tag{2.33}$$

Since e^Φ vanishes at $\tau = 0$, the spatial gradient terms become small here, which implies that stringy modes with polarizations along $\vec{y}$ are not suppressed. However, because $\partial_+\Phi$ blows up here, the polarizations along the $\vec{W}$ direction are suppressed.

2.6. Penrose limits and matrix theory. To learn a little more about the string theory in the bulk it is useful to consider the Penrose limit of our background. For this purpose it is convenient to rewrite the metric as

$$ds^2 = r^2[-dt^2 + dq^2 + e^{F(z^+)}(dx_2^2 + dx_3^2)] + \frac{dr^2}{r^2} + d\psi^2 + \sin^2\psi\, d\Omega_4^2, \tag{2.34}$$

where we have used the affine parameter z^+ defined by $z^+ = \int^{x^+} dx\ e^{f(x)}$ along a null geodesic instead of x^+, and the function $F(z^+)$ is defined by $F(z^+) = f(x^+(z^+))$. The coordinates q, t are defined by

$$z^+ = \frac{1}{\sqrt{2}}(q+t), \qquad x^- = \frac{1}{\sqrt{2}}(t-q). \tag{2.35}$$

Now we zoom in on a null geodesic given by

$$r = \sin U, \qquad t = -\cot U, \qquad \psi = U, \tag{2.36}$$

After the usual scaling associated with a Penrose limit and a complicated coordinate transformation the Einstein frame metric is given by [**2**]

$$ds^2 = 2dUdV - [H(U)\vec{X}^2 + \vec{Y}^2](dU)^2 + d\vec{X}^2 + d\vec{Y}^2. \tag{2.37}$$

In the Penrose limit, the coordinate U is related to the coordinate z^+ by $z^+ - \frac{1}{\sqrt{2}} \cot U$ and the function $H(U)$ is determined in terms of $F(z^+)$ by

$$\begin{aligned} H(U) &= 1 - \frac{[1+2(z^+)^2]^2}{4}\left[\frac{d^2F}{(dz^+)^2} + \frac{1}{2}\left(\frac{dF}{dz^+}\right)^2\right] \\ &= 1 + \frac{[1+2(z^+)^2]^2}{8}\left(\frac{d\Phi}{dz^+}\right)^2, \end{aligned} \tag{2.38}$$

In terms of these coordinates, the singularity appears at $U = \pi/2$. Near this point,

$$H(U) \sim \frac{1}{(U-\frac{\pi}{2})^2}, \qquad e^{\Phi(U)} \sim \left(U - \frac{\pi}{2}\right)^{\frac{\sqrt{8}}{3}}. \tag{2.39}$$

Thus the Penrose limit of our original spacetime is singular as well. In fact, it turns out that the pp-wave is singular if and only if the pre-Penrose limit original spacetime is singular [**28**].

The pp-wave spacetime has spacelike and null isometries. In a way similar to the null dilaton cosmologies in the previous section, one may write down a matrix membrane theory for such a background which has a compact null direction $x^- \sim x^- + 2\pi R$ and $x^8 \sim x^8 + 2\pi R_B$. The resulting $(2+1)$-dimensional Yang-Mills action is

$$\begin{aligned} \mathcal{L} = \mathrm{Tr}\frac{1}{2}\Big\{&[(D_\tau\chi^\alpha)^2 - e^{\Phi(\tau)}(D_\sigma\chi^\alpha)^2 - e^{-\Phi(\tau)}(D_\rho\chi^\alpha)^2] \\ &+ \frac{1}{G_{YM}^2}[e^{\Phi(\tau)}F_{\sigma\tau}^2 + e^{-\Phi(\tau)}F_{\rho\tau}^2 - F_{\rho\sigma}^2] \\ &- H(\tau)[(\chi^1)^2 + (\chi^2)^2] - (\chi^3)^2\cdots(\chi^6)^2 - 4(\chi^7)^2 \\ &+ \frac{G_{YM}^2}{2}[\chi^\alpha,\chi^\beta]^2 + 2iG_{YM}\chi^7[\chi^5,\chi^6] + \frac{4}{G_{YM}}\chi^7F_{\sigma\rho}\Big\}, \end{aligned} \tag{2.40}$$

Unlike the matrix membrane in the linear null dilaton discussed in the previous section (i) the Yang-Mills coupling of this model is independent of τ, and (ii) both ∂_ρ and ∂_σ have time-dependent factors. In the IR, the fields in the theory become commuting and may be chosen to be diagonal and the extent of the ρ direction shrinks to zero size. The Lagrangian then reduces to the light-cone gauge Green-Schwarz worldsheet Lagrangian for the fundamental string in the relevant pp-wave background. An analysis similar to that in section 2.1 now shows that excited modes of both D-strings and fundamental strings are now produced by the time-dependent background.

It would be interesting to analyze worldsheet string theory in this time-dependent pp-wave. Backgrounds with similar singularities in the *string frame* metric have been studied earlier [**29, 28**]. In that case the worldsheet equations of motion are solvable in terms of special functions and certain statements about the validity of string theory could be made. In our background the worldsheet action is quadratic in the fields, but the equations of motion are not readily solvable. Our analysis of the matrix theory seems to indicate that nonperturbative physics becomes important. Nevertheless some insight from the worldsheet theory could be valuable.

3. Time dependent solution

In [**1**] we also obtained solutions with spacelike singularities. These backgrounds are not so well understood as the null backgrounds. However, it is of interest to examine some general features.

3.1. Kasner solutions. A particularly interesting class of solutions is of the Kasner form

$$\begin{aligned} ds^2 &= \frac{1}{z^2}\left[dz^2 - dt^2 + \sum_{i=1}^{3} t^{2p_i} dx^i dx^i\right], \qquad \sum_i p_i = 1 \\ e^{\Phi(t)} &= |t|^{\sqrt{2(1-\sum p_i^2)}} \end{aligned} \tag{3.1}$$

The string coupling – and therefore the Yang-Mills coupling – still goes to zero at the spacelike singularity at $t = 0$, but diverges at early or late times. As we shall see below it still turns out that the Weyl anomaly of the boundary theory in these coordinates vanishes. However, because of the divergence of the Yang-Mills coupling, it is unclear whether the gauge theory makes sense.

We will concentrate on solutions with $p_1 = p_2 = p_3 = \frac{1}{3}$. Redefining the time coordinate, this solution may be written in the form

$$\begin{aligned} ds^2 &= \frac{1}{z^2}\left[dz^2 + \frac{2t}{3}\left(-dt^2 + (dx^1)^2 + (dx^3)^2\right)\right] \\ e^{\Phi(t)} &= |t|^{\sqrt{3}} \end{aligned} \tag{3.2}$$

This solution has a spacelike singularity at $t = 0$.

Since the boundary metric on $z = 0$ is conformally flat, there should be a PBH transformation which leads to a foliation with a flat boundary. This is indeed true. The solution for $t > 0$ becomes

$$\begin{aligned} ds^2 = \frac{1}{\rho^2}\Bigg[d\rho^2 &- \frac{(16T^2 - 5\rho^2)^2}{256T^4} dT^2 \\ &+ \frac{(16T^2 - \rho^2)^{\frac{4}{3}}(16T^2 + 5\rho^2)^{\frac{2}{3}}}{256T^4}\left((dx^1)^2 + (dx^3)^2\right)\Bigg] \end{aligned} \tag{3.3}$$

where the new coordinates (ρ, T) are related to the coordinates (z, t) in the region $\rho < 4T$ by the transformations

$$z = \frac{32\rho T^{\frac{5}{2}}}{\sqrt{6}} \frac{1}{16T^2 - \rho^2}$$

$$t = T \left(\frac{16T^2 + 5\rho^2}{16T^2 - \rho^2} \right)^{\frac{2}{3}} \tag{3.4}$$

The dilaton may be written down in new coordinates by substituting (3.4) in (3.2).

It is clear that in this new foliation defined by slices of constant ρ, the boundary $\rho = 0$ has a flat metric. However, this coordinate system has a coordinate singularity at $\rho = 4T$, but may be extended beyond this point.

The arguments in the previous section then indicate that there is a dual field theory which lives in a flat spacetime, but with a time-dependent coupling which vanishes at $T = 0$. Unike the null solutions, the coupling *diverges* at early or late times – and we cannot make any careful argument about the behavior of this dual theory.

As noted in the introduction, these solutions have a curvature singularity at any finite time, though the singularity goes away at early and late times. The bulk Ricci scalar is given by

$$R_5 = -\left(\frac{9z^2}{4t^3} + 20 \right) \tag{3.5}$$

In the global geometry the Poincaré horizon is a product of a null plane times an S^2. This singularity appears at *one* point on this null plane. The rest of the Poincaré horizon is non-singular.

3.2. New class of time-dependent solutions. A necessary condition for a well defined dual theory is that the coupling should be bounded at all times. This motivates us to search for new solutions which have spacelike singularities of this type. We will present such solutions in this section.

These solutions are special cases of a class of time-dependent solutions whose boundary metrics are FRW universes. The Einstein frame metric[5] is given by

$$ds^2 = \frac{1}{z^2} \left[dz^2 + A(t) \left[-dt^2 + \frac{dr^2}{1 - k\, r^2} + r^2 \left(d\theta^2 + \sin\theta^2\, d\phi^2 \right) \right] \right] \tag{3.6}$$

with $k = 0, \pm 1$, and

$$\Phi(t) = \pm\sqrt{3} \int \frac{dt}{A(t)} \tag{3.7}$$

[5]These solutions can be derived from a generic ansatz with diagonal metric, and imposing that the dilaton is a function $\Phi(t)$ of time t alone.

where

$$A(t) = C_1 \sin(2\sqrt{k}\,t) + C_2 \cos(2\sqrt{k}\,t). \tag{3.8}$$

The solutions with $k=-1$ are particularly interesting. If we choose $A(t) = |\sinh(2t)|$, the dilaton becomes

$$e^{\Phi(t)} = g_s |\tanh\ t|^{\sqrt{3}} \tag{3.9}$$

so that the coupling is bounded and vanishes at $t = 0$. There is a spacelike singularity at $t = 0$. In the following we will restrict our attention to the "big crunch" part of the space-time, i.e. for $t < 0$. In this case we use $A(t) = |\sinh(2t)|$.

The boundary metric is in fact conformal to parts of Minkowski space. This is seen by defining new coordinates (for $t < 0$)

$$r = \frac{R}{\sqrt{\eta^2 - R^2}} \quad e^{-t} = \sqrt{\eta^2 - R^2} \tag{3.10}$$

The solution now becomes

$$\begin{aligned} ds^2 &= \frac{1}{z^2}\left[dz^2 + \left|1 - \frac{1}{(\eta^2 - R^2)^2}\right| [-d\eta^2 + dR^2 + R^2 d\Omega_2^2]\right] \\ e^{\Phi} &= \left|\frac{\eta^2 - R^2 - 1}{\eta^2 - R^2 + 1}\right|^{\sqrt{3}} \end{aligned} \tag{3.11}$$

The $t > 0$ part of the solution also becomes this metric after a coordinate transformation obtained by reversing the sign of t in (3.10). In these coordinates it is clear that as $t \to -\infty$, i.e. $\eta^2 - R^2 \to \infty$, the spacetime is AdS and e^{Φ} asymptotes to a constant.

The coordinate transformation (3.10) is valid in the region $\eta^2 - R^2 > 0$, and $\eta^2 - R^2 = 1$ are the two spacelike singularities. Even though we started with the form of the metric (3.6) we could extend the solution beyond this part of Minkowski space in the standard manner. In this extended solution, there are timelike singularities at $R^2 - \eta^2 = 1$. As is evident, the dilaton shows a singular behavior at the location of these singularities, even though the value of e^{Φ} goes to zero. In the following we will be interested in the solution in the regions $(\eta^2 - R^2) > 1$, i.e. the spacetime described by the metric (3.6).

Like the Kasner solutions, these solutions generically have curvature singularities at $z = \infty$. In the big crunch region, The bulk Ricci scalar is given by

$$R_5 = -\left(20 - 3\frac{z^2}{(\sinh(2t))^3}\right), \tag{3.12}$$

where t is as in (3.6) with $A(t) = -\sinh(2t)$ in this $t < 0$ region of the spacetime. The global nature of this singularity is similar to the Kasner type solution. In particular, at early times $t \to -\infty$ there is no such singularity.

We can, therefore, view these backgrounds in the same way as the null backgrounds. For $t < 0$ the space-time is pure $AdS_5 \times S^5$ in the infinite past. As time evolves one generates a *spacelike singularity* at $t = 0$ which extends to the boundary defined at $z = 0$. However, since the boundary metric is conformal to flat space, we can choose a different foliation by performing a PBH transformation and choose a boundary which is completely flat. (In this case, we have not been able to find the exact PBH transformations, but – as detailed in the Appendix – the PBH transformation may be found in an expansion around the boundary). The gauge theory defined on this latter boundary is on flat space with a time-dependent coupling constant which vanishes at the location of the bulk singularity. The source in the gauge theory evolves the initial vaccum state. On the supergravity side, a (timelike) singularity develops at $z = \infty$. While we do not have a clear idea of the meaning of this singularity in the gauge theory it is reasonable to presume – in view of the usual AdS/CFT duality – that this should manifest itself in the infrared behavior. Finally, as the time evolves, the gauge coupling goes to zero – this manifests itself as a spacelike singularity in the bulk in a region where supergravity itself breaks down.

The analysis of this dual gauge theory appears to be more complicated than the dual gauge theory for null backgrounds. One issue is related to the fact that the gauge theory Lagrangian has an overall factor of $e^{-\Phi}$. When Φ depends only on a null direction, it was shown in [**2**] that a choice of light-cone gauge, together with a field redefinition, converts the kinetic terms in the action into standard form for constant couplings. All factors of couplings then appear in the nonlinear terms as positive powers of $e^{\Phi(x^+)}$, which vanish at the location of the bulk singularity. This allowed us to arrive at some clean conclusions about the behavior of the gauge theory. In [**20**] a different gauge choice was used – this again made analysis of the gauge theory easier. For time-dependent backgrounds, we have not been able to find a gauge choice and a field redefinition which leads to a similar simplification. Nevertheless we expect that the theory is amenable to perturbative analysis near $t = 0$ where the gauge coupling becomes weak.

4. Issues

The toy models of cosmology described in this talk might provide an interesting way to resolve a null or spacelike singularity.

The situation is clean for null singularities. In the asymptotic past (in light-cone time) the gauge theory has a valid spacetime interpretation in terms of supergravity. As we approach the null or spacelike singularity, the 't Hooft coupling approaches zero, and the spacetime description breaks down. Our investigations *suggest* that the weakly coupled gauge theory remains controlled and it is this description which should be used to approach and even continue past the singularity. Our analysis is not detailed enough to decide whether such a continuation is indeed possible. Some of our conclusions were based on a treatment in the light-cone gauge and in light-front quantization. Sometimes light-front quantization leads to subtleties with the zero longitudinal momentum mode. These subtleties could give rise to infrared effects which have to be interpreted suitably.

We have similar solutions for spacelike singularities and possibly a similar scenario. However, the analysis of the gauge theory is a lot more involved and we cannot make any firm statement at this point.

Acknowledgements

I would like to thank Adel Awad, Jeremy Michelson, K. Narayan and Sandip Trivedi for a very enjoyable collaboration, Costas Bachas, Ben Craps, David Gross, A. Harindranath, David Kutasov, Gautam Mandal, Shiraz Minwalla, Tristan McLoughlin and Alfred Shapere for discussions at various stages of the research presented here. I am thankful to Amit Sever and Stephen Shenker for correspondence which clarified several issues and Arkady Tseytlin for a correspondence about conformal anomalies. I thank the organizers of the Sowers Worskhop at Virginia Tech for hosting an extremely stimulating meeting. The reserach reported here was supported in part by the United States National Science Foundation Grant Numbers PHY-0244811 and PHY-0555444 and Department of Energy contract No. DE-FG02-00ER45832.

References

[1] S. R. Das, J. Michelson, K. Narayan and S. P. Trivedi, Phys. Rev. D **74**, 026002 (2006) [arXiv:hep-th/0602107].

[2] S. R. Das, J. Michelson, K. Narayan and S. P. Trivedi, Phys. Rev. D **75**, 026002 (2007) [arXiv:hep-th/0610053].

[3] G. 't Hooft, arXiv:gr-qc/9310026; L. Susskind, J. Math. Phys. **36**, 6377 (1995) [arXiv:hep-th/9409089].

[4] For reviews and references see I. R. Klebanov, arXiv:hep-th/9108019; S. R. Das, arXiv:hep-th/9211085; E. J. Martinec, Class. Quant. Grav. **12**, 941 (1995) [arXiv: hep-th/9412074].

[5] T. Banks, W. Fischler, S. H. Shenker and L. Susskind, Phys. Rev. D **55**, 5112 (1997) [arXiv:hep-th/9610043]; L. Susskind, arXiv:hep-th/9704080; T. Banks and N. Seiberg, Nucl. Phys. B **497**, 41 (1997) [arXiv:hep-th/9702187]; L. Motl, arXiv:hep-th/9701025; R. Dijkgraaf, E. P. Verlinde and H. L. Verlinde, Nucl. Phys. B **500**, 43 (1997) [arXiv:hep-th/9703030].

[6] J. M. Maldacena, Adv. Theor. Math. Phys. **2**, 231 (1998) [Int. J. Theor. Phys. **38**, 1113 (1999)] [arXiv:hep-th/9711200]; E. Witten, Adv. Theor. Math. Phys. **2**, 253 (1998) [arXiv:hep-th/9802150]; S. S. Gubser, I. R. Klebanov and A. M. Polyakov, Phys. Lett. B **428**, 105 (1998) [arXiv:hep-th/9802109].

[7] S. R. Das, arXiv:hep-th/0503002.

[8] S. R. Das and L. H. Santos, Phys. Rev. D **75**, 126001 (2007) [arXiv:hep-th/0702145]; J. L. Karczmarek, arXiv:0711.2680 [hep-th].

[9] S. R. Das, arXiv:0706.4188 [hep-th].

[10] T. Hertog and G. T. Horowitz, JHEP **07**, 073 (2004) arXiv:. T. Hertog and G. T. Horowitz, "Holographic description of AdS cosmologies," JHEP **04**, 005 (2005) [arXiv:hep-th/0503071].
[11] C. S. Chu and P. M. Ho, JHEP **0604**, 013 (2006) [arXiv:hep-th/0602054].
[12] F. L. Lin and W. Y. Wen, JHEP **0605**, 013 (2006) [arXiv:hep-th/0602124].
[13] F. L. Lin and D. Tomino, JHEP **0703**, 118 (2007) [arXiv:hep-th/0611139].
[14] A. Awad, S. R. Das, K. Narayan and S. P. Trivedi, arXiv:0711.2994 [hep-th].
[15] P. Kraus, H. Ooguri and S. Shenker, Phys. Rev. D **67**, 124022 (2003) [arXiv:hep-th/0212277]; L. Fidkowski, V. Hubeny, M. Kleban and S. Shenker, JHEP **0402**, 014 (2004) [arXiv:hep-th/0306170].
[16] M. Berkooz, Z. Komargodski and D. Reichmann, arXiv:0706.0610 [hep-th].
[17] E. S. Fradkin and A. A. Tseytlin, Phys. Lett. B **134**, 187 (1984).
[18] H. Liu and A. A. Tseytlin, Nucl. Phys. B **533**, 88 (1998) [arXiv:hep-th/9804083].
[19] S. Nojiri and S. D. Odintsov, Phys. Lett. B **444**, 92 (1998) [arXiv:hep-th/9810008].
[20] C. S. Chu and P. M. Ho, arXiv:0710.2640 [hep-th].
[21] R. Penrose and W. Rindler, "*Spinors and Space-Time, vol. 2*" Cambridge University Press, UK. (1986) 501p
[22] J. D. Brown and M. Henneaux, Commun. Math. Phys. **104**, 207 (1986).
[23] C. Imbimbo, A. Schwimmer, S. Theisen and S. Yankielowicz, Class. Quant. Grav. **17**, 1129 (2000) [arXiv:hep-th/9910267].
[24] M. Henningson and K. Skenderis, JHEP **9807**, 023 (1998) [arXiv:hep-th/9806087]; S. de Haro, S. N. Solodukhin and K. Skenderis, Commun. Math. Phys. **217**, 595 (2001) [arXiv:hep-th/0002230]; K. Skenderis, Int. J. Mod. Phys. **A16**, 740 (2001) [arXiv: hep-th/0010138].
[25] M. Fukuma, S. Matsuura and T. Sakai, Prog. Theor. Phys. **109**, 489 (2003) [arXiv:hep-th/0212314].
[26] For an introduction, see A. Harindranath, arXiv:hep-ph/9612244.
[27] R. R. Metsaev and A. A. Tseytlin, Phys. Rev. D **63**, 046002 (2001) [arXiv: hep-th/0007036]; R. R. Metsaev, C. B. Thorn and A. A. Tseytlin, Nucl. Phys. B **596**, 151 (2001) [arXiv:hep-th/0009171]; J. Polchinski and L. Susskind, "String theory and the size of hadrons," [arXiv:hep-th/0112204].
[28] M. Blau, M. Borunda, M. O'Loughlin and G. Papadopoulos, "Penrose Limits and Spacetime Singularities," Class. Quant. Grav. **21**, L43 (2004) [arXiv:hep-th/0312029].
[29] G. Papadopoulos, J. G. Russo and A. A. Tseytlin, Class. Quant. Grav. **20**, 969 (2003) [arXiv:hep-th/0211289].

Department of Physics and Astronomy, University of Kentucky, Lexington, KY 40506, USA.
E-mail address: das@pa.uky.edu

AMS/IP Studies in Advanced Mathematics
Volume 44, 2008

Is SUSY Natural?

Keith R. Dienes, Michael Lennek, David Sénéchal, and Vaibhav Wasnik

ABSTRACT. Spacetime supersymmetry is widely believed to play an important role in most fundamental theories of physics, and is usually invoked in order to address problems of naturalness. In this talk, we examine the question of whether supersymmetry itself is "natural" (*i.e.*, likely to exist as a fundamental component of nature at high energy scales). Our approach to answering this question is based on a statistical examination of the heterotic string landscape, and our conclusion is that supersymmetry is an exceedingly rare phenomenon. We also find that the likelihood of supersymmetry appearing is very dependent on the gauge symmetries present at the string scale, with certain gauge groups strongly favoring the appearance of $\mathcal{N} = 1$ supersymmetry and others not. This talk summarizes results from several recent papers, yet also contains some new results. In particular, one new result is that the heterotic landscape appears to favor either the non-supersymmetric Standard Model or an $\mathcal{N} = 1$ SUSY GUT gauge group at the string scale; by contrast, both the MSSM or a non-supersymmetric GUT are disfavored.

1. Introduction

Most theoretical frameworks for physics beyond the Standard Model involve the introduction of supersymmetry (SUSY), and there are many reasons why this is so. First, supersymmetry solves the technical gauge hierarchy problem. Second, supersymmetry provides a dynamical mechanism for triggering electroweak symmetry breaking. Third, supersymmetry improves the accuracy of gauge coupling unification, and fourth, it provides a dark matter candidate. As a result, supersymmetry is truly ubiquitous in particle physics, appearing virtually everywhere – except the data.

However, through the years, lots of competing or alternative theories have been proposed. Some involve large extra dimensions and some involve

2000 *Mathematics Subject Classification.* Primary 81T60; Secondary 81V22.

Talk delivered by KRD at the Sowers Theoretical Physics Workshop (held at Virginia Tech University, May 2007); at the PLANCK 2007 Conference (held at the University of Warsaw, Poland, June 2007); and at the Eötvös-Cornell Summer Workshop on Particle Theory (held at Eötvös University, Budapest, Hungary, June 2007).

small extra dimensions. Others do not explicitly involve extra dimensions at all, yet contain new, strongly coupled sectors. Likewise, especially over the past decade, the substance of phenomenological model-building has changed dramatically. Indeed, it is now quite common that a talk introducing a new particle-physics scenario will begin with a litany of sequential assumptions that would have sounded increasingly fantastic to the ear of a physicist a mere decade ago. We are made of open strings. And we live on a brane. And the brane lives in extra dimensions. And the brane is wrapped and intersects other branes. And the extra dimensions are warped. And the warping is severe and forms a throat. And the brane is falling into a throat. And so forth. Indeed, such scenarios now often form the backbone of cutting-edge model-building.

Admittedly, all of this may sound highly unnatural, and it is excusable to yearn for the simpler days of the MSSM and their cousins, the SUSY GUTs. But is SUSY itself truly natural? What does it mean to be "natural," anyway?

There are many different notions of naturalness that have appeared in the literature. For example, Dirac naturalness stipulates that an effective field theory (EFT) is natural if the dimensionless coefficients for all operators are $\sim \mathcal{O}(1)$ – *i.e.*, no exceedingly small or large numbers are allowed. In this sense, the large electroweak gauge hierarchy is unnatural, which is one of the biggest motivations for supersymmetry. Another notion is 't Hooft naturalness: even if such a number is small, it can be viewed as "natural" if its smallness is protected by a nearly unbroken symmetry.

But neither of these addresses the question as to whether a theory, even if "natural" in the above sense, is *likely* to be right. How *likely* is SUSY to be the correct theory?

The word *likely* often causes us to shudder. Indeed, even though we constantly judge theories this way, we don't say this word aloud because the question of theoretical likelihoods seems more philosophical than scientific, especially when we have no data upon which to base our assessments. How likely relative to *what*? To all other theories that one can imagine? And who is doing the imagining? One might get very different responses depending on the identity of the unlucky proponent. Ultimately, we seem to be faced with a dead-end question. *How can one compare the likelihood of one theory against another?*

String theory provides a framework in which this question can be addressed in a mathematical way. This is because string theory provides a large set of possible vacuum solutions ("vacua", or string "models", collectively called "the landscape"), each of which corresponds to a different alternative universe with different physical laws. In this context, we can then place our likelihood question on firmer footing. *In the landscape of possible string solutions, how many of these solutions are supersymmetric? Is SUSY "natural" on this landscape, or relatively rare?*

This is the subject of this talk.

2. Our Study: the heterotic landscape

In order to approach this question in a somewhat tractable way, we shall focus on the landscape associated with perturbative four-dimensional heterotic strings which are realizable using free-field constructions (such as those based on free worldsheet bosons or fermions).

In a recent paper [**1**], we conducted a random exploration of the landscape associated with such models, using the free-fermionic construction method [**2**]. In this construction, each model can be described in terms of its left- and right-moving worldsheet conformal field theories. These conformal field theories consist of tensor products of non-interacting, free, complex fermionic fields. Different models are then achieved in this method by varying the boundary conditions of these worldsheet fermionic fields around the two non-contractible loops of the worldsheet torus. The sets of allowed boundary conditions for the fermionic fields are restricted by numerous string self-consistency conditions which also must be applied. These self-consistency conditions guarantee that the string partition function can be viewed as the trace over a Fock space corresponding to a self-consistent string model. Other perturbative self-consistency constraints (such as conformal and modular invariance) are also imposed.

This study was similar to earlier studies performed in Refs. [**3, 4**], and has some important limitations. For example, this study examined string models for which all real worldsheet fermions could be grouped together to form complex worldsheet fermions. This effectively reduces the number of consistent sets of fermion worldsheet boundary conditions, but allows for the utilization of many time-saving computational algorithms. The major phenomenological consequence of this restriction is that these string models have gauge groups with a fixed rank (namely twenty-two in four dimensions).

In our study, two models were considered distinct if their spacetime phenomenologies differ in some way. For example, two models were considered distinct if they differ in their amounts of unbroken supersymmetry, or in their gauge group or massless particle spectrum. Moreover, after a statistically significant number of such string models was surveyed, a novel technique [**5**] was utilized in order to extract the stable statistical results which do not vary as a function of sample size. Such results are therefore likely to be indicative of the corresponding landscape associated with such models as a whole. The techniques by which such results are extracted are discussed in Ref. [**5**].

3. Results

In this section, we shall describe the results of our study. As indicated above, the statistical methodology utilized in extracting these results is discussed in Ref. [**5**], and a full discussion of the results of Sects. 3.1 and 3.2 can be found in Ref. [**1**].

3.1. Prevalence of SUSY on the heterotic landscape. One of the first questions addressed in our study concerned the prevalence of spacetime supersymmetry in the heterotic landscape. In other words, what percentage of the heterotic string landscape consists of string models exhibiting an unbroken spacetime supersymmetry at the string scale? Clearly, answering this question is the first step towards addressing the overall issue of the naturalness of supersymmetry.

The results that are found [**1**] are reproduced below. Note that in quoting these results, we are explicitly focusing on only that portion of the heterotic string landscape which is stable at tree level. In other words, we are explicitly disregarding those non-supersymmetric portions of the landscape (which amount to approximately 32.1% of the total) for which tachyonic states are present at tree level.

This table represents our final partitioning of the tree-level four-dimensional heterotic landscape according to its degree of supersymmetry. There are several rather striking facts which are evident from these results.

- First, we see that more than half of the stable heterotic landscape is non-supersymmetric and yet tachyon-free. Indeed, this proportion remains near half of the total even when the non-supersymmetric tachyonic models are included.
- Second, we see that the supersymmetric portion of the heterotic landscape appears to account for less than one-quarter of the full four-dimensional heterotic landscape.
- Finally, models exhibiting extended ($\mathcal{N} \geq 2$) supersymmetries are exceedingly rare, together representing less than one percent of the full landscape.

Of course, we stress once again that these results hold only for the *tree-level* landscape, *i.e.*, models which are stable at tree level only. *A priori*, it is not clear whether these results would persist after full moduli stabilization. However, it seems likely that they would, given that most modern methods of moduli stabilization (fluxes, superpotentials, *etc.*) tend to further

TABLE 1. Classification of the four-dimensional heterotic landscape as a function of the number of unbroken spacetime supersymmetries. We are explicitly focusing on string models which are stable (and thus tachyon-free) at tree level.

SUSY class	% of heterotic landscape
$\mathcal{N}=0$ (tachyon-free)	68.48
$\mathcal{N}=1$	30.78
$\mathcal{N}=2$	0.74
$\mathcal{N}=4$	0.0044

break (rather than restore) spacetime supersymmetry. Indeed, under these assumptions, these results then lead to a number of interesting conclusions.

The first conclusion is that the properties of the tachyon-free heterotic landscape as a whole are statistically dominated by the properties of string models which do *not* have spacetime supersymmetry. Indeed, the $\mathcal{N}=0$ string models account for over three-quarters of this portion of the heterotic string landscape. The fact that the $\mathcal{N}=0$ string models dominate the tachyon-free portion of the landscape suggests that breaking supersymmetry without introducing tachyons is actually *favored* over preserving supersymmetry for this portion of the landscape. Indeed, we expect this result to hold even after full moduli stabilization (as has been argued within the context of Type I strings [**6**]), unless an unbroken supersymmetry is somehow restored by stabilization.

The second conclusion which can be drawn from these results is that the supersymmetric portion of the landscape is almost completely comprised of $\mathcal{N}=1$ string models. Indeed, only 2% of the supersymmetric portion of the heterotic landscape has more than $\mathcal{N}=1$ supersymmetry. This suggests that the correlations present for the supersymmetric portion of the landscape can be interpreted as the statistical correlations within the $\mathcal{N}=1$ string models, with the $\mathcal{N}=2$ correlations representing a correction at the level of 2% and the $\mathcal{N}=4$ correlations representing a nearly negligible correction.

In fact, the SUSY fraction of the full string landscape may be even smaller than quoted here. One reason is that free-field string constructions (such as we are employing here) probably tend to artificially favor models with unbroken supersymmetry. Second, even when stabilized string models nevertheless exhibit spacetime SUSY at the string scale, it is statistically unlikely that this SUSY will survive all the way down to the electoweak scale.

Thus, we conclude that weak-scale SUSY is rather *unnatural* from a string landscape perspective. On the one hand, this result shifts the burden of proof onto the SUSY enthusiasts, which represents a dramatic reframing of the underlying question of whether SUSY should exist at or above electroweak scale. But the conclusion that weak-scale SUSY is unnatural should not necessarily be viewed as a problem for string phenomenology. In fact, this result might even be considered good news: it implies that we will actually learn something about string theory and its preferred compactifications if/when weak-scale supersymmetry is actually discovered in upcoming collider experiments!

3.2. Correlations between SUSY and gauge groups. We shall now examine the effects of supersymmetry on the probability of realizing different gauge group factors in this landscape. What percentage of string models with a given level of supersymmetry will contain a given gauge group factor amongst its unbroken gauge group at the string scale? We shall also

TABLE 2. The percentage of heterotic string models exhibiting specific gauge group factors as functions of their spacetime supersymmetry. Here $SU_{>5}$ and $SO_{>10}$ collectively indicate gauge groups $SU(n)$ and $SO(2n)$ for any $n > 5$, while $\mathcal{N}$ refers to the number of unbroken supersymmetries at the string scale. Note that the $\mathcal{N}=0$ models are all tachyon-free. The final column of this table is derived from the earlier columns using the landscape weightings in Table 1.

gauge group	$\mathcal{N}=0$	$\mathcal{N}=1$	$\mathcal{N}=2$	$\mathcal{N}=4$	full landscape
U_1	99.9	94.5	68.4	89.6	98.0
SU_2	62.46	97.4	64.3	60.9	73.2
SU_3	99.3	98.0	93.0	45.1	98.9
SU_4	14.46	30.0	39.0	53.5	19.4
SU_5	16.78	43.5	66.3	33.8	25.4
$SU_{>5}$	0.185	1.7	10.6	73.0	0.73
SO_8	0.482	1.6	6.2	21.1	0.87
SO_{10}	0.084	0.2	1.6	18.7	0.13
$SO_{>10}$	0.005	0.038	0.77	7.5	0.021
$E_{6,7,8}$	0.0003	0.03	0.16	11.5	0.011

be interested in knowing the likelihood of realizing different gauge group factors for the full landscape as a whole.

The results [**1**] are presented in Table 2. As can clearly be seen, supersymmetry has a profound effect upon the prevalence of different gauge group factors in this landscape.

As we see from Table 2, supersymmetry can have quite sizable effects upon the probability of realizing specific groups. However, there are some general trends that hold for the full heterotic landscape. These trends include:

- A preference for $SU(n+1)$ over $SO(2n)$ groups for each rank n. Even though these two groups have the same rank, it seems that SU groups are more common than the SO groups for all levels of supersymmetry.
- Groups with smaller rank are much more common than groups with larger rank. Once again, this also appears to hold for all levels of supersymmetry.
- Finally, the gauge-group factors comprising Standard Model gauge group $G_{\rm SM} \equiv SU(3)\times SU(2)\times U(1)$ are particularly common, much more so than those of any of its grand-unified extensions.

3.3. SUSY naturalness. At this point, we have presented both the unrestricted probability of finding different levels of supersymmetry on the heterotic landscape, and the restricted probability of finding different gauge group factors when a certain degree of supersymmetry is assumed. However, a more useful quantity might be the "inverse" of this last probability, namely the probability of finding different levels of supersymmetry given a specific gauge group factor. This would give an indication of how "natural" each level of supersymmetry is for different gauge group factors.

In order to derive these probabilities, we can utilize the results presented above. First, let us recall some basic elements of probability theory. The results in the first four columns of Table 2 are necessarily conditional probabilities: each number in these columns represents the probability of finding a specific gauge group factor *given* a certain level of supersymmetry. Thus, if A represents the occurrence of a specific gauge group factor and B represents the occurrence of a specific level of supersymmetry, then the results presented in Table 2 are all of the form

$$P(A \mid B) \equiv \frac{P(A \cap B)}{P(B)}. \tag{3.1}$$

Of course, what we now seek is not $P(A \mid B)$, but the "inverse" $P(B \mid A)$. In general, the relationship between $P(B \mid A)$ and $P(A \mid B)$ is given as

$$P(B \mid A) = \frac{P(B)}{P(A)} P(A \mid B). \tag{3.2}$$

Fortunately, all of the probabilities needed within the right side of Eq. (3.2) are present in Tables 1 and 2. Specifically, $P(A)$ is given within the final column of Table 2, while $P(B)$ is given in Table 1 and $P(A \mid B)$ is given in the rest of Table 2. Thus, given the presence of a specific gauge group factor at the string scale, we can now determine the corresponding probability of finding different levels of unbroken supersymmetry at the string scale. Our results are presented in Table 3.

By comparing these probabilities to the probabilities given in Table 1, it is possible to determine which gauge group factors tend to favor different levels of supersymmetry *beyond their expected representations on the landscape as a whole*. For example, we see from Table 1 that 68.48% of the stable heterotic landscape is non-supersymmetric. Thus, if a given gauge group factor is associated with models of which fewer than 68.48% are non-supersymmetric, then this gauge group factor can be said to preferentially *favor* supersymmetry.

These results also have a number of dramatic implications:

- First, we observe that gauge group factors with large rank (greater than four) actually favor the appearance of unbroken supersymmetry.
- Second, we observe that the gauge group factors which comprise the Standard Model gauge group do *not* generally favor supersymmetry.
- Finally, we see that the $SU(n)$ gauge group factors (with $n > 5$) and the exceptional gauge groups $E_{6,7,8}$ overwhelmingly favor $\mathcal{N} = 1$ supersymmetry.

TABLE 3. The percentage of string models with different levels of supersymmetry as a function of different gauge group factors. Thus, if we know that a given string model gives rise to a specific gauge group factor at the string scale, this table lists the corresponding probabilities that this model will have various levels of unbroken supersymmetry. This table can therefore be viewed as the "inverse" of Table 2.

SUSY	U_1	SU_2	SU_3	SU_4	SU_5	$SU_{>5}$	SO_8	SO_{10}	$SO_{>10}$	$E_{6,7,8}$
$\mathcal{N}=0$	69.80	58.41	68.79	50.98	45.29	17.33	37.98	43.68	16.21	1.85
$\mathcal{N}=1$	29.68	40.94	30.51	47.53	52.78	71.56	56.66	46.75	55.38	83.00
$\mathcal{N}=2$	0.51	0.65	0.69	1.48	1.92	10.65	5.25	8.95	26.84	10.59
$\mathcal{N}=4$	0.004	0.002	0.002	0.012	0.006	0.44	0.11	0.63	1.57	4.57

We thus conclude that the heterotic string landscape appears to favor either the non-supersymmetric Standard Model gauge group or an $\mathcal{N}=1$ SUSY GUT gauge group at the string scale. However, the opposite outcomes (namely the MSSM or a non-SUSY GUT gauge group) are significantly disfavored.

One important caveat is that the gauge group factors presented in Table 3 do *not* generally specify the gauge group fully. Indeed, these gauge group factors could be part of either a hidden sector or the visible sector. However, the gauge group factors listed in these tables are necessarily among those which are explicitly present in the full gauge group of the string model at the string scale.

4. Discussion

In this talk, we have presented results concerning the prevalence of spacetime supersymmetry at the string scale and its possible statistical correlations with the unbroken gauge group which might also appear at the string scale. Since these two quantities (spacetime supersymmetry and unbroken gauge group) are completely independent in ordinary quantum field theory based on point particles, these sorts of correlations represent true predictions of string theory and thereby provide one possible route towards answering the question as to whether supersymmetry is truly "natural" as a component of Beyond-the-Standard-Model physics.

As we have seen, the results of this calculation show that spacetime supersymmetry is *not* generically a feature of the low-energy limit of string theory. However, spacetime supersymmetry is actually statistically favored for certain gauge group factors.

There are some inherent limitations to these results which must continually be borne in mind. The first major problem with these string models is that they are actually unstable: the non-supersymmetric models generically have non-zero dilaton tadpoles, and the supersymmetric string models have flat directions. Thus, one might think that these results might change if only fully-stabilized models are considered. Unfortunately, no such perturbative heterotic models have been constructed. One could even argue that if the non-supersymmetric string models were required to be as "stable" as the supersymmetric string models (*e.g.* only have some finite number of flat directions), then these results would also change. However, at this point in time, the string models considered in this study are state-of-the-art and are as stable as many of the other classes of string models which have been considered in other statistical studies. Another issue facing this study concerns the extent to which these sorts of statistical correlations can be trusted, given that the full heterotic landscape has not been surveyed. However, this issue has been discussed in Ref. [**5**], and the methods developed there have been used in order to extract each of the results quoted here. Thus, to the best of our knowledge, the results quoted here are independent of the size of our sample of heterotic string models, and thus should persist even as additional models are examined. Finally, one could argue that the construction method utilized in this study necessarily only probes certain sectors of the heterotic landscape. Nevertheless, these sectors are the ones most likely to realize non-abelian gauge symmetries, and are thus the most likely to have non-trivial low energy phenomenologies.

The interpretation of these results is also open to some debate. The probability-based definiton of naturalness used in this paper is far from universally accepted, and may only have relevance for a landscape study such as the one we are performing. However, this definition of naturalness has the advantage of being applicable in a wide variety of contexts, and does not resort to any aesthetic or theoretical prejudices concerning the parameters that appear in effective Lagrangians. As such, probability-based definitions of naturalness may have inherent advantages over other definitions.

There are several extensions to these results which are currently under investigation. For example, we would like to understand how the presence of supersymmetry affects the statistical appearance of the entire composite Standard Model gauge group $G_{\rm SM} \equiv SU(3) \times SU(2) \times U(1)$, and not merely the appearance of its individual factors. We would also like to understand how the presence or absence of supersymmetry affects other features which are equally important for the overall architecture of the Standard Model: these include the appearance of three chiral generations of quarks and leptons, along with a potentially correct set of gauge couplings and Yukawa couplings. This work will be reported elsewhere [**7**].

Acknowledgments

The work of KRD, ML, and VW was supported in part by the U.S. Department of Energy under Grant DE-FG02-04ER-41298, by the U.S. National Science Foundation under Grant PHY/0301998, and by a Research Innovation Award from Research Corporation. Since September 2007, the work of ML has been supported through the Ecole Polytechnique. We are happy to thank M. Trapletti for discussions, and P. Fox and S. Raby for suggestions leading to the new results reported in Sect. 3.3.

References

[1] K. R. Dienes, M. Lennek, D. Sénéchal and V. Wasnik, "Supersymmetry versus Gauge Symmetry on the Heterotic Landscape," Phys. Rev. D **75**, 126005 (2007) [arXiv:0704.1320 [hep-th]].

[2] H. Kawai, D. C. Lewellen and S. H. H. Tye, "Construction Of Fermionic String Models In Four-Dimensions," Nucl. Phys. B **288**, 1 (1987); I. Antoniadis, C. P. Bachas and C. Kounnas, "Four-Dimensional Superstrings," Nucl. Phys. B **289**, 87 (1987); H. Kawai, D. C. Lewellen, J. A. Schwartz and S. H. H. Tye, "The Spin Structure Construction Of String Models And Multiloop Modular Invariance," Nucl. Phys. B **299**, 431 (1988).

[3] D. Sénéchal, "Search For Four-Dimensional String Models. 1," Phys. Rev. D **39**, 3717 (1989).

[4] K. R. Dienes, "Statistics on the heterotic landscape: Gauge groups and cosmological constants of four-dimensional heterotic strings," Phys. Rev. D **73**, 106010 (2006) [arXiv:hep-th/0602286].

[5] K. R. Dienes and M. Lennek, "Fighting the floating correlations: Expectations and complications in extracting statistical correlations from the string theory landscape," Phys. Rev. D **75**, 026008 (2007) [arXiv:hep-th/0610319].

[6] For a review, see: E. Silverstein, "TASI / PiTP / ISS lectures on moduli and microphysics," arXiv:hep-th/0405068. See also: A. Saltman and E. Silverstein, "A new handle on de Sitter compactifications," JHEP **0601**, 139 (2006) [arXiv:hep-th/0411271].

[7] K. R. Dienes, M. Lennek, D. Sénéchal and V. Wasnik, to appear.

Department of Physics, University of Arizona, Tucson, AZ 85721 USA
E-mail address: `dienes@physics.arizona.edu`

Department of Physics, University of Arizona, Tucson, AZ 85721 USA; Centre de Physique Théorique, Ecole Polytechnique, F-91128 Palaiseau Cedex, France
E-mail address: `mlennek@cpht.polytechnique.fr`

Département de Physique, Université de Sherbrooke, Sherbrooke, Québec J1K 2R1 Canada
E-mail address: `david.senechal@usherbrooke.ca`

Department of Physics, University of Arizona, Tucson, AZ 85721 USA
E-mail address: `wasnik@physics.arizona.edu`

AMS/IP Studies in Advanced Mathematics
Volume 44, 2008

Brane Induced Gravity: Codimension-2

Nemanja Kaloper

ABSTRACT. We review the results of arXiv:hep-th/0703190, on brane induced gravity (BIG) in $6D$. Among a large diversity of regulated codimension-2 branes, we find that for near-critical tensions branes live inside very deep throats which efficiently compactify the angular dimension. In there, $4D$ gravity first changes to $5D$, and only later to $6D$. The crossover from $4D$ to $5D$ is independent of the tension, but the crossover from $5D$ to $6D$ is not. This shows how the vacuum energy problem manifests in BIG: instead of tuning vacuum energy to adjust the $4D$ curvature, generically one must tune it to get the desired crossover scales and the hierarchy between the scales governing the $4D \to 5D \to 6D$ transitions. In the near-critical limit, linearized perturbation theory remains under control below the crossover scale, and we find that linearized gravity around the vacuum looks like a scalar-tensor theory.

Based on the talks given at the "Sowers Workshop", Virginia Tech, May 14–18, 2007, "Cosmology and Strings" workshop at ICTP, Trieste, Italy, July 9–13, 2007, "Dark Energy In the Universe", Hakone, Japan, Sep 1–4, 2007 and "Zagreb Workshop 2007", Zagreb, Croatia, Nov 9–11, 2007.

1. Introduction

The cosmological constant problem is the deepest problem of fundamental physics. To date, the attempts to explain a small cosmological constant in effective field theory coupled to General Relativity using dynamical adjustment have not yielded an answer [**1**]. The perspective may change if our universe is a brane in extra dimensions [**2**]–[**4**]. In such setups one may divert the brane vacuum energy into the extrinsic curvature [**4**], so that the vacuum energy could remain invisible to long distance $4D$ gravity, yielding a low curvature $4D$ geometry. However, while recovering exact $4D$ General Relativity at large distances, by using additional branes in the bulk one rediscovers a fine-tuning similar to the standard $4D$ one [**4, 5**]. A program of [**6**] has been initiated to seek the means to relax these fine-tunings in codimension-2 brane setups.

2000 *Mathematics Subject Classification.* Primary 83E15.

One may try to alleviate the problem by only approximating $4D$ General Relativity at large distances, as in the brane induced gravity (BIG *a.k.a.* DGP) [**7**]. There the extra-dimensional space may have an infinite volume, so that the $4D$ graviton completely decouples. Thus $4D$ gravity must emerge from the exchange of the continuum of bulk modes. This may occur with the help of the induced curvature terms on the brane [**7**], which control the momentum transfer due to the scattering of virtual bulk gravitons and may yield a force which scales as $1/r^2$, simulating $4D$ behavior [**7**]. The exact gravitational shock waves of [**8**] confirm this nicely.

These mechanisms may operate on higher-codimension defects [**7, 9**]. In [**10**], with D. Kiley we explored in a very detailed fashion the BIG mechanism on general codimension-2 branes floating in $6D$. We presented exact background solutions with flat $4D$ geometries for nonvanishing tensions, and regulating the short distance properties by considering hollow cylindrical branes we calculated the various regimes of gravity. The most interesting case, where the theory displayed a remarkably good perturbative behavior, was for near-critical tensions, where the deficit angle around the brane was almost 2π. In this limit the bulk compactifies to a very thin sliver [**11**], inside which bulk gravity is $5D$. Because of this, gravity along the brane changes dimensions with distance according to $4D \to 5D \to 6D$. The crossover scale where gravity stops being $4D$ is independent of the tension, and can be quickly obtained by using the naive crossover formula on a codimension-1 brane, realized as a wrapped 4-brane in a $6D$ flat space with a compact circle: by Gauss formulas for Planck masses $M_{4\,eff}^2 = M_5^3 r_0$ and $M_{5\,eff}^3 = M_6^4 r_0$, we find exactly $r_c \sim \frac{M_{4\,eff}^2}{M_{5\,eff}^3} \sim \frac{M_5^3}{M_6^4}$, which is our crossover formula from the exact shock waves, and which is the same as the see-saw scale $r_c \sim \frac{M_4^2}{M_6^4 r_0}$ of [**9**]. Thus we see precisely *how* the see-saw mechanism emerges. However, BIG is not free of the vacuum energy problem: although the brane is flat, when we fix M_4, we must finely tune the tension of the brane to get the bulk sliver geometry, that yields the separation of the $4D$ and $6D$ regimes by a $5D$ one.

From the shock waves we have seen that the theory contains $4D$ General Relativity. We unveil the rest in the linearized perturbation theory, which shows that there are scalar modes with gravitational couplings to matter. In the near-critical limit linearized perturbation theory around the vacuum remains under control in the $4D$ regime below the crossover scale. While the radion is suppressed, the helicity-0 mode mediates an extra long range force around the $4D$ vacuum backgrounds, and where the linearized theory is $4D$ it is a Brans-Dicke gravity with $\omega = 0$. This would disagree with the classic tests of General Relativity [**12**], but perhaps non-linearities or dynamics on vacua of more complex structure could come to the rescue. As the tension decreases towards sub-critical values the scalars start to couple strongly, and the perturbativity of the $4D$ regime is lost exactly at the $4D$ crossover scale

r_c, which is the Vainshtein scale [**13**] of the vacuum itself. This happens since the scalar gravitons probe the regulating axion flux, which sources brane bending. On the other hand, we do not find any instabilities in the leading order of perturbation theory.

In this note, we will review the near-critical limits for the sake of clarity. The detailed analysis of the general sub-critical setups with flat $4D$ geometries can be found in [**10**]. We should note that very recently in [**14**] similar features of BIG theory in $6D$ environments were argued to emerge in a framework with more branes.

2. Regulated near-critical branes

We model a 3-brane in $6D$ (codimension-2) by a 4-brane wrapped as a cylinder [**15, 16**], in order to regulate the short distance singularities which will appear even in the classical theory. When the 4-brane has nonzero tension, with the vacuum action $S_{\text{vacuum}} = -\int d^5x\sqrt{g_5}\lambda_5$, its stress-energy tensor will be $T^A_B = -\lambda_5\delta(\rho - r_0)\,\mathrm{diag}(1,1,1,1,1,0)$, where we wind it around the circle of radius r_0, with ρ along the normal to the brane. Thus, to wrap the 4-brane into a cylinder, we must cancel the pressure $\propto \lambda_5$ in the compact direction. A simple way to do it is to put an axion-like field Σ on the 4-brane, with a vacuum action $S_{\text{vacuum}} = -\int d^5x\sqrt{g_5}\big(\lambda_5 + \frac{1}{2}g^{ab}\partial_a\Sigma\partial_b\Sigma\big)$. Taking the fourth coordinate to be the angle on the compact circle ϕ and substituting the Scherk-Schwarz ansatz $\Sigma = q\phi$ we choose q to precisely cancel T^ϕ_ϕ. This requires

$$\lambda_5 = \frac{1}{2}q^2 g^{\phi\phi}\,, \tag{2.1}$$

where $g^{\phi\phi} = \frac{1}{r_0^2}$ is the inverse radius squared of the compact dimensions. Then the remaining components of the 4-brane stress-energy become precisely $T^\mu_\nu = -2\lambda_5\delta^\mu_\nu$. Thus the brane source now reads

$$T^A_B = -2\lambda_5\delta(\rho - r_0)\,\mathrm{diag}(1,1,1,1,0,0),$$

and the tensor structure is precisely the same as in the stress-energy tensor of a thin 3-brane. By inspection, we see that if we shift the argument of the remaining radial δ-function to $\rho - r_0$, it becomes $\delta^{(2)}_{\text{thick}}(\vec{y}) = \frac{1}{2\pi r_0}\delta(\rho - r_0)$, and so the effective $4D$ tension is

$$\lambda = 4\pi r_0\lambda_5, \tag{2.2}$$

relating the $5D$ and the effective $4D$ vacuum energies. The effective $4D$ vacuum energy λ contains the contributions from the classical axion field Σ, doubling up its value, because of the cancellation condition of Eq. (2.1).

The field equations for a 4-brane in a $6D$ bulk which include brane localized gravity terms are [**7, 9**], in Gaussian-normal gauge, and with the

4-brane residing at $\rho = r_0$,

$$M_6^4 G_{6\,B}^{\,A} + M_5^3 G_{5\,b}^{\,a} \delta_a^A \delta_B^b \delta(\rho - r_0) = T_b^a \delta_a^A \delta_B^b \delta(\rho - r_0)\,, \tag{2.3}$$

which we apply to our cylindrical brane vacuum, imposing that the stress-energy tensor is covariantly conserved, which requires $\partial^a \partial_a \Sigma = 0$. It is easy to check that by axial symmetry the ansatz $\Sigma = q\phi$ trivially solves the latter equation. Then, the tensor structure of the source,

$$T_B^A = -2\lambda_5 \delta(\rho - r_0)\,\mathrm{diag}(1,1,1,1,0,0),$$

guarantees that the flat $4D$ metric is a solution if the tension is again off-loaded into the bulk, like in the thin 3-brane case [**17**]. Tracing the field equations (2.3), and using (2.1) we find that the condition for this is that the metric in the remaining two dimensions, coordinatized by the angular direction along the brane and the radial distance away from it, has curvature

$$R_2 = \frac{4\lambda_5}{M_6^4} \delta(\rho - r_0)\,. \tag{2.4}$$

It is straightforward to find the metric [**10**], which is

$$ds_6^2 = \eta_{\mu\nu} dx^\mu dx^\nu + d\rho^2 + \Big[(1 - b\Theta(\rho - r_0))\rho + b r_0 \Theta(\rho - r_0)\Big]^2 d\phi^2\,, \tag{2.5}$$

where $b = \frac{2\lambda_5 r_0}{M_6^4}$. For $\rho < r_0$ the $2D$ part of the metric is $ds_2^2 = d\rho^2 + \rho^2 d\phi^2$, i.e. a flat disk, while for $\rho > r_0$ it becomes $ds_2^2 = d\rho^2 + \big((1-b)\rho + b r_0\big)^2 d\phi^2 = d\rho^2 + (1-b)^2\big(\rho + \frac{b}{1-b} r_0\big)^2 d\phi^2$, precisely the metric on a cone, which we see by comparing it to the thin 3-brane metric from the previous section. Thus the combined solution represents a truncated cone, with a flat mesa of radius r_0 on top, as depicted in Fig. (1). The geometry of the brane is $4D$ Minkowski $\times$ circle.

Looking at Eq. (2.5) for $\rho > r_0$ we see that the critical string tension, where the deficit angle becomes 2π, is again given by $b = 1$, which now corresponds to the tension $\lambda_{5\,cr} = \frac{M_6^4}{2r_0}$. Comparing to the relation between the $5D$ tension λ_5 and the effective $4D$ tension λ given by Eq. (2.2), we find that the critical value of the effective $4D$ tension is, not surprisingly, $\lambda_{cr} = 2\pi M_6^4$. Now, when $b = 1 - \epsilon$, it's easy to see that the bulk cone looks like a sliver. In this limit, the exterior bulk metric is approximately $ds_6{}^2 = \eta_{\mu\nu} dx^\mu dx^\nu + d\rho^2 + \big(\epsilon(\rho - r_0) + r_0\big)^2 d\phi^2$. This looks like the metric on a cylinder for $r_0 \le \rho \lesssim r_0/\epsilon$, because the variation of the radius of the sliver, due to the change of radial distance ρ, is very small compared to the radius of the brane r_0. Near-critical branes reside inside deep bulk throats that only asymptotically open into conical geometries [**11**]; see Fig. (2).

Note that our construction is completely covariant from the bulk point of view, since we can define the regulator by a variational principle from a

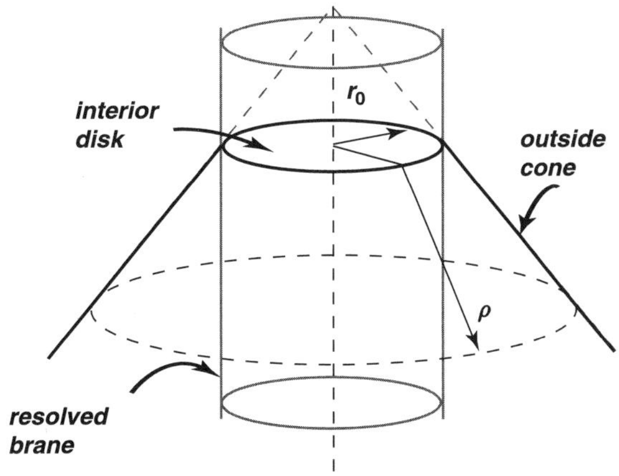

FIGURE 1. $2D$ bulk geometry of the resolved brane.

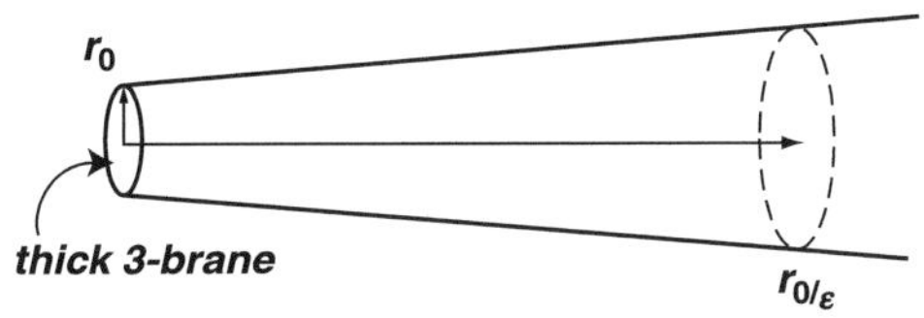

FIGURE 2. $2D$ bulk geometry of the near-critical resolved 3-brane.

well defined 4-brane action. Therefore we can explicitly check the properties of brane-localized gravity, and calculate the crossover scale inferred in [**9**] exactly.

3. Matter on thick branes

We can take the matter to be localized on the 4-brane, and described by the action $S_{\rm matter} = -\int d^5x\sqrt{g_5}\,\mathcal{L}_{\rm matter}$. Matter couples to the induced metric on the 4-brane $g_{5\,ab}$. Since we are interested in the behavior of the theory at length scales greater than the radius of the brane, $r \gg r_0$, we can dimensionally reduce the brane-localized sector on this direction, representing all brane fields by the standard Fourier expansion $\Phi_{\mathcal{N}}(x^\mu, r_0\phi) = \sum_{n=-\infty}^{\infty} \Phi_{\mathcal{N},n}(x^\mu)\, e^{in\phi}$, where $\mathcal{N}$ represents the quantum numbers of any particular representation on the brane. The Fourier coefficients then give rise to the $4D$ KK towers of fields, with masses $M^2 = m^2 + n^2/r_0^2$, where m are the explicit $5D$ masses and the mass gap is $1/r_0$. This expansion also applies to the brane-induced metric and its curvature. Thus, at distances $r \gg r_0$ we can always model the angular dependence of any configuration as a small,

Yukawa-suppressed correction from a KK momentum mode stretching along the compact circle.

To determine the stress-energy tensors of the thin loops of matter on the wrapped brane we can model them by a Nambu-Goto action for a string, $S_{\rm matter} = -\mu \int d^2\sigma \sqrt{\gamma}$, where μ is the mass per unit length of the string, σ^α are its two worldsheet coordinates, which we will gauge fix to be t and $r_0\phi$, and $\gamma_{\alpha\beta}$ is the induced worldsheet metric. We then vary this action with respect to the metric of the target space on which the string moves, in this case the metric of the background thick brane. Using the standard definition of the stress-energy tensor τ^{ab}, given by the functional derivative equation $\delta S_{\rm matter} = \frac{1}{2}\int d^5x\,\sqrt{g_5}\delta g_{ab}\tau^{ab}$, to read off the stress-energy tensor of the loop. Rewriting then the action as $S_m = -\mu \int d^2\sigma d^5x\,\sqrt{\gamma}\,\delta^{(5)}(x^M - x^M(\sigma))$, where the δ-function puts the string on its shell, with the gauge choice $\sigma^0 = x^0$ and $\sigma^1 = r_0\phi$, and using the vacuum solution (2.5) as the target space, so that $\gamma_{\alpha\beta} = \eta_{\alpha\beta}$, we find

$$\tau^{ab} = -\mu \int d^2\sigma \delta^{(5)}(x^c - x^c(\sigma))\eta^{\alpha\beta}\frac{\partial x^a}{\partial\sigma^\alpha}\frac{\partial x^b}{\partial\sigma^\beta}. \tag{3.1}$$

This string stress-energy is conserved, $\nabla_a\tau^{ab} = 0$, by virtue of the string equations of motion. For static sources, the string representing only the lightest modes must be translationally invariant in the compact direction, because by the discussion above any inhomogeneities along the string can be viewed as heavy KK states at distances $r \gg r_0$, which decouple. For such axially symmetric sources, the stress-energy boils down to

$$\tau^a{}_b = -\mu\,\delta^{(3)}(\vec{x})\,{\rm diag}(1,0,0,0,1)\,. \tag{3.2}$$

The parameter μ is the total rest mass per unit length of the configuration. For a relativistic string, we can simply boost (3.2) along the direction of motion $x_\|$, go to the light cone coordinates $x_\| = v+u$, $t = v-u$ and take the limit of infinite boost parameter while sending $\mu \to 0$ [**18, 19**]. Since the momentum of the string p is the length of the string $2\pi r_0$ multiplied by the finite limit of $\mu\cosh\beta$ as β diverges, for sources composed of the lightest modes as in (3.2) this yields, thanks to the properties of the vacuum (2.5),

$$\tau^\mu_\nu = \frac{2p}{2\pi r_0\sqrt{g_4}}g_{4uv}\delta(u)\delta^{(2)}(\vec{x}_\perp)\delta^\mu_v\delta^u_\nu. \quad \tau^\phi{}_\phi = 0, \tag{3.3}$$

To find the gravitational fields of these sources, we need to solve the BIG field equations (2.3), in Gaussian-normal gauge with the brane at $y = \rho - r_0 = 0$ and with a specified T^a_b, which includes the background tension and axion flux, and τ^a_b as given in (3.2) or (3.3). Rewriting the field equations as conditions on the Ricci tensor, we need to solve

$$M_6^4 R_{6B}^{\,A} = \Big[\Big(T_b^a - M_5^3\Big(R_{5b}^{\,a} - \frac{1}{2}\delta_b^a R_5\Big)\Big)\delta_a^A \delta_B^b - \frac{1}{4}\delta_B^A\Big(T_a^a + \frac{3}{2}M_5^3 R_5\Big)\Big]\delta(\rho - r_0). \tag{3.4}$$

4. Shocks

We determine the shock wave solution by introducing a discontinuity in the metric of Eq. (2.5) according to the prescription of [**19, 8**], and demanding that the wave profile solves Eq. (3.4). By axial symmetry of the source, we can only look for shock wave profiles $f(\vec{x}_\perp, \rho)\delta(u)$ that depend on the transverse coordinates along the brane $\vec{x}_\perp$ and the transverse distance ρ orthogonal to the brane. Thus the shock wave metric is

$$ds_6^2 = 4dudv - 4\delta(u) f du^2 + d\vec{x}_\perp^2 + d\rho^2 + \Big[(1 - b\,\Theta(\rho - r_0))\rho + b\,r_0\Theta(\rho - r_0)\Big]^2 d\phi^2. \tag{4.1}$$

Substituting this and (3.3) in Eq. (3.4), we find the shock wave equation

$$\nabla_4^2 f + \frac{M_5^3}{M_6^4}\nabla_2^2 f\,\delta(\rho - r_0) = \frac{2p}{2\pi r_0 M_6^4}\,\delta^2(\vec{x}_\perp)\,\delta(\rho - r_0)\,. \tag{4.2}$$

The Laplacian ∇_4^2 is defined on the truncated cone (2.5), such that for axially symmetric shocks, $\nabla_4^2 f = \nabla_{\vec{x}_\perp}^2 f + f'' + \frac{1 - b\Theta(\rho - r_0)}{(1 - b\Theta(\rho - r_0))\rho + br_0\Theta(\rho - r_0)} f'$, where the prime denotes the ρ-derivative. The latter term contains a jump across the 4-brane.

To solve (4.2), we Fourier transform in the brane transverse space. This yields [**10**]

$$(\nabla_{\vec{y}}^2 - k^2)\varphi_k = \Big(\frac{2p}{2\pi r_0 M_6^4} + \frac{M_5^3}{M_6^4}k^2\varphi_k\Big)\delta(\rho - r_0)\,. \tag{4.3}$$

We can use the standard methods for finding bulk field wave functions in the codimension-1 brane setups: we solve Eq. (4.3) inside ($-$) and outside ($+$) the brane, and then use the boundary conditions, which are the continuity of φ_k, the jump of φ_k' as prescribed by Gaussian pillbox integration of (4.3) around the brane, and the regularity of the solution in the center of (2.5) and at infinity. Then using the axial symmetry in the $\vec{x}_\perp$ plane to integrate over the transverse spatial angle in the Fourier transform formula, which gives $f = \frac{1}{2\pi}\int_0^\infty dk\, k\varphi_k J_0(k|\vec{x}_\perp|)$, we finally obtain the expression for the shock wave profile [**10**]:

$$f(\vec{x}_\perp, \rho) = -\frac{p}{2\pi^2 r_0}\int_0^\infty dk\,\frac{I_0(k\rho_<)K_0(k\rho_>)J_0(k|\vec{x}_\perp|)}{\Upsilon_1}, \tag{4.4}$$

where

$$\Upsilon_1 = M_6^4\left(I_0(kr_0)K_1\left(k\frac{r_0}{1-b}\right) + I_1(kr_0)K_0\left(k\frac{r_0}{1-b}\right)\right) + M_5^3 k I_0(kr_0)K_0\left(k\frac{r_0}{1-b}\right)$$

and

$$I_0(k\rho_<)K_0(k\rho_>) = \begin{cases} I_0\Big(k\rho\Big)K_0\Big(k\frac{r_0}{1-b}\Big), & \rho \le r_0 \\ I_0\Big(kr_0\Big)K_0\Big(k\Big(\rho + \frac{br_0}{1-b}\Big)\Big), & \rho \ge r_0 \end{cases} \tag{4.5}$$

This solution is clearly regular on and off the brane, as long as $r_0 > 0$. On the brane, $\rho = r_0$, it reduces to

$$f(\vec{x}_\perp, r_0) = -\frac{p}{2\pi^2 r_0}\int_0^\infty dk\, \frac{I_0(kr_0)K_0\left(k\frac{r_0}{1-b}\right)J_0(k|\vec{x}_\perp|)}{\Upsilon_2}, \tag{4.6}$$

where

$$\Upsilon_2 = M_6^4\left(I_0(kr_0)K_1\left(k\frac{r_0}{1-b}\right) + I_1(kr_0)K_0\left(k\frac{r_0}{1-b}\right)\right) + M_5^3 k I_0(kr_0)K_0\left(k\frac{r_0}{1-b}\right),$$

and where all the individual contributions remain finite.

5. Scales of gravity

Now we can look at the field along the brane, (4.6). The solution is dominated by the modes with the momenta $k \sim 1/|\vec{x}_\perp|$, due to the oscillatory nature of J_0. The contributions of the modes with momenta k far from $1/|\vec{x}_\perp|$ will interfere destructively. At transverse distances $|\vec{x}_\perp| \gg r_0$, we need to focus on the momenta for which $kr_0 \ll 1$. Thus we can always replace the terms $\propto I_n(kr_0)$ by their small argument expansion. For sub-critical branes, $1 - b \sim \mathcal{O}(1)$, and we can likewise replace any $K_n\left(k\frac{r_0}{1-b}\right)$ by their small argument expansion too. However, in the near-critical limit the deficit angle approaches 2π and so $|1-b| \ll 1$. Hence the argument of $K_0\left(k\frac{r_0}{1-b}\right)$ may be very large even when $kr_0 \ll 1$. Thus we must consider the near-critical branes very carefully.

For sub-critical branes, $1 - b \lesssim \mathcal{O}(1)$, $kr_0 \ll 1 - b$, and we can approximate all Bessel functions by their small argument form. The shock along the brane then becomes

$$f(\vec{x}_\perp, r_0) \simeq -\frac{p}{2\pi^2 M_6^4}\int_0^\infty dk\, k\, \frac{\ln\left[\frac{2(1-b)}{kr_0}\right]J_0(k|\vec{x}_\perp|)}{(1-b) + \frac{M_5^3 r_0}{M_6^4}\, k^2 \ln\left[\frac{2(1-b)}{kr_0}\right]}. \tag{5.1}$$

When $k \sim 1/|\vec{x}_\perp|$ is large enough so that the second term in the denominator dominates, which happens when $k^2 > k_c^2(k) = \frac{(1-b)M_6^4}{M_5^3 r_0 \ln\left[\frac{2(1-b)}{kr_0}\right]}$, the shock wave profile is

$$(5.2)\quad f(\vec{x}_\perp, r_0) \simeq -\frac{p}{2\pi^2 M_5^3 r_0}\int_{k_c}^{\infty} dk \frac{J_0(k|\vec{x}_\perp|)}{k} = \frac{p}{\pi M_4^2}\ln\Big(k_c|\vec{x}_\perp|\Big) + \cdots .$$

where we have used Gauss's law to normalize the lightest KK modes in the expansion of the brane-induced curvature on the circle of radius r_0 to $M_4^2 = 2\pi M_5^3 r_0$. This is precisely the $4D$ Aichelburg-Sexl solution [**18**], with the log profile normalized to the inverse of the critical momentum k_c. So for sub-critical branes the gravitational field looks $4D$, like (5.2), only when $r < r_c = 1/k_c$. At distances larger than r_c, gravity eventually changes to $6D$. In this limit (4.5) or (4.6) yield

(5.3)
$$f \simeq -\frac{p}{2\pi^2(1-b)M_6^4}\int_0^{\infty} dk\, k\, I_0(kr_0)K_0\left(k\left(\rho + \frac{r_0}{1-b}\right)\right) J_0(k|\vec{x}_\perp|) + \cdots ,$$

where the integral can be done in closed form, leading to [**21**]

$$(5.4)\qquad f \simeq -\frac{p}{2\pi^2(1-b)M_6^4}\frac{1}{|\vec{x}_\perp|^2 + \rho^2} + \cdots .$$

This is the $6D$ gravitational shock wave, constructed in [**20**], with the 'lightning rod' term accounting for the brane tension as found in [**21**]. The distance scale which controls the transition, $r_c^2(k) = \frac{M_4^2}{2\pi(1-b)M_6^4}\ln\left[\frac{2(1-b)}{kr_0}\right]$, is *very different* from the conjectured see-saw scale of [**9**]. So for sub-critical branes the $4D \to 6D$ transition is controlled by the 'naive' ratio of brane and effective bulk Planck scales, and depends on the UV cutoff r_0 of the brane core only through the logarithmic correction reminiscent of 'running' in real space, which was discussed for locally localized gravity in [**22**].

In the near-critical limit, $b \to 1$, the story is a lot more interesting. In this case the bulk around the brane compactifies to a long cylinder, which opens up into a cone only very far from the brane [**11**]. Bulk gravity in the throat is in the $5D$ regime, which separates the $4D$ and $6D$ ones. Thus the crossover scale beyond which gravity is not $4D$ determines where it changes to $5D$ gravity which lives inside the throat. In the shock wave solution (4.6), we can always use the small argument expansion for $I_n(z)$ at large distances from the source, but in the functions $K_n(z)$ we must use the large argument expansion for the momenta $1 - b \ll kr_0 \ll 1$, which correspond to length scales $\ell \lesssim \frac{r_0}{1-b}$ inside the conical throat. In this case gravity is $4D$ for momenta $k > k_c = \frac{M_6^4}{M_5^3}\frac{K_1(k\frac{r_0}{1-b})}{K_0(k\frac{r_0}{1-b})}$, which thanks to the limit $K_1 \to K_0$ saturates to $k_c = \frac{M_6^4}{M_5^3}$. Therefore the scale below which gravity is $4D$ is

$r_c = \frac{M_5^3}{M_6^4}$, which is valid as long as $r_c < \frac{r_0}{1-b}$. This scale is exactly the see-saw scale of [**9**]: using Gauss's law $M_4^2 = 2\pi M_5^3 r_0$, we find

$$r_c = \frac{M_4^2}{2\pi M_6^4 r_0}\,. \tag{5.5}$$

This explains the see-saw effect, showing that at least with our regularization, it does happen naturally but only in the near-critical limit.

To confirm that inside the conical throat gravity is $5D$, we can look at our solution for a fixed distance $|\vec{x}_\perp|$ and take the limit $b \to 1$ while holding r_0 fixed. Replacing K_n's by their large argument expansion for any finite momentum, cancelling the like terms, and taking the thin brane limit, we find [**10**]

$$\begin{aligned} f(\vec{x}_\perp, \rho) &= -\frac{p}{2\pi^2}\int_0^\infty dk\, \frac{J_0(k|\vec{x}_\perp|)}{M_6^4 r_0 + M_5^3 r_0 k} \\ &= -2p\int_{-\infty}^{\infty} \frac{d^2\vec{k}}{(2\pi)^2}\, \frac{e^{i\vec{k}\cdot\vec{x}_\perp}}{2\pi M_6^4 r_0\, k + 2\pi M_5^3 r_0\, k^2}\,. \end{aligned} \tag{5.6}$$

This is *precisely* the shock wave profile on a flat thin 3-brane in $5D$, on the normal branch, with brane and bulk Planck masses $M_{4\,eff}^2 = 2\pi M_5^3 r_0$ and $M_{5\,eff}^3 = 2\pi M_6^4 r_0$ [**8, 10**].

So the crossover scale for sub-critical and near-critical branes, beyond which gravity is not $4D$ anymore, is given by

$$r_c^2 = \begin{cases} \dfrac{M_4^2}{2\pi(1-b)M_6^4}\ln\left[\frac{2(1-b)}{kr_0}\right], & \text{subcritical tension} \\[2ex] \dfrac{M_4^4}{4\pi^2 M_6^8 r_0^2}\,, & \text{near-critical tension,} \end{cases} \tag{5.7}$$

only in the latter case agreeing with the see-saw scale discussed in [**9**].

6. Gravity of slow sources

The shock wave analysis can only discover the presence of helicity-2 modes, since, in the limit of relativistic sources when the rest masses vanish, the source stress energy tensor becomes traceless. Such sources will not excite the additional helicities in the graviton spectrum in the linearized theory. Thus to see what else is in the spectrum, one eventually needs to engage in the laborious procedure of gravitational perturbation theory around slow, nonrelativistic, sources. This analysis has been done in full detail in [**10**]. Here we will merely recall a few basic results.

The perturbed metric in the brane-fixed Gaussian-normal gauge is

$$ds_6^2 = (\eta_{\mu\nu} + h_{\mu\nu})\, dx^\mu dx^\nu + d\rho^2 + (1+\tilde{\Phi})\,\alpha(\rho)\, d\phi^2\,, \tag{6.1}$$

with $\tilde{\xi} = 0$ and $\alpha = g_{\phi\phi}$ of the background. The axion perturbation obeys $\partial_4^2\sigma = 0$, and so in linear order the axion is not sourced by axially symmetric matter distributions and can be set to zero. Substituting this in the field equations (3.4) we can get the equations of perturbations. We can solve them most easily by decomposing into helicities, using the transverse-traceless tensor projection operator $K_{\mu\nu\alpha\beta}$, such that $K_{\mu\nu}^{\alpha\beta}h_{\alpha\beta} = \gamma_{\mu\nu}$. This operator must satisfy $K_{\mu\nu\alpha\beta} = K_{\nu\mu\alpha\beta} = K_{\mu\nu\beta\alpha}$ and $K^{\mu}_{\mu\alpha\beta} = \partial^\mu K_{\mu\nu\alpha\beta} = 0$. The unique solution is

$$K_{\mu\nu\alpha\beta} = -\frac{1}{3}\Bigg[\eta_{\mu\nu}\eta_{\alpha\beta} - \frac{3}{2}\Big(\eta_{\mu\alpha}\eta_{\nu\beta} + \eta_{\mu\beta}\eta_{\nu\alpha}\Big) - \eta_{\mu\nu}\frac{\partial_\alpha\partial_\beta}{\partial_4{}^2} - \eta_{\alpha\beta}\frac{\partial_\mu\partial_\nu}{\partial_4{}^2} + \frac{3}{2}\Big(\eta_{\nu\beta}\frac{\partial_\mu\partial_\alpha}{\partial_4{}^2} + \eta_{\mu\beta}\frac{\partial_\nu\partial_\alpha}{\partial_4{}^2} + \eta_{\nu\alpha}\frac{\partial_\mu\partial_\beta}{\partial_4{}^2} + \eta_{\mu\alpha}\frac{\partial_\nu\partial_\beta}{\partial_4{}^2}\Big) - 2\frac{\partial_\mu\partial_\nu\partial_\alpha\partial_\beta}{\partial_4{}^4}\Bigg]. \tag{6.2}$$

Since the vacuum (2.5) is $4D$-flat, stress-energy conservation implies $\partial_\mu\tau^\mu_\nu = 0$, and so $K_{\mu\nu}^{\alpha\beta}\tau_{\alpha\beta} = \tau^\mu_\nu - \frac{1}{3}\left(\delta^\mu_\nu - \frac{\partial^\mu\partial_\nu}{\partial_4^2}\right)\tau^\alpha_\alpha$. Hence the TT-tensor field equation is

$$\gamma^{\mu\prime\prime}_\nu + \frac{\alpha'}{2\alpha}\gamma^{\mu\prime}_\nu - k^2\gamma^\mu_\nu - \frac{M_5^3}{M_6^4}k^2\gamma^\mu_\nu\,\delta(\rho - r_0) = -\frac{2}{M_6^4}\left[\tau^\mu_\nu - \frac{1}{3}\left(\delta^\mu_\nu - \frac{k^\mu k_\nu}{k^2}\right)\tau^\alpha_\alpha\right]\delta(\rho - r_0)\,. \tag{6.3}$$

Subtracting this from the full set of equations isolates the vectors and scalars. Because the backgrounds (2.5) are Lorentz-invariant, the vector sources vanish at linear order and so the vectors decouple from the matter distribution, so they can all be set to zero. The scalars remain. They are all controlled by a single independent scalar field, the 'radion,' defined by $X = \frac{1}{4}(h_4 - \partial_4^2\Psi)$, where Ψ is fully determined by X [**10**], and ${}^s h^\mu_\nu = \partial^\mu\partial_\nu\Psi + X\delta^\mu_\nu$. A straightforward albeit lengthy calculation [**10**] then yields the field equation for it, which can be written as

$$X'' + \frac{\alpha'}{2\alpha}X' - k^2X - \frac{M_5^3}{M_6^4}\left(k^2 + \frac{3bM_6^4}{4M_5^3r_0} - \beta(k^2)\right)X\,\delta(\rho - r_0) = -\frac{1}{6M_6^4}\left(\tau^\alpha_\alpha + \frac{3M_5^3r_0k^2}{bM_6^4}\,\frac{1 - b - \frac{bM_6^4r_0}{M_5^3}}{1 - b + k^2r_0^2}\,\tau^\phi_\phi\right)\delta(\rho - r_0)\,, \tag{6.4}$$

where

$$\beta(k^2) = \frac{3M_5^3r_0}{4bM_6^4}\,\frac{1 - b}{1 - b + k^2r_0^2}\left(k^2 + \frac{bM_6^4}{M_5^3r_0}\right)^2, \tag{6.5}$$

is the correction to the brane kinetic term of the scalar X originating from the brane bending contributions. This equation is very similar to the TT-tensor equation (6.3), with the differences being in the brane-localized terms.

We can solve both of these equations using the standard Green's function techniques [**10**]. The Green's functions for the TT-tensor and radion are defined by, respectively,

$$\mathcal{G}_{\mathrm{TT}}'' + \frac{\alpha'}{2\alpha}\mathcal{G}_{\mathrm{TT}}' - k^2\mathcal{G}_{\mathrm{TT}} - \frac{M_5^3}{M_6^4}k^2\mathcal{G}_{\mathrm{TT}}\,\delta(\rho - r_0) = \frac{1}{M_6^4}\,\delta(\rho - r_0)\,, \tag{6.6}$$

$$\mathcal{G}_{\mathrm{X}}'' + \frac{\alpha'}{2\alpha}\mathcal{G}_{\mathrm{X}}' - k^2\mathcal{G}_{\mathrm{X}} - \frac{M_5^3}{M_6^4}\left(k^2 + \frac{3bM_6^4}{4M_5^3 r_0} - \beta(k^2)\right)\mathcal{G}_{\mathrm{X}}\,\delta(\rho - r_0) = \frac{1}{M_6^4}\,\delta(\rho - r_0)\,. \tag{6.7}$$

The solutions are, formally,

$$\gamma^\mu_\nu(k,\rho) = -2\,\mathcal{G}_{\mathrm{TT}}(k,\rho)\left[\tau^\mu{}_\nu - \frac{1}{3}\left(\delta^\mu{}_\nu - \frac{k^\mu k_\nu}{k^2}\right)\tau^\alpha{}_\alpha\right], \tag{6.8}$$

$$X(k,\rho) = -\frac{1}{6}\,\mathcal{G}_{\mathrm{X}}(k,\rho)\left[\tau^\alpha{}_\alpha + \frac{3M_5^3 r_0 k^2}{bM_6^4}\,\frac{1 - b - \frac{bM_6^4 r_0}{M_5^3}}{1 - b + k^2 r_0^2}\,\tau^\phi{}_\phi\right]. \tag{6.9}$$

Since the theory has scalars, in general one must worry about which frame one extracts the long range fields in. The long distance dynamics will be described by a scalar-tensor theory, where the radion is the Brans-Dicke-like scalar. To go to the effective 'Einstein' frame, we can conformally remove the radion which on the brane yields the Einstein frame metric perturbation

$$h^{\mathrm{E}}{}^\mu_\nu = \gamma^\mu_\nu - 4\left(X - \frac{r_0 k^2}{2}\xi\right)\frac{k^\mu k_\nu}{k^2} + \left(\frac{\xi}{r_0} - \frac{X}{2}\right)\delta^\mu_\nu\,, \tag{6.10}$$

where of course ξ is given by X [**10**]. However in the near-critical limit, it turns out that the radion is suppressed and so in fact the metric coincides with the TT-tensor to leading order, as we are about to see.

For static solutions, the Fourier transforms all have $k^\mu = (0, \vec{k})$, and therefore are given by $\mathcal{G}(k) = 2\pi\delta(k^0)\,\mathcal{G}(\vec{k})$. Factorizing out the energy δ-function, setting $k^0 = 0$ in the formulas of the previous section, and dropping the integration $\int \frac{dk^0}{2\pi}$ from the Fourier integral we then solve the Green's function equations (6.6) and (6.7). On the brane $\rho = r_0$ at distances $\ell \gg r_0$ we must be cautious with K_0 and K_1 since they depend on the deficit angle, as in the shock wave case. The consistent approximation for the Green's functions on the brane is [**10**]

$$\mathcal{G}_{\mathrm{TT}} = \frac{1}{M_6^4} \frac{1}{\frac{K_1\left(k\frac{r_0}{1-b}\right)}{K_0\left(k\frac{r_0}{1-b}\right)}k + \frac{M_5^3}{M_6^4}k^2}, \tag{6.11}$$

$$\mathcal{G}_{\mathrm{X}} = \frac{1}{M_6^4} \frac{1}{\frac{K_1\left(k\frac{r_0}{1-b}\right)}{K_0\left(k\frac{r_0}{1-b}\right)}k + \frac{M_5^3}{M_6^4}\left[k^2 + \frac{3bM_6^4}{4M_5^3 r_0} - \beta(k^2)\right]}, \tag{6.12}$$

For the TT-tensor, the behavior of the theory is governed by the ratio $k_c = \frac{M_6^4 K_1(k\frac{r_0}{1-b})}{M_5^3 K_0\left(k\frac{r_0}{1-b}\right)}$, which controls the denominator of $\mathcal{G}_{\mathrm{TT}}$ at low momenta. For $k < k_c$, or distances $\ell > 1/k_c$, the dominant contributions always come from $\propto \frac{K_1\left(k\frac{r_0}{1-b}\right)}{K_0\left(k\frac{r_0}{1-b}\right)}$. In this limit the long range fields manifestly reveal the extra dimensions, since the scaling of the potentials will not be $4D$. The critical value k_c is *exactly* the same as the one discovered by the shock wave analysis. Thus, specifically in the near-critical limit, the crossover scale is $r_c = \frac{M_5^3}{M_6^4}$, where gravity first changes into a $5D$ theory. Eventually, the circle opens up and the theory turns into a $6D$ gravity on a cone. The crucial dynamics which manufactures this sequence of transitions is the compactification of one bulk dimension induced by a near-critical brane [**11**].

The function $\mathcal{G}_{\mathrm{X}}$ is considerably more involved because of $\beta(k^2)$ in the denominator. It behaves differently for generic sub-critical and for near-critical branes. For generic sub-critical branes, $b \lesssim 1$, at scales $k \ll 1/r_0$, the function $\mathcal{G}_{\mathrm{X}}$ (6.12) reduces to

$$\mathcal{G}_{\mathrm{X}} = \frac{1}{M_6^4} \frac{r_0 \ln\left[\frac{2(1-b)}{kr_0}\right]}{1 - b - \frac{3M_5^6 r_0^2}{4bM_6^8}k^2\left(k^2 + \frac{2bM_6^4}{3M_5^3 r_0}\right)\ln\left[\frac{2(1-b)}{kr_0}\right]}, \tag{6.13}$$

where we have approximated $\frac{K_1(k\frac{r_0}{1-b})}{K_0\left(k\frac{r_0}{1-b}\right)}$ with the small argument expansion. At very large distances $\mathcal{G}_{\mathrm{X}} \simeq \frac{r_0 \ln\left[\frac{2(1-b)}{kr_0}\right]}{(1-b)M_6^4}$, implying that the configuration space solution for the scalar field X depends on the distance as $X \sim \mathcal{M}/|\vec{x}|^3$, *i.e.* as a field in $6D$. But as we get closer to the source, the negative momentum-dependent terms in the denominator of $\mathcal{G}_{\mathrm{X}}$ make the scalar force grow, and the scalar eventually becomes strongly coupled as k approaches the pole of (6.13). This happens when $k \sim k_*$, where

$$k_*^2 \simeq \frac{2(1-b)M_6^4}{M_5^3 r_0 \ln\left[\frac{2(1-b)}{kr_0}\right]}. \tag{6.14}$$

But this is basically the same as the crossover momentum in the TT-tensor sector! So, for sub-critical branes inside the regime of length scales, $\ell < r_*$,

where

$$r_*^2 \simeq \frac{M_4^2}{2(1-b)M_6^4} \ln\left[\frac{2(1-b)}{kr_0}\right], \tag{6.15}$$

the scalar sector is in fact strongly coupled. This makes the perturbation theory around the vacuum on sub-critical branes completely unreliable inside the regime where gravity might be $4D$. The negative signs of the momentum-dependent terms in the denominator of (6.13) might naively suggest the presence of ghosts in this regime, but at this level of the approximation we cannot conclude that decisively. We conclude that at the level of linearized perturbation theory the gravitational effects below the crossover scale are *not* calculable for generic subcritical cases.

In the near-critical limit the scalar sector behaves very differently. At distances $\ell \gg r_0$, the scalar Green's function is [**10**]

$$\mathcal{G}_{\mathrm{X}} = \frac{1}{M_6^4} \frac{1}{k + \frac{M_5^3}{M_6^4}\frac{r_0^2k^2}{1-b+r_0^2k^2}\left(k^2 + \frac{3}{4r_cr_0}\right)}, \tag{6.16}$$

where $r_c = M_3^5/M_6^4$ is the TT-tensor crossover scale. Now, it is convenient to define the scale $r_{\mathtt{vac}} = r_0/\sqrt{1-b}$. When $b \to 1$ we have $r_{\mathtt{vac}} \ll \frac{r_0}{1-b}$, and so the scale $r_{\mathtt{vac}}$ is always smaller than the size of the conical throat surrounding the brane. This is why the scale $r_{\mathtt{vac}}$ may compete with r_c for the control over the scalar sector. Both $r_{\mathtt{vac}}$ and r_c are smaller than the size of the throat, and the details of sub-crossover dynamics depend on their ratio. Skipping some details, we can check [**10**] that the scalar X is either more weakly coupled to matter than the TT-tensor (when $r_{\mathtt{vac}} < r_c$) or is Yukawa-suppressed, by a mass term $m_{\mathrm{X}}^2 = \frac{3}{4r_cr_0}$ (when $r_{\mathtt{vac}} > r_c$). Either way, the long range forces generated by scalars in the linearized approximation on near-critical branes turn out to be small compared to those arising from the TT-tensors.

Therefore the effective $4D$ theory below the crossover scale r_c, in the regime of distances $r_0 \ll \ell \ll r_c$ is completely controlled by the TT-tensors. In this regime their Green's function reduces to $\mathcal{G}_{\mathrm{TT}} \to \frac{1}{M_5^3k^2}$. For non-relativistic matter sources $\tau^a{}_b = -\mu\,\mathrm{diag}(1,0,0,0,1)$ the linearized solution is

$$\gamma^\mu_\nu(k,\rho) = \frac{2\mu}{3M_5^3k^2}\begin{cases} 2\,, & \mu=\nu=0\,,\\ -\left(\delta^j{}_k - \frac{k^jk_k}{k^2}\right), & \mu=j,\ \nu=k\,. \end{cases} \tag{6.17}$$

Fourier-transforming to coordinate space and recalling that μ is the mass per unit length of string, so that $\mu = \frac{\mathcal{M}}{2\pi r_0}$, and using as before the Gauss law $M_5^3 = \frac{M_4^2}{2\pi r_0}$, and introducing the effective $4D$ Newton's constant $G_{N\,eff} =$

$\frac{1}{8\pi M_4^2}$, we finally obtain the TT-tensor in the $4D$ regime:

$$\gamma^0{}_0(\vec{x}) = \frac{8}{3}\, G_{N\,eff}\, \frac{\mathcal{M}}{|\vec{x}|}\,,$$
$$\gamma^j{}_k(\vec{x}) = -\frac{2}{3}\, G_{N\,eff}\, \frac{\mathcal{M}}{|\vec{x}|}\left(\delta^j{}_k + \frac{x^j x_k}{\vec{x}^2}\right). \tag{6.18}$$

The Newtonian potential $V_N = -h^0_0/2 = -\frac{4}{3}\, G_{N\,eff}\, \frac{\mathcal{M}}{|\vec{x}|}$, is a factor of 4/3 larger than the usual formula of General Relativity, which a manifestation of the Iwasaki-van Dam-Veltman-Zakharov discontinuity [**12**]. This signals the presence of the helicity-0 modes in the graviton multiplet. The factor 4/3 enhancement is precisely what one expects based on other examples [**7**], and in this case shows that the extra helicity-0 modes in the spin-2 multiplet mediate an attractive force, just as the helicity-2 modes. In fact the solution (6.18) mimics a Brans-Dicke theory with $\gamma = \frac{1}{2}$, or therefore, with the Brans-Dicke ω parameter equal to zero, where the parameter ω is defined in the usual way in the Brans-Dicke action as $S_{BD} = \int d^4x\sqrt{g}(\Phi R - \omega(\nabla\Phi)^2/\Phi)$. This theory is in conflict[1] with observational data as it stands, and hence is not a realistic description of our Universe.

However, it is conceivable (although not certain) that different strong coupling effects at higher orders in perturbation theory might improve the behavior of long range fields of masses far from the source but well below the crossover scale [**13, 24**]–[**28**]. Another possibility might be that the additional tweaks of the bulk theory, e.g. curving the bulk locally, could change how the helicity-0 mode couples to brane matter. Either way, the scalar couplings remain consistently *weak* for static sources on near-critical branes with $b \to 1$. The static solutions are governed by Euclidean momenta, and so they are always finite in linearized theory around near-critical vacua, in contrast to the fields of masses which perturb generic sub-critical vacua. Thus linearized perturbation theory remains under control on near-critical vacua. Similar arguments show that the near-critical propagating solutions also remain under control [**10**].

7. Summary

BIG models may in fact approximate $4D$ worlds between the UV cut-off that resolves the core of the gravity-localizing defect and the crossover scale, if their Scherk-Schwarz sector is carefully tuned to yield $4D$ Minkowski vacua. We have seen this explicitly in the example of regulated codimension-2 defects, which remain perturbatively calculable and have a $4D$ regime for near-critical tensions. We feel that similar conclusions may extend to any setups with codimension ≥ 2. The gravitational sector contains additional

[1]Ideas for decoupling of the Brans-Dicke scalar by using radiatively induced potential were explored in [**23**]; however the field Φ here is, as we noted, really the helicity-0 graviton.

scalars, but at least at the linearized order there are no ghosts, of the kind that plague the codimension-1 DGP models [**27, 29**]–[**33**], or were argued to arise in soft gravity resolutions of higher-codimension cases [**34**].

In BIG codimension-two setups, the cosmological constant problem is very different. Putting brane localized terms and changing gravity in the IR by itself does not solve it. Instead, one still has the vacuum energy problem, since the tension of the brane needs to be tuned to get a desired low energy theory. But now vacuum energy does not need to curve the $4D$ geometry at all. Instead, we must tune the vacuum energy to get a weak $4D$ gravity over a large spatial region and maintain perturbativity in the scalar graviton sector. This might offer a different strategy to help with the vacuum energy problem: perhaps the fact that graviscalars go into the strong coupling for detuned tensions can be used as a new tool. One might try to look, for example, for theories where the tuned value of the tension, which ensures perturbativity, marks an IR fixed point of the combined matter and graviscalar theory. In this case, at least some of the premises of Weinberg's no-go theorem for cosmological constant adjustment [**1**] would be violated, and it would be interesting to see what the consequences are.

The numerology of the near-critical codimension-2 BIG is rather intriguing. Using the formula for the $4D \to 5D$ crossover scale $r_c = M_5^3/M_6^4$, we see that to get r_c to be of the order of the current horizon size, $H_0^{-1} \sim 10^{28}$cm, and ensure that matter on the brane looks pointlike at energies below a TeV, we need $M_6 \gtrsim$ TeV, and therefore $M_5 \lesssim 10^{19}$ GeV. Further, to reproduce the $4D$ Newton's constant below the crossover scale, one needs to pick $r_0 \gtrsim 10^{-19}\,\mathrm{GeV}^{-1}$. So we must tune the theory to make sure that the brane is very thin, with many hidden sector fields, to get such hierarchies. We don't have much to add to the discussion of how to do it, but merely note that this is in line with the current philosophy of the BIG models [**7**]. While some work along the lines of embedding the theory into string theory, that would provide the framework for its full UV completion, has been pursued [**35, 36, 37**], the existing constructions are really semiclassical models where gravity is treated classically. At the least, we can explore the low energy consequences of such models.

Many questions remain open. Can the helicity-0 modes decouple? Are there dangerous instabilities beyond the linear order? Can BIG theory be UV-completed? Most interestingly, could the new guise of the cosmological constant problem open some new directions to circumvent the Weinberg no-go theorem [**1**]? While we can't address these issues here, we hope to return to them soon.

Acknowledgements

We would like to thank Derrick Kiley for collaboration and Alberto Iglesias, Robert Myers, Minjoon Park, Oriol Pujolas, Lorenzo Sorbo and John Terning for interesting discussions. N.K. is grateful to Galileo Galilei

Institute, Florence, Italy, for kind hospitality in the course of this work, and to the organizers of the "Sowers Workshop", Virginia Tech, May 14–18, 2007, "Cosmology and Strings" workshop at ICTP, Trieste, Italy, July 9–13, 2007, "Dark Energy In the Universe", Hakone, Japan Sep 1–4, 2007 [**38**] and "Zagreb Workshop 2007", Zagreb, Croatia, Nov 9–11, 2007, for the opportunity to present the results and the kind hospitality. This work has been supported in part by the DOE Grant DE-FG03-91ER40674, in part by the NSF Grant PHY-0332258 and in part by a Research Innovation Award from the Research Corporation.

References

[1] S. Weinberg, Rev. Mod. Phys. **61**, 1 (1989).
[2] V. A. Rubakov and M. E. Shaposhnikov, Phys. Lett. B **125**, 139 (1983).
[3] B. Holdom, ITP-744-STANFORD preprint, 1983.
[4] N. Arkani-Hamed, S. Dimopoulos, N. Kaloper and R. Sundrum, Phys. Lett. B **480**, 193 (2000); S. Kachru, M. B. Schulz and E. Silverstein, Phys. Rev. D **62**, 045021 (2000).
[5] S. Forste, Z. Lalak, S. Lavignac and H. P. Nilles, Phys. Lett. B **481**, 360 (2000).
[6] Y. Aghababaie *et al.*, JHEP **0309**, 037 (2003); C. P. Burgess, J. Matias and F. Quevedo, Nucl. Phys. B **706**, 71 (2005).
[7] G. R. Dvali, G. Gabadadze and M. Porrati, Phys. Lett. B **485**, 208 (2000); G. R. Dvali and G. Gabadadze, Phys. Rev. D **63**, 065007 (2001).
[8] N. Kaloper, Phys. Rev. Lett. **94**, 181601 (2005) [Erratum-ibid. **95**, 059901 (2005)]; Phys. Rev. D **71**, 086003 (2005) [Erratum-ibid: **D71** (2005) 086003].
[9] G. Dvali, G. Gabadadze, X. r. Hou and E. Sefusatti, Phys. Rev. D **67**, 044019 (2003).
[10] N. Kaloper and D. Kiley, JHEP **0705**, 045 (2007).
[11] N. Kaloper, Phys. Lett. B **652**, 92 (2007).
[12] Y. Iwasaki, Phys. Rev. D **2**, 2255 (1970); H. van Dam and M. J. G. Veltman, Nucl. Phys. B **22**, 397 (1970); V. I. Zakharov, JETP Lett. **12** 312, (1970).
[13] A. I. Vainshtein, Phys. Lett. B **39** 393, (1972).
[14] C. de Rham, G. Dvali, S. Hofmann, J. Khoury, O. Pujolas, M. Redi and A. J. Tolley, arXiv:0711.2072 [hep-th].
[15] C. P. Burgess, F. Quevedo, G. Tasinato and I. Zavala, JHEP **0411**, 069 (2004); C. P. Burgess, D. Hoover and G. Tasinato, JHEP **0709**, 124 (2007).
[16] M. Peloso, L. Sorbo and G. Tasinato, Phys. Rev. D **73**, 104025 (2006).
[17] R. Sundrum, Phys. Rev. D **59**, 085010 (1999).
[18] P. C. Aichelburg and R. U. Sexl, Gen. Rel. Grav. **2**, 303 (1971).
[19] T. Dray and G. 't Hooft, Nucl. Phys. B **253**, 173 (1985); Class. Quant. Grav. **3**, 825 (1986).
[20] V. Ferrari, P. Pendenza and G. Veneziano, Gen. Rel. Grav. **20**, 1185 (1988); H. de Vega and N. Sanchez, Nucl. Ph. B **317**, 706 (1989).
[21] N. Kaloper and D. Kiley, JHEP **0603**, 077 (2006).
[22] N. Kaloper and L. Sorbo, JHEP **0508**, 070 (2005).
[23] A. Albrecht, C. P. Burgess, F. Ravndal and C. Skordis, Phys. Rev. D **65**, 123507 (2002).
[24] C. Deffayet, G. R. Dvali, G. Gabadadze and A. I. Vainshtein, Phys. Rev. D **65** 044026, (2002).
[25] N. Arkani-Hamed, H. Georgi and M. D. Schwartz, Annals Phys. **305** 96, (2003).
[26] V. A. Rubakov, `arXiv:hep-th/0303125`.
[27] M. A. Luty, M. Porrati and R. Rattazzi, JHEP **0309**, 029 (2003).

[28] A. Nicolis and R. Rattazzi, JHEP **0406**, 059 (2004).
[29] K. Koyama, Phys. Rev. D **72**, 123511 (2005); D. Gorbunov, K. Koyama and S. Sibiryakov, Phys. Rev. D **73**, 044016 (2006).
[30] C. Charmousis, R. Gregory, N. Kaloper and A. Padilla, JHEP **0610**, 066 (2006); R. Gregory, N. Kaloper, R. C. Myers and A. Padilla, JHEP **0710**, 069 (2007).
[31] C. Deffayet, G. Gabadadze and A. Iglesias, JCAP **0608**, 012 (2006).
[32] K. Izumi, K. Koyama and T. Tanaka, JHEP **0704** (2007) 053.
[33] M. Carena, J. Lykken, M. Park and J. Santiago, Phys. Rev. D **75**, 026009 (2007).
[34] S. L. Dubovsky and V. A. Rubakov, Phys. Rev. D **67**, 104014 (2003).
[35] E. Kiritsis, N. Tetradis and T. N. Tomaras, JHEP **0108**, 012 (2001).
[36] S. Corley, D. A. Lowe and S. Ramgoolam, JHEP **0107**, 030 (2001).
[37] I. Antoniadis, R. Minasian and P. Vanhove, Nucl. Phys. B **648**, 69 (2003).
[38] http://www.resceu.s.u-tokyo.ac.jp/workshops/resceu07s/files/Kaloper3.pdf

Department of Physics, University of California, Davis, CA 95616
E-mail address: `kaloper@physics.ucdavis.edu`

AMS/IP Studies in Advanced Mathematics
Volume 44, 2008

Local Bulk Operators in AdS/CFT and the Fate of the BTZ Singularity

Alex Hamilton, Daniel Kabat, Gilad Lifschytz, and David A. Lowe

Presented by DK at the 2007 Sowers workshop and GL at the pre-strings 2007 workshop

ABSTRACT. This paper has two parts. First we review the description of local bulk operators in Lorentzian AdS in terms of non-local operators in the boundary CFT. We discuss how bulk locality arises in pure AdS backgrounds and how it is modified at finite N. Next we present some new results on BTZ black holes: local operators can be defined inside the horizon of a finite N BTZ black hole, in a way that suggests the BTZ geometry describes an average over black hole microstates, but with finite N effects resolving the singularity.

1. Introduction

Quantum gravity in asymptotically anti-de Sitter space is dual to a conformal field theory on the boundary of AdS [**Mal98**]. One of the most interesting questions raised by this duality is: how does approximately local bulk gravitational physics emerge from the CFT?

To address this one needs some way of probing local physics in the bulk. We are mostly interested in the semiclassical limit of small Planck length. In this limit we should be able to recover the traditional results of quantum field theory in curved space [**BD82**]. So it's natural to ask: how can a local quantum field in the bulk of AdS be represented in the boundary CFT?

2000 *Mathematics Subject Classification.* Primary 81T20, 81T30; Secondary 81T40, 83C57, 83E30.

Key words and phrases. AdS/CFT correspondence, BTZ black hole.

AH was supported by a Columbia University Initiatives in Science and Engineering grant and by DOE grant DE-FG02-92ER40699.

DK was supported by DOE grant DE-FG02-92ER40699.

GL is supported in part by Israeli science foundation grant number 568/05.

DL was supported by DOE grant DE-FG02-91ER40688-Task A.

This question was addressed in [**BDHM98, BGL99, Ben00**] and was further developed by the authors in [**HKLL06a, HKLL06b, HKLL07**]. In the latter works it was shown that local operators in the bulk could be represented as non-local operators in the CFT. The CFT operators turn out to have support on a compact region of the complexified boundary. This representation makes several properties manifest. It makes it clear why bulk locality is exact at large N, but breaks down at finite N, in exactly the manner required by holography. It also provides a simple CFT description of the horizon and singularity of a BTZ black hole in the large N limit.

An outline of this paper is as follows. In section 2 we review the representation of local bulk operators in terms of operators on the complexified boundary. In section 3 we use these boundary operators to discuss bulk locality and holography from the point of view of the CFT. In section 4 we extend the construction to the BTZ black hole and discuss the horizon and singularity in the large N limit. We conclude in section 5 with some speculation on the fate of the horizon and singularity at finite N. Sections 2–4 are a review; the results in section 5 are new.

2. Local operators in the semiclassical limit

In Poincaré coordinates the metric on Lorentzian AdS_D is

$$ds^2 = \frac{R^2}{Z^2}\left(-dT^2 + |d\mathbf{X}|^2 + dZ^2\right). \tag{2.1}$$

Here R is the AdS radius. The Poincaré horizon is at $Z = \infty$, while the $\mathrm{CFT}_{d=D-1}$ lives on the boundary at $Z = 0$. Consider a scalar field of mass m in AdS, with normalizable fall-off near the boundary.

$$\phi(T, \mathbf{X}, Z) \sim Z^\Delta \phi_0(T, \mathbf{X}) \quad \text{as } Z \to 0$$

The parameter Δ is related to the mass of the field by

$$\Delta = \frac{d}{2} + \sqrt{\frac{d^2}{4} + m^2R^2}\,.$$

We will refer to ϕ_0 as the boundary field. It's dual to an operator of dimension Δ in the CFT.

$$\phi_0(T, \mathbf{X})_{\mathrm{SUGRA}} \leftrightarrow \mathcal{O}(T, \mathbf{X})_{\mathrm{CFT}} \tag{2.2}$$

The question is, can we express ϕ in terms of ϕ_0? If so, then we can use (2.2) to find the CFT operator dual to a local operator in the bulk.

For now we'll study this in the semiclassical limit

$$\begin{aligned} &\ell_S, \ell_P \to 0 && \text{in the bulk} \\ &N, \lambda \to \infty && \text{on the boundary} \end{aligned}$$

Here ℓ_S and ℓ_P are the bulk string and Planck lengths, while N and λ are parameters for some kind of 't Hooft large N expansion in the CFT whose details won't matter for us. The basic idea is to represent

$$\phi(T, \mathbf{X}, Z) = \int dT' d^{d-1}X' \, K(T', \mathbf{X}'|T, \mathbf{X}, Z) \, \phi_0(T', \mathbf{X}')$$

using a kernel or smearing function K. Since AdS has a timelike boundary, this is not a standard Cauchy problem, and neither existence nor uniqueness of K is guaranteed. Indeed in [**HKLL06a**] we discuss examples where both existence and uniqueness are violated.

A cure for these problems, at least in a pure semiclassical AdS background, is to make a Wick rotation to de Sitter space. Define a new set of boundary spatial coordinates by setting $\mathbf{X} = i\mathbf{Y}$. This turns the AdS metric (2.1) into

$$ds^2 = \frac{R^2}{Z^2}\left(-dT^2 - |d\mathbf{Y}|^2 + dZ^2\right). \tag{2.3}$$

This is de Sitter space in flat FRW coordinates, with Z playing the role of conformal time. The past boundary of de Sitter space is at $Z = 0$.

In de Sitter space we have a standard Cauchy problem. As shown in Fig. 1 we can use a retarded de Sitter Green's function to solve for the bulk field in terms of data on the past boundary. The explicit analytic expressions

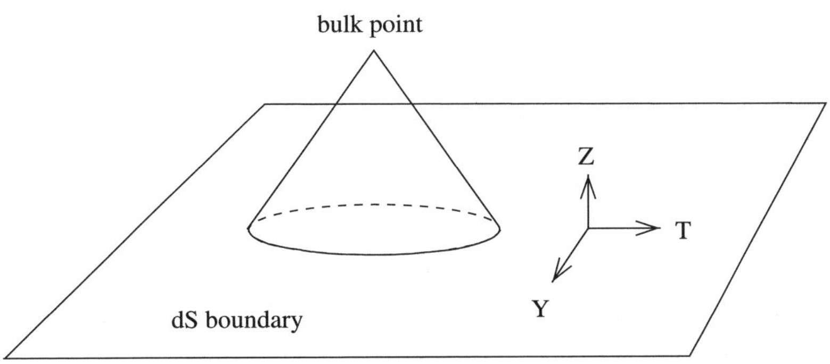

FIGURE 1. The field at a bulk point in de Sitter space can be expressed in terms of data on the past de Sitter boundary. The slice $Y = 0$ also describes a region in AdS. So we can also regard this as expressing the field in AdS in terms of data on the complexified AdS boundary.

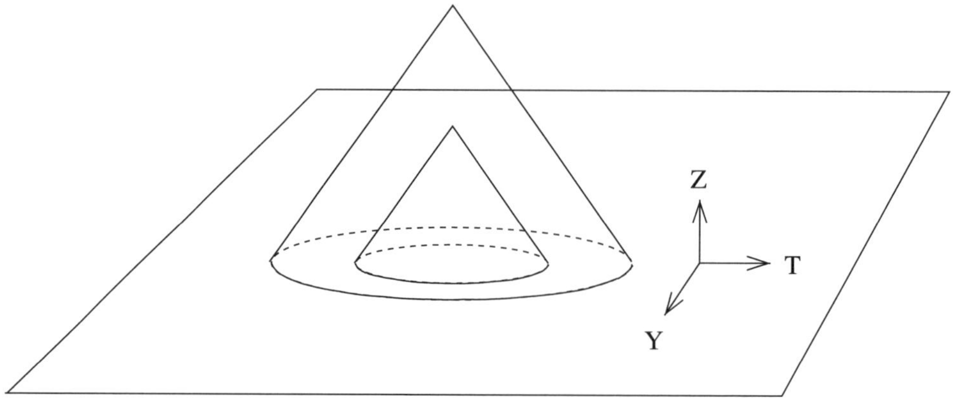

FIGURE 2. Smearing functions for two bulk points separated only in the Z direction. The smearing functions overlap on the boundary; nonetheless the smeared operators commute at infinite N.

are pretty simple: the field at a point in AdS can be expressed as [**HKLL07**]

$$\phi(T, \mathbf{X}, Z) = \frac{\Gamma\left(\Delta - \frac{d}{2} + 1\right)}{\pi^{d/2}\,\Gamma(\Delta - d + 1)} \int_{T'^2 + |\mathbf{Y}'|^2 < Z^2} dT' d^{d-1}Y' \cdot \left(\frac{Z^2 - T'^2 - |\mathbf{Y}'|^2}{Z}\right)^{\Delta - d} \phi_0(T + T', \mathbf{X} + i\mathbf{Y}'). \tag{2.4}$$

Note that we have to integrate over a compact region of the de Sitter boundary (the region inside the past light-cone of the bulk point). Equivalently we integrate over a compact region of the complexified AdS boundary (the region spacelike separated from the bulk point).

By construction this lets us reproduce bulk correlation functions in the semiclassical limit.

$$\langle \phi(x_1, Z_1)\phi(x_2, Z_2)\rangle_{SUGRA} = \int dx_1' dx_2'\, K(x_1'|x_1, Z_1)K(x_2'|x_2, Z_2)\langle \mathcal{O}(x_1')\mathcal{O}(x_2')\rangle_{CFT} \tag{2.5}$$

This is guaranteed to work, just because ϕ_0 and $\mathcal{O}$ have identical correlators. Although somewhat trivial, this result has an interesting corollary. In the semiclassical limit of vanishing Planck length, bulk operators will commute at spacelike separation. Therefore the corresponding smeared boundary operators will commute with each other. This is true even though, as shown in Fig. 2, the smearing functions might overlap on the boundary.

3. Bulk locality and holography at finite N

What we've done so far is exact in the semiclassical limit; it can be regarded as a set of statements about free wave equations in a pure AdS background. In this section we'll remain in a pure AdS background, but ask what happens at finite N.

First we need to decide what smearing functions to use. One possibility is to use the same smearing functions at finite N. For example, in $\mathcal{N} = 4$ Yang-Mills the operator

$$\Phi(T, \mathbf{X}, Z) = \int dT' d^3X' \, K(T', \mathbf{X}'|T, \mathbf{X}, Z) \operatorname{Tr} F^2 \tag{3.1}$$

can be defined at any N, where K is the kernel appearing in (2.4). In a pure AdS background we believe these must be the right operators to use at finite N, just because the construction is singled out by the symmetries. To see this, introduce a distance function on AdS.

$$\begin{aligned} \sigma(T, \mathbf{X}, Z|T', \mathbf{X}', Z') &= \cosh\left(\frac{\text{geodesic distance}}{R}\right) \\ &= \frac{Z^2 + Z'^2 + |\mathbf{X} - \mathbf{X}'|^2 - (T - T')^2}{2ZZ'} \end{aligned} \tag{3.2}$$

σ is invariant under AdS isometries. In Poincaré coordinates the isometry

$$(T, \mathbf{X}, Z) \to \lambda(T, \mathbf{X}, Z)$$

acts as a scale transformation on the boundary. Z has conformal weight -1, and since $K \sim \lim_{Z' \to 0} (\sigma Z')^{\Delta - d}$, we see that K transforms covariantly under AdS isometries with conformal weight $d - \Delta$. But given an operator of dimension Δ, this is exactly what we need for $\int d^d x \, K\mathcal{O}$ to behave like a scalar field in the bulk! That is, *the smearing functions we have defined provide the unique covariant way to map a primary field in the CFT to a scalar field in the bulk.*[1,2]

This might seem very strange from the point of view of holography. In (3.1) we've defined a continuous infinity of bulk operators. How can this be compatible with the holographic bound [**SW98**], which should only allow a finite number of degrees of freedom in any given region in the bulk?

We believe the resolution is that at finite N not all the operators defined in (3.1) commute at spacelike separation. To see this, consider a fixed-T hypersurface in the bulk, with operators placed at some radial position Z. The smearing functions we have defined have an extent in the time direction $\Delta T = Z$.[3] As shown in Fig. 3 the operators will be spacelike separated on the

---

[1]Thus the smearing functions should have a purely group-theoretic interpretation in terms of representations of the (complexified) isometry group $SO(d, 2)$, along the lines of [**VK91**]. We are grateful to Djordje Minic for discussions on this point.

[2]To avoid a possible confusion: by construction, the operators defined in (3.1) satisfy a free wave equation. So they do not have the right interactions to be identified with the bulk dilaton field at finite N. Nevertheless, they should serve as good probes of local physics in the bulk.

[3]They also have an extent in the imaginary spatial directions, but as shown in [**HKLL07**] it's only the extent in time that matters here.

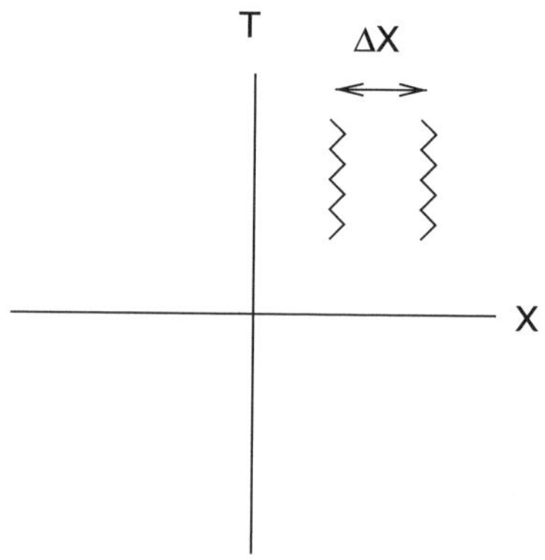

FIGURE 3. The smearing functions have support on the two jagged lines. For $|\Delta X| > Z$ they are spacelike separated.

boundary provided the bulk operators have a spatial separation $|\Delta X| > Z$. In this case the operators are guaranteed to commute, just by locality of the boundary theory. At finite N we do not expect operators separated by $|\Delta X| < Z$ to commute. (This is unlike the semiclassical situation discussed in section 2, where operators could commute even though they overlapped.) So we expect $1/Z^{d-1}$ commuting operators per coordinate area on the boundary. Using the AdS metric (2.1), this means we expect $1/R^{d-1}$ commuting operators per proper area in the bulk. Equivalently we expect one commuting operator per unit cell, where the cell volume $\sim R^{d-1}$. At finite N, bulk locality breaks down at distances set by the AdS radius of curvature!

This is a bit disturbing, in that it seems the commuting operators we can build from $\mathrm{Tr}\, F^2$ aren't sufficient to describe a local bulk dilaton at distances less than an AdS radius. We don't have a complete resolution of this puzzle. It's probably too strong a condition to require that all operators describing the dilaton commute exactly. Take the set of commuting operators built from $\mathrm{Tr}\, F^2$. Perhaps there are additional operators which can be used to describe the dilaton, which do not all commute, but whose commutators are so small at low energies that they can be ignored.[4] The operators built from $\mathrm{Tr}\, F^2$ using our smearing functions are good first candidates for the job, since their commutators do vanish in the large N limit. But it could also be that operator mixing is important, so that operators besides $\mathrm{Tr}\, F^2$ can contribute.

One might also wonder about the holographic bound. Here things work out very nicely. The holographic bound in AdS states that the entropy per longitudinal coordinate area is bounded by $s < N^2/Z^{d-1}$. Each operator in the CFT gives an entropy density $\sim 1/Z^{d-1}$, so the bound is saturated if we

[4]More precisely, we expect matrix elements of the commutators between low-energy states to be small.

have N^2 commuting operators in the CFT. This seems quite reasonable, as $\mathcal{N} = 4$ Yang-Mills involves $N \times N$ matrices and has a central charge $\sim N^2$.

4. Semiclassical BTZ black hole

We now turn to excitations of AdS_3, in particular we will study non-extremal BTZ black holes. But first let's consider AdS_3 in accelerating or Rindler-like coordinates.

$$ds^2 = -\frac{r^2 - r_0^2}{R^2} dt^2 + \frac{R^2}{r^2 - r_0^2} dr^2 + r^2 d\phi^2 \tag{4.1}$$
$$-\infty < t, \phi < \infty \qquad 0 < r < \infty$$

Here r_0 is an arbitrary parameter with units of length.[5] We'll frequently work in terms of the rescaled coordinates

$$\hat{t} = r_0 t / R^2 \qquad \hat{\phi} = r_0 \phi / R\,.$$

It's straightforward to construct smearing functions in these coordinates. The Wick rotation $\hat{\phi} = iy$ turns the AdS_3 metric (4.1) into de Sitter space, now expressed in static coordinates.

$$ds^2 = -\frac{r^2 - r_0^2}{R^2} dt^2 + \frac{R^2}{r^2 - r_0^2} dr^2 - \frac{r^2 R^2}{r_0^2} dy^2 \tag{4.2}$$
$$-\infty < t < \infty \qquad y \approx y + 2\pi \qquad 0 < r < \infty$$

(The periodicity in y is necessary to avoid a singularity at $r = 0$.) One can use a retarded de Sitter Green's function to construct smearing functions in AdS_3. Alternatively, one can just translate our previous result (2.4) into Rindler coordinates, to find[6]

$$\phi(t, r, \phi) = \int dt' d\phi' \, K_{\text{Rindler}}(t, r, \phi | t', \phi') \, \phi_0(t', \phi') \tag{4.3}$$
$$K_{\text{Rindler}}(t, r, \phi | t', \phi') = \frac{(\Delta - 1)\, 2^{\Delta - 2}}{\pi R^3} \lim_{r' \to \infty} \left(\frac{\sigma}{r'}\right)^{\Delta - 2}.$$

A BTZ black hole can be obtained from AdS_3 by identifying $\hat{\phi} \approx \hat{\phi} + 2\pi$. This produces an orbifold singularity at $r = 0$. But making this identification doesn't change the smearing functions at all: if the boundary field has the necessary periodicity, so will the bulk field. So we can use the same smearing functions (4.3) in a BTZ background.

In the semiclassical limit this gives a clear picture of the BTZ horizon and singularity. First, the horizon. The integral in (4.3) is over points on the complexified boundary that are spacelike separated from the bulk point. As

---

[5]We denoted this parameter r_+ in our previous work.

[6]The boundary field ϕ_0 is defined slightly differently in Rindler coordinates: $\phi(t, r, \phi) \sim \phi_0(t, \phi)/r^{\Delta}$ as $r \to \infty$.

the bulk point approaches the (future, past) horizon the integration region extends to ($t = +\infty$, $t = -\infty$). Thus to probe the horizon requires an infinite time from the boundary point of view. This fits nicely with the bulk definition of a horizon, as bounding the region where light rays cannot escape to infinity.[7]

What about the BTZ singularity? With scalar fields as probes we cannot directly study the bulk geometry. However it turns out that the orbifold singularity generates a divergence in scalar correlators. To see this, we use the fact that in the semiclassical limit we can make the bulk correlator periodic with an image sum [**LO94**].

$$\langle \phi(\hat{t}, r, \hat{\phi}) \phi(\hat{t}', r', \hat{\phi}') \rangle_{\mathrm{BTZ}} = \sum_{n=-\infty}^{\infty} \langle \phi(\hat{t}, r, \hat{\phi}) \phi(\hat{t}', r', \hat{\phi}' + 2\pi n) \rangle_{\mathrm{AdS}} \tag{4.4}$$

But $r = 0$ is a fixed point of the isometry $\phi \to \phi + \text{const.}$, so correlators in AdS are ϕ-independent at $r = 0$. If we compute a bulk correlator in a BTZ background, the image sum diverges when one of the bulk points is located at the singularity.

The same divergence arises from the boundary point of view. The CFT dual to AdS_3 in Rindler coordinates lives on $\mathbb{R}^{1,1}$, while the CFT dual to BTZ lives on $\mathbb{R} \times S^1$. In the semiclassical limit the BTZ boundary correlator can be given the necessary periodicity with an image sum.

$$\langle \phi_0(\hat{t}, \hat{\phi}) \phi_0(\hat{t}', \hat{\phi}') \rangle_{\mathrm{BTZ}} = \sum_{n=-\infty}^{\infty} \langle \phi_0(\hat{t}, \hat{\phi}) \phi_0(\hat{t}', \hat{\phi}' + 2\pi n) \rangle_{\mathrm{AdS}}$$

To recover a bulk correlator we convolve this with our smearing functions as in (2.5). Again the image sum diverges when one of the bulk points is located at the singularity. So we also get a divergent correlator at $r = 0$ from the boundary point of view.

For future reference it's useful to study the divergence in a little more detail. Consider a point in AdS located near $r = 0$, at

$$(t, r, \phi) \quad \text{with } t = 0,\ r \to 0,\ \phi = 0$$

and a second point in AdS located near the boundary, at

$$(t', r', \phi') \quad \text{with } r' \to \infty.$$

[7]One can also describe bulk points that are located inside the horizon [**HKLL07**]. However this requires the use of smearing functions with support on both the left and right boundaries of the extended Kruskal diagram. For a bulk point inside the future horizon the smearing function extends to $t = +\infty$ on the right boundary and $t = -\infty$ on the left boundary, where time has the same orientation on the left and right. From the boundary point of view this means we are using operators that act on both copies of the thermofield-doubled CFT.

The invariant distance (3.2) between these points is approximately

$$\sigma \approx \frac{r'}{r_0}\left(\frac{r}{r_0}\cosh\hat{\phi}' + \sinh\hat{t}'\right).$$

The AdS bulk correlation function [**IS95**] decays exponentially at large ϕ'.

$$\begin{aligned} \langle\phi(t,r,\phi)\phi(t',r',\phi')\rangle_{\text{AdS}} &= \frac{1}{4\pi R\sqrt{\sigma^2-1}}\frac{1}{\left(\sigma+\sqrt{\sigma^2-1}\right)^{\Delta-1}} \\ &\sim e^{-\Delta\hat{\phi}'} \quad \text{for} \quad |\hat{\phi}'| > \hat{\phi}_{\max} \sim \log(r_0/r) \end{aligned} \tag{4.5}$$

This behavior means the BTZ image sum (4.4) is cut off at $|n| \sim \frac{1}{2\pi}\log(r_0/r)$, which in turn means the BTZ correlator diverges logarithmically as $r \to 0$.

$$\langle\phi(t,r,\phi)\phi(t',r',\phi')\rangle_{\text{BTZ}} \sim \frac{\log(r_0/r)}{2\pi^2 R}\left(\frac{r_0}{2r'\sinh\hat{t}'}\right)^{\Delta} \quad \text{as } r \to 0 \tag{4.6}$$

5. BTZ at finite N

What happens to the BTZ black hole at finite N? Although definitive statements are hard to come by, there are a few interesting observations to make. First, there's the issue of what smearing functions to use. In a pure AdS background we gave a symmetry argument that the same smearing functions should be used at any value of N. For excited states such as BTZ, where the symmetries are broken, this argument is not valid.[8] Nonetheless one might be tempted to use the same semiclassical smearing functions for BTZ even at finite N. This is indeed a reasonable prescription for bulk points outside the horizon. However as the bulk point approaches the (future) horizon the semiclassical smearing functions extend to $t = +\infty$ on the boundary, and for points inside the horizon they grow exponentially with time.[9]

$$K(\hat{t},\hat{\phi}|\cdot) \sim e^{(\Delta-d)\hat{t}} \quad \text{as } t \to \infty$$

In the semiclassical limit this causes no problems. CFT correlators at infinite N decay exponentially [**BSS02**], making the convolution of a boundary correlator with a smearing function well-defined.

$$\langle\phi_0(\hat{t},\hat{\phi})\phi_0(0,0)\rangle_{BTZ} \sim e^{-\Delta\hat{t}}$$

But when N is finite such behavior cannot persist indefinitely. At finite N, the CFT on $\mathbb{R}\times S^1$ has a finite thermal entropy. Eventually the discrete

[8] From the boundary point of view BTZ is related to a CFT on a Euclidean 2-torus, and the bulk-to-boundary map could depend on the modular parameter of the torus in a way that cannot be determined from modular invariance.

[9] This is the behavior on the right boundary. For points inside the future horizon the smearing function also extends to $t = -\infty$ on the left boundary, and grows as $e^{-(\Delta-d)\hat{t}}$ in the far past.

spectrum of the CFT becomes important and causes correlators to oscillate quasi-periodically rather than decay exponentially [**DLS02**]. So we could not hope to use our semiclassical smearing functions to reproduce sensible bulk correlators. It seems there are a couple of options.

(1) Perhaps this supports the fuzzball picture of Mathur and collaborators [**Mat05**], in which the geometry inside the horizon differs radically from what one would expect based on the traditional black hole metric. In this case one would need to know the exact microstate of the black hole to make sense of the interior of the horizon.
(2) Maybe there is some modification to the smearing functions which gives sensible bulk correlators even for points inside the horizon. "Sensible" means with small $1/N$ corrections to the semiclassical result, except near $r = 0$, where the divergence (4.6) should be smoothed out. This would support the picture that the semiclassical geometry is a good description, even inside the horizon, but with quantum gravity effects resolving the singularity.
(3) It could be that options 1 and 2 are compatible, if one is able to recover the semiclassical BTZ metric from the fuzzball picture by a suitable averaging procedure [**AdBM06, BCL$^+$07**].

We conclude by presenting a prototype construction to show how option 3 might be realized. The difficulty is that for points inside the horizon the semiclassical smearing functions grow exponentially with time. A simple cure is to define a modified smearing function $\widetilde{K}$ which vanishes if $\hat{t}$ is larger than some cutoff time $\hat{t}_{\text{max}}$:

$$\widetilde{K}(\hat{t},\hat{\phi}|\cdot) = \begin{cases} K(\hat{t},\hat{\phi}|\cdot) & \text{if } |\hat{t}| < \hat{t}_{\text{max}} \\ 0 & \text{otherwise} \end{cases}$$

(We impose the cutoff on both the left and right boundaries of the black hole.) We can use this modified smearing function to define – purely within the CFT – a set of "bulk operators."

$$\widetilde{\phi}(t,r,\phi) \stackrel{\text{def}}{=} \int dt'd\phi'\, \widetilde{K}(t,r,\phi|t',\phi')\,\mathcal{O}(t',\phi') \tag{5.1}$$

With the cutoff in place, the operators $\widetilde{\phi}$ have finite correlation functions.

Imposing a cutoff in this way might seem very arbitrary. But in fact there is a good physical motivation for the cutoff which sets an upper bound on t_{max}. With

$$n(E) = (\#\text{ CFT states with energy } < E) = e^{S(E)}$$

the density of states

$$\frac{dn}{dE} = \beta e^{S}$$

implies a spacing between energy levels

$$\Delta E = \frac{1}{\beta e^S}\,. \tag{5.2}$$

This spacing corresponds to a time (the Heisenberg time [**Sre99**])

$$t_H = \beta e^S.$$

By this time CFT correlators begin to oscillate quasi-periodically rather than decay exponentially [**BR03, BR04**]. So imposing a cutoff on the smearing functions at $t_{\max} \lesssim t_H$ is the minimal change necessary to obtain well-defined bulk correlators.[10] Moreover, such a cutoff has a nice physical interpretation. From the CFT point of view measurements with a duration exceeding t_H can resolve individual microstates of the CFT. So putting a cutoff at $t_{\max} \lesssim t_H$ implies an average over microstates. This means the smearing functions we constructed based on the classical BTZ geometry break down at the horizon when N is finite, unless one averages over microstates. This suggests that the region inside the horizon of the classical BTZ geometry isn't a good description of any individual microstate of the black hole. Rather the region inside the horizon only provides a good description of ensemble averages over black hole microstates.

With a cutoff at $t_{\max} \lesssim t_H$, $\widetilde{K}$ represents the minimal modification to the smearing functions necessary to plausibly represent BTZ correlators at finite N. In the semiclassical limit we expect $t_{\max} \to \infty$, so the bulk operators we have defined have the correct semiclassical limit.

Ideally at this point we would compute correlation functions of the operators $\widetilde{\phi}$ in the CFT at finite N. Such a calculation might be within reach [**Wit07**]. But for the time being we will regard $t_{\max}$ as a fixed *ad hoc* cutoff and study correlators of the operators $\widetilde{\phi}$ in the large N limit of the CFT. Up to small $1/N$ corrections, this should be a good guide to behavior at finite N.

We first work in Rindler coordinates on AdS_3 and consider the correlation function

$$\langle\, \widetilde{\phi}(t,r,\phi)\, \widetilde{\phi}(t',r',\phi')\,\rangle_{\mathrm{AdS}} \tag{5.3}$$

between a bulk operator located at $(t=0, r, \phi=0)$ and an operator near the boundary at (t', r', ϕ') with $r' \to \infty$. In the semiclassical limit the boundary correlator

$$\langle \phi_0(\hat{t},\hat{\phi})\phi_0(0,0)\rangle_{\mathrm{AdS}} = \frac{(r_0^2/2)^{\Delta}}{2\pi R\left(\cosh\hat{\phi} - \cosh\hat{t}\right)^{\Delta}} \tag{5.4}$$

---

[10] In general $t_{\max}$ should be set by the time at which CFT correlators begin to behave quasi-periodically. This could occur before t_H, so really t_H is an upper bound on $t_{\max}$. The main point is that at finite N the upper bound is finite. We are grateful to Hong Liu and Massimo Porrati for discussions on this topic.

decays exponentially at spacelike separation. So the bulk-boundary correlator (5.3) will be exponentially small provided the boundary point is spacelike separated from the support of the smearing function. As shown in Fig. 4, with the modified smearing functions the cutoff at $t_{\rm max}$ means that – even for a bulk point inside the Rindler horizon – the correlator will decay exponentially at large ϕ'.

To extend this discussion to BTZ we use the image charge construction (4.4). For points inside the horizon $\hat{t}_{\rm max}$ serves to cut off the AdS correlator at $|\hat{\phi}'| \approx \hat{t}_{\rm max}$. However as shown in (4.5) the AdS correlator is already exponentially small when $|\hat{\phi}'| > \hat{\phi}_{\rm max}$. Thus when we perform the image sum there are two possible regimes.

(1) For $\hat{t}_{\rm max} > \hat{\phi}_{\rm max}$, or equivalently for $r > r_0 e^{-\hat{t}_{\rm max}}$, the additional cutoff at $\hat{t}_{\rm max}$ isn't important. So using $\widetilde{K}$ rather than K makes a negligible change to the BTZ correlator away from $r = 0$. We can therefore probe a large region inside the BTZ horizon, roughly the region

$$r_0 e^{-\hat{t}_{\rm max}} < r < r_0 \,,$$

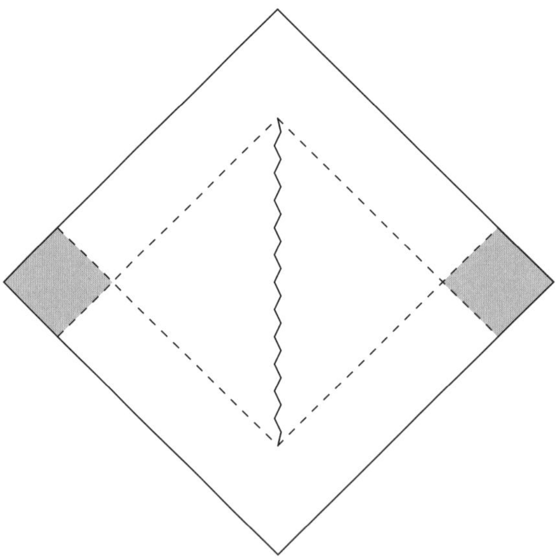

FIGURE 4. Penrose diagram of the (t, ϕ) plane. The support of $\widetilde{K}$ is indicated by the jagged line. Points in the shaded region are spacelike separated from the support of $\widetilde{K}$. When the smearing function extends from $-\hat{t}_{\rm max}$ to $+\hat{t}_{\rm max}$ the shaded region is characterized by $|\hat{\phi}| > \hat{t}_{\rm max} + |\hat{t}|$.

using only a finite time interval on the boundary, and without seeing significant deviations from the semiclassical result. Strictly speaking this means there is no horizon, at least not in the sense of section 4 where the horizon corresponded to integration over infinite time.

(2) However for $\hat{t}_{\rm max} < \hat{\phi}_{\rm max}$, or equivalently for

$$0 < r < r_0 e^{-\hat{t}_{\rm max}},$$

the additional cutoff at $\hat{t}_{\rm max}$ is crucial. It serves to regulate the image sum, cutting it off at $|n| \approx \frac{1}{2\pi}\hat{t}_{\rm max}$ so that

$$\langle\, \widetilde{\phi}(t,r,\phi)\, \widetilde{\phi}(t',r',\phi')\,\rangle_{\rm BTZ} \approx \frac{\hat{t}_{\rm max}}{2\pi^2 R}\left(\frac{r_0}{2r'\sinh \hat{t}'}\right)^{\Delta}.$$

Note that the correlator is independent of r, and the divergence at $r = 0$ has been eliminated!

We find it appealing that the same effect that eliminates the horizon also gets rid of the divergence. Note that the effects we have discussed are very robust: they are independent of any details of the CFT and only rely on the generic thermal behavior (5.2). Our results are compatible with the detailed study of extremal BTZ black holes in [**B**$^+$**07**].

To summarize: we've defined a set of operators in the CFT (5.1) which should have well-defined correlation functions even at finite N. As a guide to the behavior of these operators we studied their correlation functions in the large N limit. For bulk points well outside the horizon the cutoff at $t_{\rm max}$ has no effect on the smearing functions. For bulk points inside the horizon but well away from the singularity the cutoff at $t_{\rm max}$ makes an exponentially small change in correlators. But for points very near $r = 0$ the cutoff at $t_{\rm max}$ becomes important and renders correlation functions finite. It seems reasonable that working with the true finite-N correlation functions of the CFT, rather than their semiclassical large N limit, should only make a small change in these results. If so, this would support option 3: after a suitable average over microstates, enforced by a cutoff at $t_{\rm max} \lesssim t_H$, the semiclassical BTZ geometry becomes a good description, even inside the horizon, but with quantum gravity effects resolving the singularity.

Acknowledgements

DK is grateful to Mark Sowers and the organizers of the 2007 Sowers workshop for a delightful and stimulating conference. GL would like to thank Vijay Balasubramanian and especially Masaki Shigemori for discussions and the organizers of pre-strings 2007 for a stimulating conference.

References

[AdBM06] Luis F. Alday, Jan de Boer, and Ilies Messamah, *The gravitational description of coarse grained microstates*, JHEP **12** (2006), 063.

[B+07] Vijay Balasubramanian et al., *Typicality versus thermality: An analytic distinction*, eprint (2007), hep–th/0701122.

[BCL+07] Vijay Balasubramanian, Bartlomiej Czech, Klaus Larjo, Donald Marolf, and Joan Simon, *Quantum geometry and gravitational entropy*, eprint (2007), arXiv:0705.4431 [hep–th].

[BD82] N. D. Birrell and P. C. W. Davies, *Quantum fields in curved space*, Cambridge Univ. Press, 1982.

[BDHM98] Tom Banks, Michael R. Douglas, Gary T. Horowitz, and Emil J. Martinec, *AdS dynamics from conformal field theory*, eprint (1998), arXiv:hep–th/9808016.

[Ben00] Iosif Bena, *On the construction of local fields in the bulk of AdS(5) and other spaces*, Phys. Rev. **D62** (2000), 066007.

[BGL99] Vijay Balasubramanian, Steven B. Giddings, and Albion E. Lawrence, *What do CFTs tell us about anti-de Sitter spacetimes?*, JHEP **03** (1999), 001.

[BR03] J. L. F. Barbon and E. Rabinovici, *Very long time scales and black hole thermal equilibrium*, JHEP **11** (2003), 047.

[BR04] ———, *Long time scales and eternal black holes*, Fortsch. Phys. **52** (2004), 642–649.

[BSS02] Danny Birmingham, Ivo Sachs, and Sergey N. Solodukhin, *Conformal field theory interpretation of black hole quasi-normal modes*, Phys. Rev. Lett. **88** (2002), 151301.

[DLS02] Lisa Dyson, James Lindesay, and Leonard Susskind, *Is there really a de Sitter/CFT duality*, JHEP **08** (2002), 045.

[HKLL06a] Alex Hamilton, Daniel Kabat, Gilad Lifschytz, and David A. Lowe, *Holographic representation of local bulk operators*, Phys. Rev. **D74** (2006), 066009.

[HKLL06b] ———, *Local bulk operators in AdS/CFT: A boundary view of horizons and locality*, Phys. Rev. **D73** (2006), 086003.

[HKLL07] ———, *Local bulk operators in AdS/CFT: A holographic description of the black hole interior*, Phys. Rev. **D75** (2007), 106001.

[IS95] Ikuo Ichinose and Yuji Satoh, *Entropies of scalar fields on three-dimensional black holes*, Nucl. Phys. **B447** (1995), 340–372.

[LO94] Gilad Lifschytz and Miguel Ortiz, *Scalar field quantization on the (2+1)-dimensional black hole background*, Phys. Rev. **D49** (1994), 1929–1943.

[Mal98] Juan M. Maldacena, *The large N limit of superconformal field theories and supergravity*, Adv. Theor. Math. Phys. **2** (1998), 231–252.

[Mat05] Samir D. Mathur, *The fuzzball proposal for black holes: An elementary review*, Fortsch. Phys. **53** (2005), 793–827.

[Sre99] Mark Srednicki, *The approach to thermal equilibrium in quantized chaotic systems*, J. Phys. **A32** (1999), 1163.

[SW98] Leonard Susskind and Edward Witten, *The holographic bound in anti-de Sitter space*, eprint (1998), arXiv:hep–th/9805114.

[VK91] N.Ja. Vilenkin and A.U. Klimyk, *Representation of Lie groups and special functions*, Kluwer Academic Publishers, 1991.

[Wit07] Edward Witten, *Three-dimensional gravity revisited*, eprint (2007), arXiv:0706.3359 [hep–th].

DEPARTMENT OF PHYSICS, COLUMBIA UNIVERSITY, NEW YORK NY 10027 USA
Current address: Department of Mathematics and Applied Mathematics, University of Cape Town, Rondebosch 7700, South Africa
E-mail address: `ahamil@gmail.com`

DEPARTMENT OF PHYSICS, COLUMBIA UNIVERSITY, NEW YORK NY 10027 USA
E-mail address: `kabat@phys.columbia.edu`

DEPARTMENT OF MATHEMATICS AND PHYSICS AND CCMSC, UNIVERSITY OF HAIFA AT ORANIM, TIVON 36006 ISRAEL
E-mail address: `giladl@research.haifa.ac.il`

DEPARTMENT OF PHYSICS, BROWN UNIVERSITY, PROVIDENCE RI 02912 USA
E-mail address: `lowe@brown.edu`

AMS/IP Studies in Advanced Mathematics
Volume 44, 2008

Heterotic Geometry and Fluxes

Li-Sheng Tseng

ABSTRACT. We begin by discussing the question, "What is string geometry?" We then proceed to discuss six-dimensional compactification geometry in heterotic string theory with fluxes. A class of smooth non-Kähler compact solutions is presented and relations between Kähler and non-Kähler solutions are pointed out.

1. String geometry

Geometry plays an important role in many aspects of string theory. For spacetime compactifications, a six- (or seven-) dimensional internal compact geometry is often used to wrap the extra dimensions of string theory in order to connect with our low-energy four-dimensional world. The characteristics of the internal compact geometry then determine the physical observables we see in four dimensions. For the two-dimensional worldsheet (or higher-dimensional worldvolume), its dynamics is determined by a path integral over all possible embeddings in spacetime. The saddle points of the integration correspond to classical objects that typically have geometrically interesting embedding conditions.

An interesting question is what is the geometry string theory "sees"? Or more precisely, what geometries are identical or different from the perspective of string theory? That this differs from that of a point particle theory can already be seen in the closed bosonic string theory compactified in one dimension. Let us consider the moduli space of such compactifications. This is equivalent to the moduli space of $c = 1$ two-dimensional CFTs worked out by Dijkgraaf *et al.* [**DVV**] and Ginsparg [**G**]. This moduli space is interesting and relatively simple. Let us briefly describe it.

There are three types of $c = 1$ compactifications: the circle S^1 ($x \sim x + 2\pi R$), the interval $I = S^1/\mathbb{Z}_2$ (where $\mathbb{Z}_2 : x \to -x$), and three non-geometric

2000 *Mathematics Subject Classification.* Primary 81T30; Secondary 32J81, 83E30.

This talk is based on work done in collaboration with K. Becker, M. Becker, J.-X. Fu, and S.-T. Yau. The author is supported in part by NSF grants DMS-0306600 and DMS-0714648.

ones. Both the circle and the interval have a radius parameter, R_S and R_I, respectively. As is well-known, string theory sometimes cannot distinguish what we would consider as different spaces. For example, T-duality equates R_S and $1/R_S$ theories. And even more unexpectedly, the circle theory at $R_S = 2$ is also identical to the interval theory at $R_I = 1$. This latter identification in fact connects the moduli space of the circle compactifications with that of the interval compactifications. The three non-geometrical compactifications are also interesting. They can all be described as orbifolds of the circle theory at the self-dual radius $R_S = 1$. But unlike the geometrical compactifications, they are three isolated points on the $c = 1$ moduli space.

The one-dimensional moduli space of string compactifications was certainly a surprise when first discovered in the mid-1980s. One would like to classify all string compactifications and describe their connectedness (or disconnectedness) for higher-dimensional cases. Studying the moduli space of compactifications will certainly help us understand better how string theory relates to geometrical and also "non-geometrical" spaces. Of course we expect that the higher-dimensional cases are likely very challenging to classify and certainly much richer as well. Below, we shall describe a step in this direction in the more phenomenologically relevant case of six-dimensional compactifications in heterotic string theory.

2. Heterotic fluxes with supersymmetry

For many years, the focus on six-dimensional compactification was on Calabi-Yau (CY) geometries, mainly because a CY preserves minimal four-dimensional supersymmetry without turning on any additional background fields. However, a CY geometry typically has many scalar moduli which result in unwanted massless scalars in four dimensions. Thus, in the past several years, there has been much work in incorporating fluxes, i.e. non-zero p-form background fields, into compactification models. Adding fluxes while preserving supersymmetry typically requires the geometry to be non-Kähler and hence non-CY.

I shall describe flux compactification in heterotic theory and present in the next section a class of smooth geometrical solutions that preserve supersymmetry in four dimensions. Before proceeding, let me mention some of the advantages of studying flux compactification in heterotic theory as compared to type II theories:

(1) The heterotic background fields are $\{g_{mn}, H_{mnp}, \phi, F_{mn}\}$. The three-form flux field, H, is the only additional field not present in standard Calabi-Yau compactification. The first three fields are the NS-NS fields of the type II theories. There are no additional R-R fields, and thus, in heterotic theory, the supersymmetry conditions incorporating the fluxes are relatively "simpler" and the possible types of geometry are more constrained.

(2) The internal six-dimensional geometry can be smooth since no sources or branes need to be present. In type II theories, supergravity no-go theorems for compactification to four-dimensional Minkowski spacetime stipulate that fluxes can only be non-zero if the compact geometry has singularities (which can arise from the presence of branes). The heterotic theory bypasses the no-go theorem with the anomaly cancellation condition which modifies the Bianchi identity for the three-form H.

(3) Gauge fields are naturally present. This allows for the possibility of constructing models with interesting phenomenology and possibly reproducing the Standard Model.

2.1. Review of $N = 1$ supersymmetry constraints. In the string frame, preserving supersymmetry requires the ten-dimensional geometry to be the product space $M = M^{3,1} \times X_6$. $N = 1$ supersymmetry requires the existence of a nowhere vanishing spinor η on X_6 that satisfies the following equations:

$$\begin{aligned} \delta\psi_m &= \nabla_m \eta + \frac{1}{8} H_{mnp}\, \gamma^{np}\, \eta = 0, \\ \delta\lambda &= \gamma^m \partial_m \phi\, \eta + \frac{1}{12} H_{mnp}\, \gamma^{mnp}\, \eta = 0, \\ \delta\chi &= \gamma^{mn} F_{mn} \eta = 0. \end{aligned}$$

With a non-vanishing spinor η, we can write down a complex structure $J_m{}^n = -i\eta^\dagger \gamma_m{}^n \eta$ such that $J^2 = -1$ and the Nijenhuis tensor $N_{mn}{}^p = 0$. Furthermore, the Hermitian $(1,1)$-form and the holomorphic $(3,0)$-form which together define the geometry of X_6 can also be simply expressed as fermion bilinears as follows.

$$\begin{aligned} J^{(1,1)}_{mn} &= -i\eta^\dagger \gamma_{mn} \eta, \\ \Omega^{(3,0)}_{mnp} &= e^{-2\phi} \bar{\eta}^\dagger \gamma_{mnp} \eta. \end{aligned}$$

We now list the supersymmetry constraints on the background fields.

2.1.1. *Geometry.* The constraints on the six-dimensional geometry can be expressed as differential equations on the nowhere vanishing pair (J, Ω). We compare the case of Calabi-Yau versus the non-Kähler case in the following.

Calabi-Yau	Non-Kähler
$dJ = 0$ (Kähler condition)	$d(\|\Omega\| * J) = 0$ (conformally balanced)
$d\Omega^{(3,0)} = 0$	$d\Omega^{(3,0)} = 0$
$H = 0$	$H = i(\bar{\partial} - \partial)J \neq 0$
$\phi = \phi_0 = \text{constant}$	$e^{-2(\phi - \phi_0)} = \|\Omega\|$

Note that since $H^{(1,2)} = i\bar{\partial}J$, if $H \neq 0$, then the metric $J_{a\bar{b}} = ig_{a\bar{b}}$ must be non-Kähler. However, not any non-Kähler geometry is allowed; the metric with a non-zero H-flux turned on must be conformally balanced. The balanced condition is defined to be

$$d(*J) = \frac{1}{2}d(J \wedge J) = 0.$$

Note that Kähler metrics are also balanced since $dJ = 0$ implies $d(J \wedge J) = 0$. Indeed, the balanced condition can be thought of as a relaxation of the Kähler condition. Explicitly, consider the number of constraint equations for three complex dimensions. For the Kähler condition, $dJ = 0$ gives 9 complex constraint equations. For the balanced condition, $*d(*J) = 0$ gives only 3 complex constraint equations.

Heuristically, the additional "degrees of freedom" in a conformally balanced geometry can be thought of as being parametrized by (H, ϕ) – H parametrizes the non-Kählerity and ϕ the conformal factor or the norm of Ω. Conversely, one can just forget about the H-flux and ϕ and simply consider the compactification geometry as being defined by J and Ω which are required to satisfy the above constraints.

Perhaps an analogy to closed curves in $\mathbb{R}^2$ would be useful here. The Calabi-Yau geometry is rigid and is the most constrained and we can compare it to a circle in $\mathbb{R}^2$. If we add flux and relax to non-Kähler geometry in heterotic theory, this can be said to correspond to an ellipse or a more curvy closed curve in $\mathbb{R}^2$. The additional degrees of freedom in the geometry, say for example the eccentricity of the ellipse or the curvature of a more general curve, would correspond physically to what we call fluxes. Flux geometries are clearly much more numerous and, as we also know, nature does not always choose the most symmetrical configurations, *e.g.* the elliptical shapes of planetary motion.

2.1.2. *Gauge Bundle.* The background gauge field satisfies the Hermitian-Yang-Mills condition:

$$F_{(2,0)} = F_{(0,2)} = 0, \quad F_{mn}J^{mn} = 0.$$

The first part is the condition that the gauge bundle is holomorphic. The second is a primitivity condition. Together, the Hermitian-Yang-Mills condition is known from the work J. Li and Yau [**LY**] to correspond to the gauge bundle being "stable."

2.1.3. *Anomaly Condition.* In the heterotic theory, this is the modified Bianchi identity

$$dH = 2i\partial\bar{\partial}J = \frac{\alpha'}{4}(\mathrm{tr}R \wedge R - \mathrm{tr}F \wedge F)\ .$$

This is an important four-form equation which further relates the gauge bundle with the geometry.

3. A class of smooth compact solutions

Now we present a compactification model that satisfies all of the above supersymmetry conditions. We shall proceed in two steps. We first compactify from ten to six dimensions and then compactify another two dimensions to four dimensions.

3.1. Four compact dimensions to 6D. For four real compact dimensions, the required geometry was studied by Strominger [**St**] in 1986. The balanced condition $d(*J) = dJ = 0$ in four dimensions is just the Kähler condition. So the required geometry is that of a conformal Calabi-Yau with

$$J = e^{2\phi} J_{K3}, \qquad \Omega^{(2,0)} = \Omega_{K3}.$$

For gauge fields, one can have $U(1)$ gauge fields taking the field strengths to be the primitive (1,1) forms on $K3$:

$$U(1): \qquad F^{(1,1)} \wedge J_{K3} = 0, \qquad F^{(1,1)} \in H^{(1,1)}(K3, \mathbb{Z}).$$

Alternatively, one can look for stable $SU(r)$ bundles on $K3$. These have been studied by Mukai [**M**] and are known to exist as long as

$$SU(r): \qquad c_2(F) = -\frac{p_1}{2} \geq r - \frac{1}{r}.$$

The anomaly condition being a four-form equation on a four-manifold becomes a single Laplacian equation with added sources. It can be solved as long as the source, proportional to $*(\operatorname{tr} R \wedge R - \operatorname{tr} F \wedge F)$, integrates to zero. This gives the condition that

$$C_2(F) = \frac{1}{16\pi^2} \int_{K3} \operatorname{tr} F \wedge F = C_2(R_{K3}) = 24.$$

There are certainly many possible stable bundles on $K3$ that satisfy the 2nd Chern class condition. Note that the anomaly condition clearly shows that a conformal T^4 geometry would not be possible since $c_2(R_{T^4}) = 0$.

3.2. Six compact dimensions to 4D. We now take the conformal $K3$ solution and compactify two additional directions to get to four dimensions. We can compactify straightforwardly on a T^2 with complex coordinate $z = x + iy$. The metric and holomorphic three-form take the form

$$J = e^{2\phi} J_{K3} + \frac{i}{2} dz \wedge d\bar{z}, \qquad \Omega^{(3,0)} = \Omega_{K3} \wedge dz.$$

With the gauge field strength specified as above, we have a simple six-dimensional compactification on a Kähler $K3 \times T^2$ manifold.

But we can also compactify the two dimensions in a more sophisticated manner [**FY, BBFTY**]. Instead of a trivial T^2 metric, we can apply a Kaluza-Klein (KK) compactification on the T^2. The global one-form on the T^2 is now

$$\theta = (dz + \alpha_1 dz_1 + \alpha_2 dz_2)$$

with (z_1, z_2) the local complex coordinates on the $K3$ surface and α_i the KK gauge field. If the KK field strength $\omega = d\alpha \neq 0$, then the manifold is no longer $K3 \times T^2$. In fact, it is non-Kähler and is mathematically a T^2 bundle over a $K3$ surface. We can take the metric and holomorphic three-form to be

$$J = e^{2\phi} J_{K3} + \frac{i}{2}\theta \wedge \bar{\theta}, \qquad \Omega^{(3,0)} = \Omega_{K3} \wedge \theta. \tag{3.1}$$

In order that the manifold be complex and balanced, we need to require that $\omega = d\alpha \in H^{(1,1)}(K3, \mathbb{Z})$ and that ω is primitive, *i.e.* $\omega \wedge J_{K3} = 0$.[1] For the gauge field strengths, we can take as before the Hermitian-Yang-Mills field strengths on stable bundles of the $K3$. The only remaining question is whether the anomaly condition can be satisfied. This turns out to be an important question, for if the anomaly condition is not present, it can be shown that the above non-Kähler ansatz (3.1) would have an infinite number of scalar moduli [**BTY1**]. But inserting the non-Kähler ansatz into the anomaly equation gives a highly non-linear second order differential equation for ϕ on $K3$. Demonstrating that there exists a solution for ϕ is highly non-trivial and this was proved by Fu and Yau [**FY**] when the following topological condition is satisfied:

$$c_2(F) + \int_{K3} \|\omega\|^2 = 24.$$

The expectation that there exists a non-trivial non-Kähler solution in heterotic theory was first motivated via duality from M-theory on $K3 \times K3$ with G-flux. This is the work of Dasgupta *et al.* [**DRS**] (see also [**BD**]) where they wrote down the form of the Hermitian metric. However, whether the anomaly condition can be satisfied in the heterotic theory on a $K3$ surface was not known until the proof of Fu and Yau [**FY**]. Hence, this non-Kähler geometry has been called in the literature [**CL**] the FSY geometry (after Fu, Strominger, and Yau). It is also worth pointing out that there is now a gauged linear sigma model construction of this non-Kähler solution that is valid to all orders in α' [**AEL**].

4. Connectedness of Kähler and non-Kähler solutions

With a large class of six-dimensional compact solutions, which includes both Kähler and non-Kähler manifolds, *i.e.* $K3 \times T^2$ and T^2 bundles over

[1] More generally, ω can have a component in $H^{(2,0)}(K3, \mathbb{Z})$ (see [**BBFTY**]).

$K3$, it is interesting to ask whether any of the solutions can be related. It turns out that in fact some of the non-Kähler solutions can be continuously deformed to Kähler solutions with $U(1)$ gauge bundles [**BTY2**] (see also [**A, Se**]). The easiest way to see this is via duality to M-theory compactification on $K3 \times K3$ with G-flux. The heterotic geometry that is dual to the M-theory solution is the above six-dimensional geometry (3.1) times a circle. We can recover the above six-dimensional solution by decompactifying the circle, *i.e.* taking the limit of the circle radius to infinity. It turns out that the trivial duality of exchanging the two $K3$'s in the M-theory solution on $K3 \times K3$ can correspond in the heterotic theory to exchanging the T^2 twist ω with the $U(1)$ gauge field strength F. So for example, the heterotic Kähler solution with $(\omega = 0, F \neq 0)$ can be dualized to a non-Kähler solution with $(\omega \neq 0, F = 0)$.

Notice that the information of the T^2 twisting ω is encoded in the metric. The Kähler/non-Kähler duality of exchanging ω and F thus suggests that we should treat the the metric g and the field strength F (flux) on an equal footing when considering heterotic flux compactification. This is an important difference between Calabi-Yau and non-Kähler compactifications. In the Calabi-Yau case, all information is encoded in the geometry or specifically the metric. But in general string compactifications, we shall describe the "string" geometry with both metric and fluxes.

References

[A] A. Adams, *Conformal field theory and the Reid conjecture*, [arXiv:hep-th/0703048].

[AEL] A. Adams, M. Ernebjerg and J. M. Lapan, *Linear models for flux vacua*, [arXiv:hep-th/0611084].

[BBFTY] K. Becker, M. Becker, J.-X. Fu, L.-S. Tseng and S.-T. Yau, *Anomaly cancellation and smooth non-Kähler solutions in heterotic string theory*, Nucl. Phys. B **751** (2006), no. 1-2, 108–128. [arXiv:hep-th/0604137].

[BD] K. Becker and K. Dasgupta, *Heterotic strings with torsion*, J. High Energy Phys. **0211** (2002) 006, 34 pp [arXiv:hep-th/0209077].

[BTY1] M. Becker, L.-S. Tseng and S.-T. Yau, *Moduli space of torsional manifolds*, Nucl. Phys. B **786**, 119 (2007) [arXiv:hep-th/0612290].

[BTY2] M. Becker, L.-S. Tseng and S.-T. Yau, *Heterotic Kähler/non-Kähler Transitions*, [arXiv:0706.4290].

[CL] M. Cyrier and J. M. Lapan, *Towards the massless spectrum of non-Kähler heterotic compactifications*, Adv. Theor. Math. Phys. **10** (2006), no. 6, 853–877. [arXiv:hep-th/0605131].

[DRS] K. Dasgupta, G. Rajesh and S. Sethi, *M-theory, orientifolds and G-flux*, J. High Energy Phys. **9908** (1999), 23, 26 pp [arXiv:hep-th/9908088].

[DVV] R. Dijkgraaf, E. P. Verlinde and H. L. Verlinde, *$c = 1$ conformal field theories on Riemann surfaces*, Commun. Math. Phys. **115** (1988) 649–690.

[FY] J.-X. Fu and S.-T. Yau, *The theory of superstring with flux on non-Kaehler manifolds and the complex Monge-Ampère equation*, [arXiv:hep-th/0604063].

[G] P. H. Ginsparg, *Curiosities at $c = 1$*, Nucl. Phys. B **295** (1988) 153–170.

[LY] J. Li and S.-T. Yau, *Hermitian-Yang-Mills connection on non-Kähler manifolds* in *Mathematical aspects of string theory*, World Scientific Publ., S.-T. Yau, editor; London (1987), 560–573.

[M] S. Mukai, *Moduli of vector bundles on K3 surfaces, and symplectic manifolds*, Sugaku Expositions **1** (1988), no. 2, 139–174.
[Se] S. Sethi, *A note on heterotic dualities via M-theory*, [arXiv:0707.0295].
[St] A. Strominger, *Superstrings with torsion*, Nucl. Phys. B **274** (1986), no. 2, 253–284.

Department of Mathematics and Jefferson Physical Laboratory, Harvard University, Cambridge, MA 02138
E-mail address: tseng@math.harvard.edu

AMS/IP Studies in Advanced Mathematics
Volume 44, 2008

On the Spectrum of Pure Yang-Mills Theory

Laurent Freidel, Robert G. Leigh, Djordje Minic, and Alexandr Yelnikov

ABSTRACT. In this note we discuss the wave functional approach to the spectrum of pure Yang-Mills theory in $2+1$ and $3+1$ dimensions by highlighting the issues of dynamical mass generation and the role played by the kinetic term. We extrapolate our recent analysis of $2+1$ Yang-Mills theory to $3+1$ dimensions, and under certain heuristic assumptions use a simple quasi-Gaussian vacuum wave functional for $3+1$ Yang-Mills theory for a preliminary study of the spectrum of glueballs which nicely fits the available lattice simulations.

1. Introduction and summary

In this note we address the question of how to understand the spectrum of pure Yang-Mills (YM) theory using the formalism of wave functionals. In particular we review and clarify our previous work on $2+1$ YM theory [**1**] and discuss some preliminary results regarding the $(3+1)$-dimensional problem.

The wave functional approach we have recently used in the context of $2+1$ YM theory [**1**] turns out to fit very nicely the existing lattice simulations [**2**]. After reviewing this work we discuss a particularly simple quasi-Gaussian ansatz for the vacuum wave functional which can be used for a study of the spectrum of $3+1$ Yang-Mills theory. This ansatz is based on some heuristic postulates motivated by existing lattice simulations. Note that even though we cannot argue from first principles that we understand the large N limit, we do compare our results to the large N lattice data. It turns out that the comparison with finite N results does not alter the final results too much. We follow the general strategy towards a wave functional approach to the spectrum of $3+1$ YM that was outlined in [**3**]. The lattice results we compare to are collected, for example, in [**4, 5, 6, 7, 8, 9**].

2000 *Mathematics Subject Classification.* Primary 81T13; Secondary 81T25.

The main physics issues to be addressed in $3+1$ YM theory, as opposed to its $(2+1)$-dimensional counterpart are:

1) dynamical generation of a scale independent mass: the so-called gap equation. We discuss this issue by analyzing the origin of the mass gap from the geometry of the configuration space in the context of $2+1$ YM. We then present some very heuristic arguments about the gap origin in the more difficult $(3+1)$-dimensional problem.

2) the IR asymptotics of the quasi-Gaussian vacuum wave functional and the issue of the evaluation of the string tension of large planar Wilson loops.

3) finally, there is the question of how to recover the main feature of the UV physics, i.e. asymptotic freedom, the hallmark universal feature of QCD perturbation theory (as exemplified in the famous $11/3$ one-loop β function coefficient). This turns out to follow from the proper UV asymptotics of the vacuum wave functional.

By contrast, one does not have to worry about these issues in the analysis of $2+1$ YM. In that case the theory has a natural dimensionful parameter, the YM coupling. So, in our attempt to treat some phenomenological features of the $3+1$ YM spectrum by using the experience with $2+1$ YM theory, we implicitly assume that the above difficult issues can be addressed in a self-consistent manner in the context of simple quasi-Gaussian vacuum wave functionals. For example, we simply assume that the mass gap is self-consistently generated in the wave functional approach to the $3+1$ YM theory. Also, the form of the quasi-Gaussian ansatz for the vacuum wave functional in the $(3+1)$-dimensional context is essentially motivated by the $(2+1)$-dimensional counterpart. The IR asymptotics of the $3+1$ YM theory is very difficult to analyze from first principles: see, for example, the discussion in [**10**]. In the $3+1$ problem, the asymptotic freedom issue is dealt with by appealing to the correct UV limit of the vacuum wave functional, the requirement of gauge symmetry (so that all interaction vertices are correctly captured) and the standard RG analysis of wave functionals [**11**].

Our appeal for a self-consistent generation of the mass gap and the IR asymptotics on the level of the vacuum wave functional does have some resonance with the philosophy pursued in [**12**] as well as the AdS/QCD approach [**13**]. The details appear to be very different though. The essential physics ingredient in our case is the constituent picture which seems to work so nicely in $2+1$ YM [**1**] (for previous attempts at the constituent picture in the $3+1$ YM consult, for example, [**14**]).

This note is organized as follows:

1) We start with a general discussion of how to understand the spectrum of pure Yang-Mills theory in the formalism of vacuum wave functionals. We concentrate here on $(2+1)$-dimensional YM theory. The crucial heuristic points in this approach are the issues of a dynamical mass gap generation and the role of the kinetic term.

2) We then present a more phenomenologically minded discussion of the spectrum of $(3+1)$-dimensional Yang-Mills theory based on a quasi-Gaussian vacuum wave functional. This phenomenologically minded ansatz is on the one hand motivated by the results in $2+1$ dimensions and on the other hand by good numerical comparison with the lattice data. We think that the results presented in this paper justify the simple quasi-Gaussian vacuum wave functional approach in spite of the somewhat tentative nature of our theoretical analysis.

2. Wave functionals and the kinetic term

Here we outline a general procedure regarding the derivation of a vacuum wave functional for pure YM theory, which in the case of $2+1$ YM theory happens to be in excellent agreement with the corresponding lattice data concerning the spectrum of the vacuum states of the theory [**1**].

We contrast the abelian case, which can be exactly solved, and the more difficult non-abelian case. We do not utilize any special variables. The approach is, in principle, very general and transparent. For concreteness, in what follows, we concentrate on the $(2+1)$-dimensional physics. In later sections we will discuss the $(3+1)$-dimensional YM theory.

Given the gauge theory Hamiltonian

$$\mathcal{H} = \frac{1}{2}\int (g^2 E_i^2 + \frac{1}{g^2} B^2) \tag{2.1}$$

where E_i and B are the electric and magnetic components of $F_{\mu\nu} \equiv \partial_\mu A_\nu - \partial_\nu A_\mu + [A_\mu, A_\nu]$, we wish to reason out the vacuum wave functional using the ordinary connection variables A^a. In the Hamiltonian gauge $A_0 = 0$

$$\mathcal{H} = \frac{1}{2}\int \left(-g^2\left(\frac{\delta}{\delta A_i}\right)^2 + \frac{1}{g^2}B^2\right) \tag{2.2}$$

We are looking for Ψ_0 such that $\mathcal{H}\Psi_0 = 0$, after subtracting the infinite vacuum energy term.

In the abelian case the answer is obviously Gaussian. In the non-abelian case this is not so, for reasons of non-abelian gauge symmetry. So, in $2+1$ dimensions we consider a quasi-Gaussian ansatz, which is compatible with the Gauss law

$$\Psi_0 = \exp\left(-c\int B(K(L))B\right), \tag{2.3}$$

where c is the appropriate constant $1/(2g^2 m)$, B is the magnetic field and $L \equiv \frac{\Delta}{4m^2}$, where Δ is the covariant Laplacian, and m is the appropriate mass parameter. Note that the parameter m can be explicitly evaluated using the approach of “corner variables” [**15**] of Karabali and Nair [**16**]. In that case there is a very precise relation (for large N)

$$m \equiv \frac{g^2 N}{2\pi} \tag{2.4}$$

It suffices to say here that the parameter m is ultimately related to the value of the string tension for large Wilson loops. Also, at least formally

$$K(L) = \sum_{n=0}^{\infty} c_n L^n \tag{2.5}$$

and thus more generally

$$\mathcal{H}\Psi_0 = E_0\Psi_0 = \left[E_0 + \int B^a\,(\mathcal{R})\,B^a + \cdots\right]\Psi_0. \tag{2.6}$$

The (divergent) vacuum energy E_0 can be isolated and as expected, the leading divergence in the UV is cubic. Next, what we need to do is compute the expression that we have labeled by $\mathcal{R}$ and set it to zero. This will constitute an equation for the kernel K.

Let us consider the various terms in $\mathcal{R}$. First, the potential term $B^a B^a$ clearly contributes a fixed constant to $\mathcal{R}$. The remainder of $\mathcal{R}$ will come from the action of the kinetic energy operator T_{KN}. Given the form of the vacuum wave functional $\Psi_0 \sim e^P$, it is elementary to derive

$$\frac{\delta\Psi_0}{\delta A^a(z)} = \frac{\delta P}{\delta A^a(z)}\Psi_0 \tag{2.7}$$

and thus

$$\frac{\delta^2\Psi_0}{\delta A^a(z) A^b(w)} = \left[\frac{\delta^2 P}{\delta A^a(z) A^b(w)} + \frac{\delta P}{\delta A^a(z)}\frac{\delta P}{\delta A^b(w)}\right]\Psi_0 \tag{2.8}$$

from which we find

$$\mathcal{H}\Psi_0 = \left[TP + m\int_{z,w}\frac{\delta P}{\delta A^a(z)}\frac{\delta P}{\delta A^b(w)} + \frac{1}{m}\int B^a B^a\right]\Psi_0. \tag{2.9}$$

The second term in brackets is easy to compute. Since we want to solve the Schrödinger equation only to quadratic in B^a order,

$$P \sim \int B^a\ K\left(\frac{\Delta}{m^2}\right) B^a + \cdots \tag{2.10}$$

By evaluating functional derivatives $\delta P/\delta A$ we see that the second term in brackets in (2.9) is equal to

$$\int B^a \left[\frac{\Delta}{m^2} K^2\left(\frac{\Delta}{m^2}\right)\right] B^a \tag{2.11}$$

Therefore, we deduce that this term contributes $LK^2(L)$ to $\mathcal{R}$. This result also holds in the pure abelian case.

Let us finally proceed to the TP term in (2.9). In the abelian case this term does not depend on field variables and gives (infinite) vacuum energy only. In the non-abelian case the kinetic term plays a very important role. In $2+1$ dimensions the simplest way to proceed is to assume that the kinetic energy operator T acts homogeneously on any local operator-valued function of B. This is natural in the holomorphic corner variable approach of Karabali, Kim and Nair. This assumption can be somewhat justified by explicit computations involving the action of T on the potential term B^2. In that particular case the action of T would simply produce $2mP$, since P is quadratic in B. Now, by extrapolating this statement into [**1**]

$$T\mathcal{O}_n = (2+n)m\,\mathcal{O}_n + \cdots \tag{2.12}$$

where $\mathcal{O}_n = BL^nB$, one gets much more information[1]. Notice that here ellipsis stand for terms of higher order in B, but the same mass dimension as $\mathcal{O}_n$, and which mix with $\mathcal{O}_n$ under the action of T. Finally, note that this statement in some sense incorporates the role of topology (i.e. compactness of the configuration space) in the non-abelian case, and encapsulates the importance of the kinetic term in the dynamical origin of the mass gap, which is intuitively very much in line with [**17**].

Returning to our discussion of the Schrödinger equation we note that (2.12) implies

$$TP = -c\int B\left\{\frac{1}{2}\sum_n c_n(2+n)L^n\right\}B. \tag{2.13}$$

It is convenient to write the factor in braces formally as

$$\frac{1}{2L}\frac{d}{dL}[L^2K(L)]. \tag{2.14}$$

Assembling all of these results, we then find the following Riccati equation [**1**]:

$$\mathcal{R}\sim\left[-K-\frac{L}{2}\frac{d}{dL}[K(L)]+LK^2+1\right]=0. \tag{2.15}$$

Note that the abelian answer is given by the last algebraic part of the equation, which does not have anything to do with the spectrum of the kinetic term, i.e. $LK^2+1=0$. This of course, has to be reproduced even in the non-abelian situation at UV. Now, the discrete spectrum of T is essential for the emergence of the gap. So the first part of the equation $\frac{1}{2L}\frac{d}{dL}\left[L^2K(L)\right]$ is crucial. In the IR limit only the factor $K-1=0$ survives. The derivative

[1]A possible calculational justification for this statement might be obtained by following the strategy proposed in the recent work by Fukuma et al. [**18**].

term $\frac{L}{2}\frac{d}{dL}[K(L)]$ interpolates between the UV and the IR limits. The other essential ingredient is that there is only one normalizable K that solves the whole differential equation, as discussed in what follows.

Finally, note that the whole discussion is in principle valid for any rank of the gauge group. Even though we do not have a very good reason to rely on large N, we will perform the actual comparison of analytic results for the spectrum with large N lattice data (which are already very close to the $SU(3)$ case.)

2.1. The Gluonic Kernel. Although the above Riccati equation is non-linear, it is easily transformed into a linear second-order equation of the Bessel type, and one finds a general solution of the form

$$K(L) = \frac{1}{\sqrt{L}} \frac{CA_2(4\sqrt{L}) + Y_2(4\sqrt{L})}{CA_1(4\sqrt{L}) + Y_1(4\sqrt{L})} \tag{2.16}$$

where C is a constant and J_n (Y_n) denote the Bessel functions of the first (second) kind. As explained in [**1**] it is remarkable that the only normalizable wave functional is obtained for $C \to \infty$, which is also the only case that has both the correct UV behavior appropriate to asymptotic freedom, as well as the correct IR behavior appropriate to confinement and a mass gap! This solution is of the form

$$K(L) = \frac{1}{\sqrt{L}} \frac{J_2(4\sqrt{L})}{J_1(4\sqrt{L})} \tag{2.17}$$

This remarkable formula is reminiscent of similar results in related contexts [**12**]; here, it encodes information on the spectrum of the theory. We note that this kernel has the following asymptotics (where $L \sim -\vec{p}^2/4m^2$)

$$p \to 0, \quad K \to 1; \qquad p \to \infty, \quad K \to 2m/p \tag{2.18}$$

consistent with confinement and asymptotic freedom, respectively (Note that the argument of Bessel functions is imaginary; so instead of J_n we have I_n Bessel functions.)

Now using standard Bessel function identities we may expand

$$\frac{J_1(u)}{J_2(u)} = \frac{4}{u} + 2u \sum_{n=1}^{\infty} \frac{1}{u^2 - \gamma_{2,n}^2} \tag{2.19}$$

where the $\gamma_{2,n}$ are the ordered zeros of $J_2(u)$ [**1**]. The inverse kernel is thus

$$K^{-1}(L) = \sqrt{L}\frac{J_1(4\sqrt{L})}{J_2(4\sqrt{L})} = 1 + 8L \sum_{n=1}^{\infty} \frac{1}{16L - \gamma_{2,n}^2}. \tag{2.20}$$

Now, if we regard $L \simeq \partial\bar{\partial}/m^2$, in terms of momentum p we find

$$K^{-1}(p) = 1 + \frac{1}{2}\sum_{n=1}^{\infty} \frac{\vec{p}^2}{\vec{p}^2 + M_n^2}. \tag{2.21}$$

Here

$$M_n = \frac{\gamma_{2,n} m}{2}. \tag{2.22}$$

M_n's can be interpreted as constituents out of which glueball masses are constructed. It is not difficult now to find a Fourier transform of inverse kernel $K^{-1}(k)$. By rewriting (2.21) as

$$K^{-1}(p) = 1 + \frac{1}{2}\sum_{n=1}^{\infty} \left(1 - \frac{M_n^2}{\vec{p}^2 + M_n^2}\right) \tag{2.23}$$

we immediately obtain

$$K^{-1}(x-y) = \delta^{(2)}(x-y) + \frac{1}{2}\sum_{n=1}^{\infty} \left(\delta^{(2)}(x-y) - \frac{M_n^2}{2\pi} K_0(M_n|x-y|)\right) \tag{2.24}$$

where $K_0(x)$ is the modified Bessel function of the third kind. At asymptotically large spatial separations $|x-y| \to \infty$ this takes the form

$$K^{-1}(|x-y|) \approx -\frac{1}{4\sqrt{2\pi|x-y|}} \sum_{n=1}^{\infty} (M_n)^{\frac{3}{2}} e^{-M_n|x-y|}. \tag{2.25}$$

As shown in [**1**] we can treat B as a free field as well and therefore

$$\langle B(x)\ B(y)\rangle \sim K^{-1}(x-y). \tag{2.26}$$

The constituent masses are controlled by the zeros of J_2:

$$M_n = \frac{\gamma_{2,n} m}{2}. \tag{2.27}$$

The glueball masses then follow by evaluating the correlators of the relevant operators that probe the glueball states. For example, the mass of the 0^{++} glueball may be probed by the operator TrB^2. We have [**1**]

$$\langle \mathrm{Tr}\,(B^2)_x\ \mathrm{Tr}\,(B^2)_y\rangle \sim \left(K^{-1}(|x-y|)\right)^2. \tag{2.28}$$

and thus the 0^{++} mass is [**1**]

$$M_{0^{++}} = M_1 + M_1 \tag{2.29}$$

A more complete discussion of other glueball masses can be found in [**1**].

Now it is also not difficult to derive an expression for the vacuum expectation of the large Wilson loop of area A. At large N

$$\langle \Phi(C) \rangle = \left\langle \mathrm{Tr}\, P \exp\left(\oint_C A \right) \right\rangle \stackrel{N\to\infty}{\longrightarrow} N \exp\left(-N \int d^2x d^2y \, \langle B(x) B(y) \rangle \right) \tag{2.30}$$

or

$$\frac{1}{N} \langle \Phi(C) \rangle = \exp\left(-\frac{g_{YM}^4 N^2}{8\pi} \int d^2x d^2y \, K^{-1}(x-y) \right) \tag{2.31}$$

and from this expression we see that the leading δ-function gives the area law with a string tension that fits beautifully the lattice simulations. As for the rest of the terms that appear in (2.24), we may notice that

$$\int_A d^2x \left(\delta^{(2)}(x-y) - \frac{M_n^2}{2\pi} K_0(M_n|x-y|) \right) \;\to\; 0 \quad \text{as} \quad A \to \infty \tag{2.32}$$

and therefore these terms will give corrections to the area law behavior which vanish for asymptotically large loops. The resulting string tension is then [**16**]

$$\sqrt{\sigma} = \frac{g^2 N}{\sqrt{8\pi}} = \sqrt{\frac{\pi}{2}}\, m \tag{2.33}$$

Given this expression we can easily convert glueball masses (2.29) into units of the square root of the string tension, which makes comparison to lattice data straightforward.

Finally, a couple of comments. Note that the gluonic wave functional leads to an effective partition function of a pure 2d YM theory in the IR limit. Also, the purely Gaussian vacuum wave functionals suffice for self-consistent calculations of both the glueball masses and the string tension. Similarly, note that the J constituents do not appear as asymptotic states. Thus, it might be plausible that the glueball constituents are "seeds" for constituent quarks once the fermionic degrees of freedom are included. If so, quark confinement would be "seeded" by the confinement mechanism in the pure glue sector.

2.2. Generalizing the kernel equation. One of the most intriguing results of [**1**] is the prediction of the 0^{++} glueball mass which agrees extremely well with the lattice simulations. The derivation of that result crucially depends on equation (12) which leads to the nontrivial kernel equation (2.15). In the absence of a rigorous derivation of that equation from first principles, we want to show that equation (12) is well-motivated phenomenologically. To this end we may consider a modification of equation (12). We will also use the following results based on such a modified kernel equation to motivate our phenomenologically minded study of the spectrum of $3+1$ YM theory.

Suppose that the action of the kinetic term is as follows:[2]

$$T\mathcal{O}_n = (2 + bn)m\mathcal{O}_n + \cdots \tag{2.34}$$

where b is some free parameter.

Then the kernel $K(L)$ satisfies the following equation of Riccati type:

$$-K - b\frac{L}{2}\frac{d}{dL}[K(L)] + LK^2 + 1 = 0 \tag{2.35}$$

After changing variables as

$$K = -b\frac{y'}{2y} \tag{2.36}$$

where the prime denotes the derivative with respect to L, we arrive at

$$y'' + \frac{2}{bL}y' + \frac{4}{b^2L}y = 0 \tag{2.37}$$

which is readily solved in terms of Bessel functions. The general solution is

$$y = c_1 L^{\frac{1-2/b}{2}} J_{2/b-1}(4/b\sqrt{L}) + c_2 L^{\frac{1-2/b}{2}} Y_{2/b-1}(4/b\sqrt{L}). \tag{2.38}$$

The physically relevant solution will be

$$y = L^{\frac{1-2/b}{2}} J_{2/b-1}(4/b\sqrt{L}) \tag{2.39}$$

Using the standard recursive formulae

$$\begin{aligned} J_{\nu-1}(u) + J_{\nu+1}(u) &= \frac{2\nu}{u} J_\nu(u) \\ J_{\nu-1}(u) - J_{\nu+1}(u) &= 2J'_\nu(u) \end{aligned} \tag{2.40}$$

and analogous formulae for Y_ν, we arrive at the general solution

$$K(L) = \frac{1}{\sqrt{L}} \frac{CJ_{2/b}(4/b\sqrt{L}) + Y_{2/b}(4/b\sqrt{L})}{CJ_{2/b-1}(4/b\sqrt{L}) + Y_{2/b-1}(4/b\sqrt{L})}. \tag{2.41}$$

Once again, the only solution which is normalizable in the UV and IR is given by

$$K(L) = \frac{1}{\sqrt{L}} \frac{J_{2/b}(4/b\sqrt{L})}{J_{2/b-1}(4/b\sqrt{L})}. \tag{2.42}$$

[2]In principle we could have taken an even more general expression

$$T\mathcal{O}_n = (a + bn)m\mathcal{O}_n + \cdots$$

This modification, however, does not lead to any new conclusions compared to (2.34).

We see that the argument (and order) of the relevant Bessel functions depends on b, and we recover the standard $2+1$ result if $b = 1$. Note that in general if $TO_n = (a+bn)O_n$ we have

$$K(L) = \frac{a}{2}\frac{1}{\sqrt{L}}\frac{J_{a/b}(4/b\sqrt{L})}{J_{a/b-1}(4/b\sqrt{L})}. \tag{2.43}$$

For $a = 2$, the kernel has the following asymptotics (where $L \sim -p^2/4m^2$):

$$p \to 0, \quad K \to 1; \quad p \to \infty, \quad K \to 2m/p \tag{2.44}$$

consistent with confinement and asymptotic freedom, respectively.

Now using standard Bessel function identities (see [**1**]) or more generally

$$\frac{J_{\nu+1}(u)}{J_\nu(u)} = 2u \sum_{n=1}^{\infty} \frac{1}{u^2 - \gamma_{\nu,n}^2} \tag{2.45}$$

where the $\gamma_{\nu,n}$ are the ordered zeros of $J_\nu(u)$. The inverse kernel is thus in general

$$K^{-1}(L) = \sqrt{L}\frac{J_{2/b-1}(4/b\sqrt{L})}{J_{2/b}(4/b\sqrt{L})}. \tag{2.46}$$

Then the constituent masses in this more general case are given by a simple formula, again in parallel with [**1**]:

$$M_n = \frac{b\gamma_{2/b,n}m}{2}. \tag{2.47}$$

In the case of $2+1$ YM the ratio of the glueball masses that follows from this constituent mass and the corresponding expression for the square root for the string tension best fits the lattice data for $b = 1$, as discussed in the previous subsection.

2.3. A recapitulation. The effect of the kinetic term seems to be of utmost physical importance (at least in the context of pure 2+1 YM theory.) In the abelian case the kinetic term just contributes to the vacuum energy and is not essential for the determination of the kernel. In a topologically non-trivial case, i.e. non-abelian theory, the effect of the kinetic term is crucial. Heuristically, the compactness of the configuration space seems to imply a discrete "spectrum" for the kinetic term, and this drastically changes the nature of the kernel and the resulting mass spectrum of the theory. The mass spectrum is in turn very nicely captured by a simple quasi-Gaussian vacuum wave functional. Presumably the compact $U(1)$ could also be treated in this manner, but it is not clear whether a Gaussian ansatz captures the whole physics. It is possible that instead of $P \sim \int BK(L)B$ one should consider something non-quadratic yet consistent with the compactness of the configuration space, say $P \sim \int \cos(c_1 BK(L)B)$, where c_1 has the appropriate dimensions for the cosine to make sense.

3. On the origin of the mass gap in pure YM theory

Here we present a summary of an intuitive argument of Karabali, Kim and Nair concerning the mass gap in $2+1$ YM theory [**16**]. We then discuss possible generalizations of this argument for the case of $3+1$ YM theory.

In the Karabali-Nair formalism [**16**] one obtains that the configuration space measure is given by the exponent of the 2d WZW action. This is the volume element which then also defines a normalizable inner product of gauge invariant wave functionals,

$$\langle 1|2\rangle_{2+1} = \int d\mu(H) e^{2c_A S_{WZW}(H)} \Psi_1^*(H)\Psi_2(H) \tag{3.1}$$

This configuration space measure is crucial in the argument concerning the dynamical origin of the mass gap. (The emphasis here is on the term *dynamical*; that the gap should be proportional to the dimensionful coupling constant is obvious on dimensional grounds.)

First note that the *quadratic* part of $S_{WZW}(H)$ is given, in terms of the magnetic field, by

$$S_{WZW} \sim \frac{1}{4\pi}\int dx^2 B \frac{1}{\nabla^2} B + \cdots \tag{3.2}$$

where the higher order terms in B are dropped in the exact expression.

Now, the intuitive argument of [**16**] runs as follows. Denote by ΔE and ΔB the root mean square fluctuations of the electric and magnetic fields. By utilizing the canonical commutation relations $[E_i^a, A_j^b] = -i\delta_{ij}\delta^{ab}$, one gets $\Delta E \Delta B \sim k$, where k denotes the momentum. The estimate for the energy is then

$$\epsilon = \frac{1}{2}\left[\frac{g^2k^2}{\Delta B^2} + \frac{\Delta B^2}{g^2}\right] \tag{3.3}$$

g being the gauge coupling. Note that this would give, by minimizing over ΔB^2, $\Delta B^2_{min} \sim g^2 k$ which then implies $\epsilon \sim k$, i.e. no mass gap. This is indeed the correct answer for the abelian theory, but incorrect for its non-abelian cousin.

In the non-abelian case we need to take care of the non-trivial configuration space measure (because the physical meaning of the uncertainty relation hinges on implicitly knowing the inner product), which we have in $2+1$ thanks to [**16**]. The equation for the Gaussian part of this measure implies that $\Delta B^2 \sim \pi k^2/C_A$ for small values of the momenta, that is, in the IR limit. Thus, as pointed out in [**16**], even though the energy ϵ is minimized around $\Delta B \sim k$, the probability, which is dominated by the measure factor, is centered around $\Delta B^2 \sim k^2\pi/c_A$. Consequently, the energy scales as $\epsilon \sim g^2 c_A/2\pi + O(k^2)$ and indicates the presence of a mass gap in the spectrum, proportional to $g^2 c_A$. Again, note the crucial role played by the kinetic term in this intuitive argument.

We wish to emphasize the dynamical nature of this heuristic argument, as opposed to the final result which could be expected on dimensional grounds. Obviously the dimensional argument fails in $3+1$, so we discuss some options regarding the generalization of the dynamical part of the argument.

For example, one option would be to consider the natural generalization of the above configuration space measure for $2+1$ YM in the $(3+1)$-dimensional context,

$$\langle 1|2\rangle_{3+1} = \int d\mu(\vec{B}) e^{2c_A S_{3+1}(\vec{B})} \Psi_1^*(\vec{B})\Psi_2(\vec{B}) \tag{3.4}$$

where to quadratic order in $\vec{B}$, by generalizing the quadratic result from $2+1$,

$$S_{3+1} \sim \frac{m}{g^2}\int dx^3 \vec{B}\frac{1}{\nabla^2}\vec{B} + \cdots \tag{3.5}$$

The parameter m should represent an RG invariant mass scale and it should be dynamically determined from a gap equation, perhaps along the lines of [**10**]. (This formula is natural from the point of view of the formalism of corner variables discussed in [**19**]. A different expression for this measure was found in [**20**]. This might indicate the sensitivity of the $(3+1)$-dimensional problem to a choice of variables and the fact that there are no obvious counterparts to the WZW functional in three dimensions.) By repeating the above argument based on the uncertainty principle we get that in this case the gap is proportional to m, as expected.

Another option would be to distinguish the $(2+1)$- and $(3+1)$-dimensional problems by appealing to a more dynamical role of the potential term in the $(3+1)$-dimensional context. In this case the intuition about the infrared dynamics of $3+1$ YM theory based solely on the experience with $2+1$ YM theory would be most probably incomplete.

Whatever the precise dynamical mechanism of mass generation is in $3+1$ YM theory, the crucial question is to determine the mass gap dynamically (via a gap equation), and relate it to the universal features of perturbation theory (i.e. asymptotic freedom and Λ_{QCD}). We will not explicitly address these difficult issue here. In the next section, we will simply postulate that the wave functional "knows" about a stable mass parameter m, which is in turn related to the mass gap as well as Λ_{QCD}.

4. A preliminary study of the spectrum of $3+1$ YM

In this section, we want to approach the spectrum of $3+1$ YM theory in the manner used in the analysis of $2+1$ YM, as summarized in the preceding part of the paper. Of course, the two theories are radically different, as outlined in the introduction. Our discussion is necessarily phenomenologically minded, given the difficult issues concerning the dynamical origin of the mass gap and the IR asymptotics of the vacuum wave functional. We will be mainly motivated by the existing lattice simulations in $3+1$ YM and the success of the simple quasi-Gaussian ansatz for the vacuum wave functional for $2+1$ YM theory.

As in the case of $2+1$ YM we assume the following ansatz for the vacuum wave functional:

$$\Psi_0 = \exp\left(-\frac{1}{2g^2 m}\int B_i\left(K(L)\right)B_i\right) \tag{4.1}$$

with a non-trivial kernel K and $L \equiv \frac{\Delta}{4m^2}$, where Δ is the 3d covariant Laplacian. Note, that in contrast to $2+1$ YM here it is crucial to determine the dynamical scale m (via a gap equation), which is not explicitly present in the bare Hamiltonian of $3+1$ YM. We will model the kernel $K(L)$ motivated by the discussion from section 2.2 of a generalized kernel equation for the $(2+1)$-dimensional problem. As pointed out in section 2.2, the UV behavior of such a kernel will be just what we need in $3+1$ YM theory

$$\Psi_0^{\text{UV}} = \exp\left(-\frac{1}{2g^2}\int B_i\left(\frac{1}{\sqrt{\Delta}}\right)B_i + \cdots\right) \tag{4.2}$$

In that limit the mass m cancels and we are left with the correct UV result known from QCD perturbation theory. By applying the standard background field methods to such a UV wave functional (following, for example, the nice work of Zarembo in [**11**]) we can readily recover the universal features of perturbation theory, such as the famous $11/3$ coefficient in the one-loop beta function. The corrections to this UV result (indicated by "..." in (4.2)) can be organized in powers of the mass m over the powers of momenta and as such they encapsulate non-perturbative effects. Given the usual decoupling between the perturbative and non-perturbative effects[3], such terms will not spoil the standard results of the QCD perturbation theory. In essence, what we propose is to *assume* that such terms can be summed up to give a very particular form for the kernel, motivated by our results regarding the $2+1$ YM theory as well as the existing lattice simulations concerning the spectrum of $3+1$ YM theory.

We now proceed with a very particular kernel discussed in section 2.2 for the case of $b = 2/3$. To motivate this choice for the parameter b we may

[3]In (4.2) g should be understood as the running coupling constant $g(\Lambda)$ at some very high scale $\Lambda \gg \Lambda_{\text{QCD}}$. The non-perturbative mass parameter m should be related to $g(\Lambda)$ and Λ via $m \sim \Lambda e^{-8\pi^2/b_0 g^2}$.

naively argue as follows: In $2+1$ dimensions the magnetic field B that features in the vacuum wave functional is a scalar. In $3+1$ dimensions the magnetic field B_i is a vector, so we naturally have to contract this vector index in the vacuum wave functional above. By averaging the scalar product of 3 vectors v^i in 3d over all directions (a fixed direction being defined by the usual Euler angles θ and ϕ) one gets $\frac{4\pi}{3}v^2$, where $\frac{4\pi}{3}$ comes from the angular integral $\int \cos^2\theta \sin\theta d\theta d\phi$. Per 2π range of the ϕ angle we also get the factor $2/3$. Granting its very naive nature, this might be a heuristic way to understand why the $2/3$ factor seems to be "natural" in the discussion of the vacuum wave functional for $3+1$ YM given its $(2+1)$-dimensional counterpart, but we admit that this simplistic argument is too loose and therefore only a phenomenological justification is possible at present.

Thus we take the relevant kernel from section 2.2 by choosing $b = 2/3$:

$$K^{-1}(L) = \sqrt{L}\frac{J_2(6\sqrt{L})}{J_3(6\sqrt{L})}. \tag{4.3}$$

The constituent masses are controlled by the zeros of J_3,

$$M_n = \frac{\gamma_{3,n} m}{3} \tag{4.4}$$

Note that this is different from the usual $2+1$ result $M_n = \frac{\gamma_{2,n} m}{2}$, which involves zeros of J_2! In order to determine the mass gap (or the mass of the lowest lying glueball) consider the 0^{++} states which may be probed by the operator $\mathrm{Tr}\,(B_i^a B_i^a)$. We have

$$\langle \mathrm{Tr}\,(B_i^a B_i^a)_x \ \mathrm{Tr}\,(B_i^a B_i^a)_y \rangle \sim \left(K^{-1}(|x-y|)\right)^2. \tag{4.5}$$

The correlation function follows by Wick's theorem given our knowledge of the quasi-Gaussian vacuum wave functional, in complete parallel with [**1**].

Now, following the general procedure outlined in section 2 the mass gap, or the mass of 0^{++}, is given as a sum of two constituents

$$M_{0^{++}} = M_1 + M_1 \tag{4.6}$$

and

$$M_{0^{++*}} = M_1 + M_2 \tag{4.7}$$

etc., as well as

$$M_{2^{++}} = M_2 + M_2 \tag{4.8}$$

etc., as in [**1**]. Other states follow as in [**1**]. Note that the relevant zeros of J_3 are [**21**]:

$$\gamma_{3,1} = 6.380, \gamma_{3,2} = 9.761, \gamma_{3,3} = 13.015, \gamma_{3,4} = 16.223, \gamma_{3,5} = 19.409 \tag{4.9}$$

Given the phenomenological nature of our discussion, we note that the zeros of the J_3 Bessel function do seem to provide the best fit to the available lattice data.

In order to be able to compare our results for glueball masses with actual lattice data, we have to re-express them in terms of some other measurable quantity[4] like $\sqrt{\sigma}$ or Λ_{QCD}. In other words, we need to establish the relationship between the string tension and the mass parameter m similar to eq. (2.33) in $2+1$ dimensions. At the moment we do not have any definite proposal on the possible generalization of that equation to $3+1$ dimensions. To circumvent this difficulty we should consider the coefficient in (2.33) as an extra fit parameter. Surprisingly enough, the best fit value for this parameter turns out to be very close to $\sqrt{\frac{\pi}{2}}$. For this reason in what follows we simply use the $(2+1)$-dimensional result (2.33) to express glueball masses in terms of the square root of the string tension.

We summarize the actual numerical results in a couple of tables. As noted in the introduction, our analysis can be done at any rank N, but we do comparisons with the large N data only. The large N data are from [**4, 5**]. There are also many more data available for $SU(3)$ and $SU(8)$. The large N extrapolation is done by multiplying these data by $\sqrt{N^2-1}/N$, as we did in $2+1$. The theoretical results are shown as well as the respective difference from the lattice data. As one can plainly see, the zeros of J_3 work very well.

4.1. Comments: lattice vs. theory. The only genuine large-N lattice estimates are available for three states only,

$$\frac{m_{0^{++}}}{\sqrt{\sigma}} = 3.307(53), \tag{4.10}$$

$$\frac{m_{2^{++}}}{\sqrt{\sigma}} = 4.80(14), \tag{4.11}$$

$$\frac{m_{0^{++*}}}{\sqrt{\sigma}} = 6.07(17). \tag{4.12}$$

This is from the most recent study by Lucini, Teper and Wenger [**4**] and is based on the extrapolation of results available for N = 2,3,4,6,8. These results are consistent with a previous study by Lucini and Teper [**5**] (based on N = 2,3,4,5). The only difference is that error bars are smaller in this new simulation (about 1.5 to 3% as opposed to 5-8% in Lucini and Teper).

The 0^{++} and 2^{++} mass estimates seem to be reliable. They show a smooth approach to the continuum limit for any given value of N as well as a smooth extrapolation from finite N values to $N=\infty$.

---

[4]However, we would like to point out that eqs (4.6)–(4.8) can readily be used to obtain definite numerical values for ratios of glueball masses.

The 0^{++*} state is more troublesome. So we should compare critically data from other sources as well. Other sources are the PhD thesis of H. Meyer [**6**] (this includes SU(3) and SU(8) data; somewhat surprisingly, Meyer's data differs from Teper et al.) and Morningstar and Peardon [**7**] (SU(3) only; this is considered, at the moment, to be the most precise lattice study for $N = 3$). First, all these authors report that the finite volume corrections are still important for this state and so its identification as a glueball is questionable (i.e. this state might be a torelon). Second, according to Teper *et al.* $m_{0^{++*}} \approx 2\, m_{0^{++}}$, which leads to a possibility that this state is actually a two-glueball state with zero relative momentum. Third, the extrapolation of finite N values to $N = \infty$ is not so good (as can be seen, for example, from the inverted slope of the large-N fit in (4.12)) which suggests the possibility that different mass eigenstates were used while performing the fit to $N = \infty$. Finally, there is a large discrepancy between Teper's $N = \infty$ result $m_{0^{++*}} \approx 2\, m_{0^{++}}$ and the SU(3) result by Morningstar and Peardon $m_{0^{++*}} = 1.54\, m_{0^{++}}$ or the $SU(8)$ result by H. Meyer $m_{0^{++*}} = 1.42\, m_{0^{++}}$.

For these reasons we think that the 0^{++*} mass estimate by Teper *et al.* might be doubted. A more reasonable estimate of the 0^{++*} mass can be based on the $SU(8)$ value found in Meyer's thesis, which upon rescaling leads to our $N = \infty$ estimate of the 0^{++*} mass

$$\frac{m_{0^{++*}}}{\sqrt{\sigma}} = 4.67 \pm 0.29 \tag{4.13}$$

This is the value we use in Table 1 and it's compatible with Morningstar and Peardon (i.e. approximately 1.5 times heavier than the 0^{++} mass). However, we would like to point once again that this estimate should be treated with caution. Similarly, the rest of the lattice values in Table 1 (for 2^{++*}, 4^{++}, 4^{++*}, 6^{++} states) were obtained from $SU(3)$ and $SU(8)$ data in [**6**] by simple rescaling and therefore should also be treated with caution.

Note that Meyer's thesis contains $SU(3)$ estimates for the masses of 0^{++**} and 0^{++***}. However, these seem to be too heavy to be absolutely

TABLE 1. Large N glueball masses in YM_{3+1}. All masses are in units of the square root of the string tension. Comments on the origin of the lattice data are presented in Section 4.1.

State	Lattice, $N \to \infty$	Theory	Difference in %
0^{++}	3.307 ± 0.053	3.394	[2.6]
0^{++*}	4.67 ± 0.29	4.294	[8]
2^{++}	4.80 ± 0.14	5.193	[8]
2^{++*}	6.17 ± 0.21	6.058	[2]
4^{++}	7.81 ± 0.20	6.924	[11]
4^{++*}	9.88 ± 0.36	7.779	[21]
6^{++}	9.34 ± 0.55	8.634	[7.5]

trusted (they are in the same ballpark as Teper's 0^{++*} estimate) and therefore we choose not to include them in our comparison.

This concludes our comparison in the $PC = ++$ sector of the theory and we want to proceed now to the $PC = -+$ sector. But before we do that we would like to make a few general comments on the qualitative properties of glueball mass spectrum in both $2+1$ and $3+1$ dimensions. It has been known for a while [**14, 8**] that the low mass glueball spectrum can be qualitatively understood in terms of local gluon interpolating operators of minimal dimension: higher dimensional operators create higher mass states. In particular, in $2+1$ dimensions there is a (mass) dimension-6 operator $\mathrm{tr}\,(B^3)$ with 0^{--} quantum numbers and this is consistent with the lattice result concerning the existence of a light (with $m_{0^{--}} \approx 1.5 m_{0^{++}}$) 0^{--} glueball state in the spectrum of pure Yang-Mills theory in $2+1$ dimensions. On the contrary, in $3+1$ dimensions modern lattice studies seem to strongly rule out the existence of light 0^{--} glueball states[5]. Qualitatively this can be explained by the fact that a dimension-6 interpolating operator with 0^{--} quantum numbers simply does not exist in $3+1$ dimensions.

What does exist in $3+1$ dimensions however is the $\mathcal{O}_E = f_{abs}(\vec{E}^a \times \vec{E}^b)\cdot \vec{E}^c$ operator with 0^{-+} quantum numbers[6]. It should be clear now that in complete analogy with [**1**], *i.e.* by considering an equal-time correlator of two such operators $\langle \mathcal{O}_E(x)\, \mathcal{O}_E(y)\rangle$, we should think of the glueball mass spectrum in the $PC = -+$ sector as being constructed out of three constituent masses. More specifically, we obtain

$$M_{0^{-+}} = M_1 + M_1 + M_1 = 5.091\,\sqrt{\sigma} \tag{4.14}$$

$$M_{0^{-+*}} = M_1 + M_1 + M_2 = 5.99\,\sqrt{\sigma} \tag{4.15}$$

$$M_{2^{-+}} = M_1 + M_2 + M_2 = 6.89\,\sqrt{\sigma} \tag{4.16}$$

$$M_{2^{-+*}} = M_1 + M_2 + M_3 = 7.76\,\sqrt{\sigma} \tag{4.17}$$

A comparison of these predictions with lattice data is presented in Table 2. No genuine large N lattice data are available for $PC = -+$ states. Therefore,as before, our large N "lattice" estimates were obtained by a simple rescaling of the actual $SU(3)$ and $SU(8)$ lattice data presented in [**6**].

Finally, we include two comments indicating potential difficulties in our scheme:

1. As can be seen from Table 2 the seeming large discrepancy for the masses of 0^{-+*} and 2^{-+} glueballs can be avoided if we swap theoretical

---

[5]The only definite estimate of 0^{--} mass, $m_{0^{--}} = (2.44 \pm 0.25) m_{0^{++}}$, can be found in [**9**] and is indeed quite heavy.

[6]In fact, in (3+1)d it is possible to construct also a dimension-4 interpolating operator with 0^{-+} quantum numbers, which is $\vec{E}^a \cdot \vec{B}^a$. Similarly, one can construct 0^{++} operators $(\vec{E}^a)^2 - (\vec{B}^a)^2$ and $f_{abc}(\vec{B}^a \times \vec{B}^b)\cdot \vec{B}^c$ of dimension four and six respectively. However, as was argued previously in [**14**], for a broad class of constituent glue models the lightest 0^{-+} glueball couples to the dimension-6 operator only. This explains our choice of $\mathcal{O}_E = f_{abs}(\vec{E}^a \times \vec{E}^b)\cdot \vec{E}^c$ as a probe operator for this state.

TABLE 2. Large N glueball masses in YM_{3+1}. All masses are in units of the square root of the string tension. Comments on the origin of the lattice data are presented in Section 4.1.

State	Lattice, $N \to \infty$	Theory	Difference in %
0^{-+}	4.81 ± 0.13	5.091	[6]
0^{-+*}	7.22 ± 0.33	5.99	[22]
2^{-+}	5.95 ± 0.10	6.89	[15]
2^{-+*}	7.46 ± 0.29	7.76	[4]

values for these states. After all, there is no particular reason why the $M_1 + M_1 + M_2$ combination should be associated with 0^{-+*} and not with 2^{-+}. The same for the $M_1 + M_2 + M_2$ combination. On the other side this is similar to the situation involving 0^{++*} and 2^{++} glueballs: here the lattice data favor $m_{0^{++*}} > m_{2^{++}}$. (However, the final status of the 0^{++*} mass on the lattice is far from clear.) Nevertheless, our model predicts otherwise.

2. The 0^{-+} mass agrees nicely with the lattice result. However, there is a claim in the lattice literature that 0^{-+} should be heavier than 2^{++}. This cannot be seen from large N extrapolated data in Tables 1 and 2 (they seem to be simply degenerate) but Morningstar and Peardon say that "the pseudoscalar is clearly resolved at 7σ level to be heavier than the tensor". Our current theoretical model (based on zeros of J_3) predicts just the opposite (compare Tables 1 and 2). It's easy to see however that the $M_{0^{-+}} > M_{2^{++}}$ requirement translates into the $\gamma_{\nu,2} < \frac{3}{2}\,\gamma_{\nu,1}$ requirement for Bessel zeros. This is possible if $\nu > 3.3$. Therefore, from this point of view $\nu = 3$ is not completely satisfactory!

In spite of these important caveats we take the above phenomenological fit to the lattice data as a positive indication that the general form of the wave functionals we have discussed in this note does capture the essential non-perturbative physics. Obviously, we are only scratching the surface in our understanding of what is truly going on here. Much more detailed work is needed to elucidate the real nature of the wave functionals discussed in this paper.

Acknowledgements

We would like to thank B. Bergoltz, D. Karabali, V. P. Nair, P. Orland, R. Pisarski, S. Rajeev and other participants of the City College workshop on non-perturbative Yang-Mills theory, for interesting conversations and questions regarding the material presented in this paper. We also thank various participants of the Virginia Tech first Sowers workshop in theoretical physics for questions and comments. RGL was supported in part by the U.S. Department of Energy under contract DE-FG02-91ER40709. DM is supported in part by the U.S. Department of Energy under contract DE-FG05-92ER40677. Many thanks to Perimeter Institute, the Kavli Institute for Theoretical Physics, and IHES at Bures sur Yvette, for their respective hospitalities.

References

[1] R. G. Leigh, D. Minic and A. Yelnikov, Phys. Rev. Lett. **96**, 222001 (2006), arXiv:hep-th/0512111; R. G. Leigh, D. Minic and A. Yelnikov, Phys. Rev. D **76**, 065018 (2007), hep-th/060406; R. G. Leigh, D. Minic and A. Yelnikov, arXiv:0704.3694 [hep-th].

[2] M. Teper, Phys. Rev. D **59**, 014512 (1999) [arXiv:hep-lat/ 9804008], B. Lucini and M. Teper, Phys. Rev. D **66**, 097502 (2002) [arXiv:hep-lat/0206027] and references therein. H. B. Meyer and M. J. Teper, Nucl. Phys. B **668**, 111 (2003) [arXiv:hep-lat/0306019].

[3] L. Freidel, R. G. Leigh and D. Minic, Phys. Lett. B **641** (2006) 105, hep-th/0604184 and L. Freidel, hep-th/0604185.

[4] B. Lucini, M. Teper and U. Wenger, JHEP **0406**, 012 (2004) [arXiv:hep-lat/0404008].

[5] B. Lucini and M. Teper, JHEP **0106**, 050 (2001) [arXiv:hep-lat/0103027].

[6] H. B. Meyer, arXiv:hep-lat/0508002.

[7] C. J. Morningstar and M. J. Peardon, Phys. Rev. D **60**, 034509 (1999) [arXiv:hep-lat/9901004].

[8] J. Kuti, Nucl. Phys. Proc. Suppl. **73**, 72 (1999) [arXiv:hep-lat/9811021].

[9] L. Hu, X. Q. Luo, Q. Z. Chen, X. Y. Fang and S. H. Guo, Commun. Theor. Phys. **28**, 327 (1997) [arXiv:hep-ph/9609435].

[10] For a review, see, A. Kovner and J. G. Milhano, arXiv:hep-ph/0406165, and references therein. Some original references are: I. I. Kogan and A. Kovner, Phys. Rev. D **52**, 3719 (1995) [arXiv:hep-th/9408081]; I. I. Kogan and A. Kovner, Phys. Rev. D **51**, 1948 (1995) [arXiv:hep-th/9410067]; W. Brown, J. P. Garrahan, I. I. Kogan and A. Kovner, Phys. Rev. D **59**, 034015 (1999) [arXiv:hep-ph/9808216].

[11] For a general discussion and references, consult, for example, D. Minic and V. P. Nair, Int. J. Mod. Phys. A **11**, 2749 (1996) [arXiv:hep-th/9406074]; K. Zarembo, Mod. Phys. Lett. A **13**, 1795 (1998) [arXiv:hep-ph/9804276]; K. Zarembo, Mod. Phys. Lett. A **13**, 1709 (1998) [arXiv:hep-th/9803237].

[12] A. Migdal, Ann. of Phys. **109**, 365, (1977); **110**, 46 (1978).

[13] S. J. Brodsky, arXiv:hep-ph/0703109 and references therein.

[14] R. L. Jaffe, K. Johnson, Z. Ryzak, Ann. of Phys. **168**, 344 (1986).

[15] I. Bars, Phys. Rev. Lett. **40**, 688 (1978) ; Nucl. Phys. B **149**, 39 (1979); also, I. Bars and F. Green, Nucl. Phys. B **148**, 445 (1979), Erratum-ibid. B **155**, 543 (1979). The large N limit was explored in I. Bars, Phys. Lett. B **116**, 57 (1982); Phys. Lett. B **245**, 35 (1990).

[16] D. Karabali and V. P. Nair, Nucl. Phys. B **464**, 135 (1996); D. Karabali and V. P. Nair, in (2+1) Phys. Lett. B **379**, 141 (1996); D. Karabali, C. J. Kim and V. P. Nair, Nucl. Phys. B **524**, 661 (1998) [arXiv:hep-th/9705087]; D. Karabali, C. J. Kim and V. P. Nair, Phys. Lett. B **434**, 103 (1998) [arXiv:hep-th/9804132]; D. Karabali, C. J. Kim and V. P. Nair, Phys. Rev. D **64**, 025011 (2001) [arXiv:hep-th/0007188] and references therein.

[17] R. P. Feynman, Nucl. Phys. B188, 479 (1981).

[18] M. Fukuma, K. I. Katayama and T. Suyama, arXiv:0711.4191 [hep-th].

[19] L. Freidel, R. G. Leigh, D. Minic and A. Yelnikov, projects in progress.

[20] V. P. Nair and A. Yelnikov, Nucl. Phys. B **691**, 182 (2004)

[21] http://webcomputing.bio.bas.bg/webMathematica/webComputing/BesselZeros.jsp

Perimeter Institute for Theoretical Physics, Caroline St. N. Waterloo, N2L 2Y5, ON, Canada; Laboratoire de Physique, Ecole Normale Supérieure de Lyon, 46 Allée d'Italie, 69364 Lyon, Cedex 07, France.
E-mail address: lfreidel@perimeterinstitute.ca

Department of Physics, University of Illinois at Urbana-Champaign, 1110 West Green Street, Urbana, IL 61801-3080, USA.
E-mail address: rgleigh@uiuc.edu

Department of Physics, Virginia Tech, Blacksburg, VA 24061, USA.
E-mail address: dminic@vt.edu

Physics Department, City College of the City University of New York, 160 Convent Avenue, New York, NY 10031, USA.
E-mail address: yelnikov@yahoo.com

AMS/IP Studies in Advanced Mathematics
Volume 44, 2008

Resolving Black Hole Microstates

Vijay Balasubramanian, Jan de Boer, Sheer El-Showk, and Ilies Messamah

ABSTRACT. We review the relation between smooth supergravity solutions and black hole microstates, and discuss the connection between ensembles of states and coarse grained geometries. Under suitable circumstances, the latter can effectively be described by a singular geometry or a geometry containing a black hole, in which case we have an explicit realization of the fuzzball proposal. We specialize our discussion to 1/2 BPS states in type II string theory on $\mathrm{AdS}_5 \times \mathrm{S}^5$, on $\mathrm{AdS}_3 \times \mathrm{S}^3 \times \mathrm{T}^4$, and in M-theory on $\mathrm{AdS}_3 \times \mathrm{S}^2 \times \mathrm{CY}$, preserving 16, 8 and 4 supersymmetries respectively.

CONTENTS

2000 *Mathematics Subject Classification.* Primary 83E30; Secondary 81T30, 83E50, 83C57, 81T60.

1. Introduction

Since the initial success of string theory in accounting for the entropy of supersymmetric black holes by counting states in a field theory [**1**] there has been an ongoing effort to understand exactly what the structure of these microstates is and how they manifest themselves in gravity. In recent years progress has been made in this direction on several fronts.

In string theory black holes can often be constructed by wrapping D-branes on cycles in a compact manifold X so they appear as point-like objects in the spatial part of the non-compact spacetime, $\mathbb{R}^{1,d-1}$. As the string coupling is increased, these objects backreact on spacetime and can form supersymmetric black holes (or more general black objects) with macroscopic horizons. The entropy associated with these objects can be determined "microscopically" by counting BPS states in a field theory living on the branes and this has been shown in many cases to match the count expected from the horizon area[1]. Although the field theory description is only valid for very small values of g_s the fact that the entropy counting in the two regimes coincides can be attributed to the protected nature of BPS states that persist in the spectrum at any value of the coupling unless a phase transition occurs or a wall of stability is crossed. The fact that the (leading) contribution to the entropy of the black hole could be reproduced from counting states in a sector of the field theory suggests that the black hole microstates dominate the entropy in this sector.

While it is fortunate that these states persist, as they allow various computations to be undertaken in the weakly coupled regime, understanding the nature of these states in (super)gravity is still an open problem. As g_s is increased the branes become massive and we expect them to start backreacting on the geometry. From our understanding of AdS/CFT and of the field theory description of D-branes we can try to determine how these states would manifest themselves in a supergravity or closed string picture.

Within the framework of the AdS/CFT correspondence black holes with near horizon geometries of the form $\mathrm{AdS}_m \times \mathrm{S}^n$ must correspond to objects in a dual conformal field theory that have an associated entropy[2]. A natural candidate is a thermal ensemble or density matrix, in the CFT, composed of individual pure states (see e.g. [**3**]). AdS/CFT then suggests that there must be corresponding pure states in the closed string (or supergravity) picture and that these would comprise the microstates of the black hole. It is not immediately obvious, however, that such states are accessible in the supergravity description. First, the dual objects should be closed string *states* and may not admit a classical description. Even if they do admit a classical description they may involve regions of high curvature and hence

---

[1]Among the large number of papers on the subject, which we will not attempt to list here, [**1**] and [**2**] are of most direct relevance for us.

[2]More generally objects with horizons, microscopic or macroscopic, are expected to have an associated entropy which should manifest itself in the CFT.

be inherently stringy. For BPS black holes[3], however, we may restrict to the BPS sector in the Hilbert space, where the protected nature of the states suggests that they should persist as we tune continuous parameters (barring phase transitions or wall crossings) allowing us to take[4] $N \to \infty$ and the 't Hooft coupling, λ, small so that stringy and quantum corrections are suppressed and we may hope for the supergravity manifestation of these states. However, the large N limit bears many similarities with the $\hbar \to 0$ limit in quantum mechanics, where we know that most states do not have a proper classical limit. As we will see, only for suitable "semiclassical" states can a reliable supergravity description be obtained.

Even if this is the case and some CFT states are dual to smooth geometries these may not be distinguishable by their profile in the non-compact directions alone and much of the interesting geometry may reside in the backreaction of the compactification manifold (which is generally quite complicated). Despite these potential problems, recently, a very fruitful program has been undertaken to explore and classify the smooth supergravity duals of the CFT states making up the black hole ensemble. Smoothness here is important because if these geometries exhibit singularities we expect these to either be resolved by string-scale effects, making them inaccessible in supergravity, or enclosed by a horizon implying that the geometry corresponds, not to a pure state, but rather to an ensemble with some associated entropy.

Large classes of such smooth supergravity solutions, asymptotically indistinguishable from black hole solutions, have indeed been found [**4**–**10**] (and related to previously known black hole composites [**11**] [**12**]). These are complete families of solutions preserving a certain amount of supersymmetry with fixed asymptotic charges[5] and with no (or very mild) singularities.

In constructing such solutions it has often been possible to start with a probe brane solution in a flat background and to generate a supergravity solution with the correct asymptotic charges by backreacting the probe [**4, 7, 13**]. The fact that these probe solutions have a CFT dual implies that they should somehow be related to black hole microstates[6]. In Lin-Lunin-Maldacena [**4**] the backreaction of such branes (as well their

---

[3]Here "BPS" can mean either 1/2, 1/4 or 1/8 BPS states or black holes in the full string theory. The degree to which states are protected depends on the amount of supersymmetry that they preserve and our general remarks should always be taken with this caveat.

[4]N measures the size of the system. For black holes it is usually related to mass in the bulk and conformal weight in the CFT.

[5]In fact the question of which asymptotic charges of the microstates should match those of the black hole is somewhat subtle and depends on which ensemble the black hole is in. In principle some of the asymptotic charges might be traded for their conjugate potentials. Moreover, the solutions will, in general, only have the same isometries asymptotically.

[6]Throughout this paper we will be discussing "microstates" of various objects in string theory but the objects will not necessarily be holes (i.e. spherical horizon topology) nor will they always have a macroscopic horizon. In fact, there is no 1/2 BPS solution in

transition to flux) was identified with a complete set of asymptotically AdS_5 supergravity solutions (as described above) suggesting that the latter should be related to 1/2 BPS states of the original $D3$ probes generating the geometry. Indeed, in [**14, 15**] it was shown that quantizing the space of such supergravity solutions as a classical phase space reproduces the spectrum of BPS operators in the dual $\mathcal{N} = 4$ superconformal Yang-Mills (at $N \to \infty$).

In a different setting Lunin and Mathur [**6**] were able to construct supergravity solutions dual to states of a $D1$-$D5$ brane in six dimensions (i.e. compactified on a T^4) by utilizing dualities that relate this system to an $F1$-P system (see also [**5**]). The latter system is nothing more than a BPS excitation of a fundamental string quantized in a flat background. The backreaction of this system can be parametrized by a profile $F^i(z)$ in $\mathbb{R}^4$ (the transverse directions). T-duality relates configurations of this system to that of the $D1$-$D5$ system.

Recall that the naive backreaction of a bound state of $D1$-$D5$ branes is a singular or "small" black hole in five dimensions. The geometries dual to the $F1$-P system, on the other hand, are smooth[7] though they have the same asymptotics as the naive solution [**5**]. Each $F1$-P curve thus defines a unique supergravity solution with the same asymptotics as the naive $D1$-$D5$ black hole but with different subleading structure. Smoothness of these geometries led Lunin and Mathur to propose that these solutions should be mapped to pure states of the $D1$-$D5$ CFT. They also conjectured that the naive black hole geometry is somehow a coarse graining over all these smooth solutions. This idea now goes under the name of *the fuzzball proposal.*

The focus of these proceedings will be [**16–20**] which attempt to elucidate the relationship between smooth supergravity solutions and black hole microstates. This will also involve understanding the nature of typical black hole microstates and how they may be resolved by probes [**16**], as this is an integral part of the information loss paradox. The discussion will involve several different systems ranging from 1/2 BPS states in $AdS_5 \times S^5$ to the 1/4 BPS states of the $D1$-$D5$ system to the least controlled 1/8 BPS case, where we will study bound multi-black hole configurations in four and five dimensions. Only in the final case will genuine macroscopic horizons emerge, but the 1/2 and 1/4 BPS cases are under more technical control and hence important to study. In all cases we will try to understand how the BPS spectrum emerges in supergravity, how it is related to the BPS spectrum of the CFT (or more generally the brane theory) and how such states might contribute to the ensembles characterizing black holes in string theory. Much of what will be discussed is a review of work by other authors [**4, 7, 9–11, 14, 21**] (see also the other references cited earlier and throughout the remainder of the text).

$AdS_5 \times S^5$ with any kind of a horizon. We will, nonetheless, somewhat carelessly continue to refer to these as "microstates" of a black hole for the sake of brevity.

[7]That is, after dualizing to the $D1$-$D5$ frame.

As the "microstate" literature is voluminous and complicated we attempt to provide, in section 7, a brief survey of the various works and which branches of the field they fit into. This survey is by no means exhaustive and no doubt neglects many important works but we feel it can, nonetheless, serve as a useful reference point for readers attempting to orient themselves within the field.

2. The relation between microstates and geometries

Throughout these proceedings we will be exploring the relationship between families of smooth supergravity solutions and the microstates of supersymmetric black holes in string theory. It is perhaps best to delineate, at the outset, the various notions involved. To this end we have provided a figure (Figure 1 below) attempting to depict the relationship between the different components involved. It should also be said that much of the subsequent discussion in this section outlines an emerging perspective on the issue of black hole microstates, but its quite possible future developments will engender significant changes in this perspective.

The first important component is a (complete) family of supergravity solutions preserving the same supersymmetry and with the same asymptotic conserved charges (see footnote 5) as a BPS black hole. These families play a dual role in our understanding of the microstates of black holes. As solutions they can be related to points in a phase space and, as a family, they define a submanifold of the full phase space. Because they are BPS they can be argued to generate a proper decoupled phase space of their own[8] [**14, 15**], at least for the purpose of enumerating states. Indeed, in the cases we consider, one can check that the restriction of the symplectic form to this space is non-degenerate, implying that the space is actually a symplectic submanifold[9]. For a thorough and detailed discussion of the subtleties involved in this "on-shell quantization" the reader is referred to [**15**, Sec. 2.7–2.10]. Quantizing the space of such solutions as a phase space yields a Hilbert space populated by putative BPS microstates of the black hole. In [**14**] this was done for 1/2 BPS geometries asymptotic to $AdS_5 \times S^5$ and found to reproduce the 1/2 BPS spectrum of the dual CFT.

Viewed another way, these geometries are simply well-defined classical solutions to supergravity that are asymptotically (in the relevant cases) indistinguishable from a black hole solution. If they can be included in a decoupled AdS throat then they are amenable to study via the AdS/CFT duality, where the CFT provides access to the quantum structure of the

[8]Though these arguments are for the 1/2 and 1/4 BPS case they should extend to 1/8 BPS bearing in mind possible discontinuities in the spectrum at walls of marginal stability.

[9]As noted in [**15**] this is related to the fact that the solutions are stationary but not static, so the momenta conjugate to the spatial components of the metric are non-vanishing.

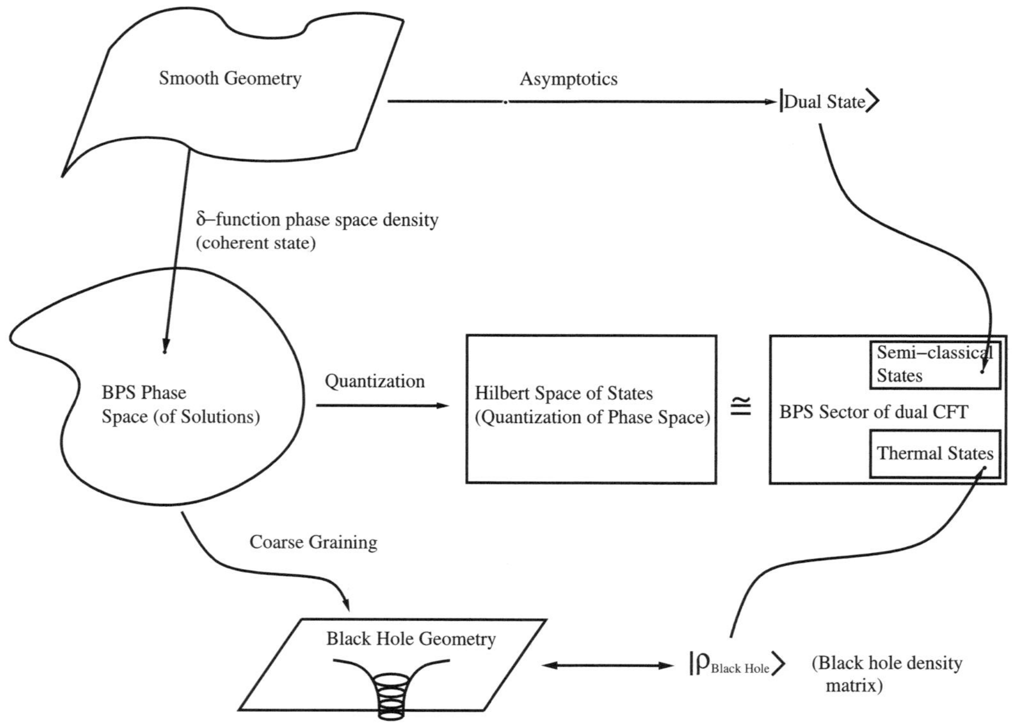

FIGURE 1. Relationship between various components appearing in the study of black hole microstates. The smooth geometries making up the phase space can be thought of either as classical solutions defining a solution space (isomorphic to a phase space) or as highly localized phase space densities corresponding to coherent states. The black hole at the bottom of the figure is thought to be generated by coarse graining (in some suitable sense) over a large number of smooth geometries; the resultant geometry need not be a black hole but may, for instance, include a naked singularity. The exact manifestation of each component and relation can differ significantly in the various cases considered in these proceedings. The quantization, for instance, is rather different for the 1/4 and 1/8 BPS case.

theory. The smooth solutions should be seen as approximating semiclassical states in the supergravity regime of the bulk (closed string) sector. Such bulk states have CFT duals that can be determined by studying the subleading asymptotics of the geometry. Because these geometries are smooth we expect the dual states to be pure states rather than generic density matrices. Unlike the direct phase space quantization, the CFT states so defined are rather special as they are dual to semiclassical "states," whereas the states arising from quantizing the phase space may be entirely quantum in nature. In the cases where we have access to a complete, decoupled set of BPS states in the CFT, however, we expect (and indeed find) that these match the states arising from the direct quantization of the BPS phase space [**14, 15**].

Furthermore, we would like to understand how these pure states, corresponding to smooth classical geometries or, more generally, phase space densities on the BPS phase space, relate to black holes. The latter have entropy so we expect them to be dual to density matrices in the CFT. The ensemble of states making up such a density matrix should be a suitable thermodynamic average over pure states in the same sector of the Hilbert space as the black hole. As our "microstate geometries" preserve the same supersymmetries and carry some of the same asymptotic charges as the black hole they provide suitable candidates for the states that form the ensemble. However, as mentioned in footnote 5 it is not generally clear what kind of ensemble corresponds to the black hole geometry, so it is not clear which sector of the Hilbert space the microstates must come from. In fact, in some cases, there is reason to believe that the Hilbert space enjoys a more refined decomposition into subsectors than just by macroscopic quantum numbers alone. In [**22–24**], for instance, a decomposition based on split attractor trees is conjectured (this will be discussed further in section 5.4). In these cases it is possible that the ensemble will only include microstates from a single subsector. It is important to note that many of the states in the ensemble may not admit a semiclassical description and are thus only accessible in the CFT or by quantizing the BPS phase space.

In [**19**] another (related) constraint on the constituents of the black hole ensemble was found. There it was argued that any microstate geometries that do not survive a near-horizon decoupling limit should not be considered in the black hole ensemble because they do not correspond to bona fide bound states of the original D-brane system generating the black hole.

Finally, in order to study this correspondence more thoroughly it is desirable to have an inverse map between the states in the CFT and classical geometries. Of course this map is not injective, as many states in the CFT are not semiclassical, so we would also like a criterion for determining which such states yield good classical geometries and which yield geometries with large quantum fluctuations. A possible criterion was discussed in [**25, 26**] and will play a role in some of the arguments that follow. The point of view that we would like to assume is based on the need for a classical observer to measure the system [**16**]. Thus we would like to identify a set of operators $\mathcal{O}_\alpha$

in the CFT dual to "macroscopic observables". The requirement that a state yield a good classical geometry can be translated into a constraint on the variance of the expectation values for these observables in the semiclassical limit.

2.1. Phase space quantization. The space of classical solutions of a theory is generally isomorphic to its classical phase space. This observation can be used to quantize the theory using a symplectic form, derived from the Lagrangian, on the space of solutions rather than on the phase space. This is an old idea [**27**] (see also [**28**] for an extensive list of references and [**29–31**] for more recent work) which was used in [**14, 15, 32, 33**] to quantize the LLM [**4**] and Lunin-Mathur [**6**] geometries. As mentioned above, an important subtlety here is that it is not the entire solution space which is being quantized, but rather a subspace of the solutions with a certain amount of supersymmetry.

In general, quantizing a subspace of the phase space will not yield the correct physics, as it is not clear that the resulting states do not couple to states coming from other sectors. It is not even clear that a given subspace will be a symplectic manifold with a non-degenerate symplectic pairing. As discussed in [**15**] we expect this to be the case only if the subspace contains dynamics; for gravitational solutions we thus expect stationary solutions, for which the canonical momenta are not trivial, to possibly yield a non-degenerate phase space. This still does not address the issue of consistency, as states in the Hilbert space derived by quantizing fluctuations along a constraint submanifold of the phase space might mix with modes transverse to the submanifold. When the submanifold corresponds to the space of BPS solutions one can argue, however, that this should not matter. The number of BPS states is invariant under continuous deformations that do not cross a wall of marginal stability or induce a phase transition. Thus, if we can quantize the solutions in a regime where the interaction with transverse fluctuations is very weak then the energy eigenstates will be given by perturbations around the states on the BPS phase space, and, although these will change character as parameters are varied the resulting space should be isomorphic to the Hilbert space obtained by quantizing the BPS sector. If a wall of marginal stability is crossed states will disappear from the spectrum, but there are tools that allow us to analyze this as it occurs (see section 5.4).

Let us emphasize that the validity of this decoupling argument depends on what questions one is asking. If we were interested in studying dynamics then we would have to worry about how modes on the BPS phase space interact with transverse modes. For the purpose of enumerating states, however, as we have argued, it should be safe to ignore these modes.

As mentioned, the LLM and Lunin-Mathur geometries have already been quantized and the resulting states were matched with states in the dual CFTs. We will have occasion to mention this briefly in the sequel but we will ultimately focus on the quantization of $\mathcal{N} = 2$ solutions in four and five dimensions. For such solutions, although a decoupling limit has been defined [**20**], the dual $\mathcal{N} = (0,4)$ CFT is rather poorly understood. Thus quantization of the supergravity solutions may yield important insights into the structure of the CFT and will be important in studying the microstates of the corresponding macroscopic black holes.

2.2. Black holes, AdS throats and dual CFT. One of the most powerful tools to study properties of black holes in string theory is the AdS/CFT correspondence [**34**]. This conjecture relates string theory on backgrounds of the form $\mathrm{AdS}_{p+1} \times \mathcal{M}$ to a CFT_p that lives on the boundary of the AdS_{p+1} space. Such backgrounds arise from taking a particular decoupling limit of geometries describing black objects such as black holes, black strings, black tubes, *etc.* This limit amounts to decoupling the physics in the near horizon region[10] of the black object from that of the asymptotically flat region by scaling the appropriate Planck length l_p to decouple the asymptotic gravitons from the bulk. At the same time one needs to scale appropriate spatial coordinates with powers of l_p to keep the energies of some excitations finite. This procedure should be equivalent to the field theory limit of the brane bound states generating the geometry under consideration.

The AdS/CFT conjecture was originally stated as an equivalence between the two theories, string theory on AdS_{p+1} on the one hand and CFT_p on the other hand [**34**] without specifying the detailed properties of the map. More progress was achieved afterwards by relying on symmetries and quantum numbers. An important role is played by supersymmetry; the more supersymmetry is preserved the better the map is understood. For this reason the BPS sector is the best studied part in this correspondence mainly because the dynamics are constrained, and also because the string theory spectrum is truncated to just the supergravity modes.

We are interested in black objects which describe normalizable deformations in the AdS_{p+1} background. These correspond to a state/density matrix on the dual CFT according to the following dictionary:

Bulk	**Boundary**
$\exp\left(-S^{\text{on shell}}_{\text{bulk}}\right)$	$\mathrm{Tr}[\rho\mathcal{O}_1 \cdots \mathcal{O}_n] = \langle \mathcal{O}_1 \cdots \mathcal{O}_n \rangle_\rho$
classical geometries	semiclassical states ?
black hole	$\rho \sim \exp\{-\sum_i \beta_i \mathcal{O}_i\}$
entropy S	$S = -\ \mathrm{Tr}(\rho \log \rho)$
bulk isometry D	$\left[\rho, \hat{D}\right] = 0$
ADM quantum numbers of D	$\mathrm{Tr}[\rho\hat{D}] = \langle \hat{D} \rangle = D_{\mathrm{ADM}}$

[10]In some of the cases treated in these proceedings the region will not be an actual near-horizon region, as the original solutions may be horizon-free but the decoupling limits are motivated by analogy with genuine black holes where the relevant region is the near horizon one.

The above table needs some clarifications. In the first line $\mathcal{O}_i$ are operators dual to sources turned on in the boundary. They are included in the bulk calculation of $\left(-S_{\text{bulk}}^{\text{on shell}}\right)$. The second line can be seen as the definition of the dual semiclassical state. Put in other words, a semiclassical state is one that has an unambiguous dual bulk geometry[11] In some ideal situations it turns out to be some coherent state. In the third line, we describe a typical form of a density matrix that we expect to describe black holes. This form is motivated by the first law of thermodynamics: the entropy as defined in the fourth line obeys $dS = \sum_i \beta_i d\langle \mathcal{O}_i \rangle$, and by matching this to the first law as derived from the bulk description of the black hole we can identify the relevant set of operators $\mathcal{O}_i$ and potentials μ_i and guess the corresponding density matrix. The fourth line simply states that we expect a relation between the bulk and boundary entropies. In the fifth and the last line, $\hat{D}$ is the current/operator dual to the bulk isometry D.

There are a lot of questions that one can try to answer about black holes in the framework of AdS/CFT correspondence. The central question that we want to shed some light on in these proceedings is; "*Given a density matrix ρ on the CFT side, is there a dual geometry in the bulk?*" There are a lot of reasons as to why one might be interested in such a question. The main one is the "fuzzball proposal": Mathur et al conjectured that what we see as a singular black hole in classical gravity is nothing but an average over a huge ensemble of many smooth geometries and the black hole entropy is nothing but the statistical entropy of this ensemble. Having such a picture will resolve two standing questions about black holes. On the one hand, a geometrical explanation of black hole entropy can be put forward. On the other hand, one resolves the information loss paradox; smoothness means information is preserved.

On general grounds one could have expected that a general density matrix ρ should be dual to a suitably weighted sum over geometries, each of which could be singular, have regions with high curvature, and probably not have proper unambiguous classical limits. As a result the dual gravitational description is not trustworthy and quite useless. However, under suitable favorable circumstances, it can happen that there is a dual "*effective*" geometry that describes the density matrix ρ very well. This procedure of finding the effective geometry is what goes under the name "*coarse graining.*" In the gravity description, this amounts to neglecting the details that a classical observer cannot access anyway due to limitations associated to the resolution of their apparatus. So, one can phrase our question in the opposite direction, "*What are the characteristics of a density matrix on the CFT side, so that there is a good dual effective geometry that describes the physics accurately?*".

---

[11]Unambiguous here means that in the classical limit ($N \to \infty$ and $\hbar \to 0$) macroscopic observables take on a fixed expectation value with vanishing variance.

One can try to construct the dual effective geometry following the usual AdS/CFT prescription. To do so, one should first calculate all the non-vanishing expectation values of all operators dual to supergravity modes (assuming one knows the detailed map between the two). On the CFT side, these vevs are simply given by

$$\langle \mathcal{O}_i \rangle = \mathrm{Tr}(\rho\, \mathcal{O}_i),$$

and they determine the boundary conditions for all the supergravity fields. The next step is to integrate the gravity equations of motion subject to these boundary conditions to get the dual geometry. This is in principle what has to be done according to the AdS/CFT prescription. The problem with this straightforward approach is that it is not terribly practical, and we will therefore revert to a different approach[12]. Before describing various examples in more detail, we first describe the main idea in general terms. We will first start by describing the connection between quantum physics and the classical phase space. After that, we are going to briefly describe the philosophy behind constructing effective geometries.

2.3. Phase space distributions. To have an idea about what it means to average over ensembles of geometries, or "coarse grain" as we will refer to it, we need to understand some general features of the bulk theory. In general, we will assume that we are dealing with a supergravity theory in the bulk[13]. Recall that solutions to the supergravity equation of motion can be associated with points in a phase space (see for example [**31**]). The boundary theory, on the other hand, is generally studied as a quantum conformal field theory. As a result we are looking for a map between quantum states (CFT) and classical objects living in a phase space (bulk). A well known example of such a map is the map between quantum states and their corresponding classical phase space densities (see the review [**35**] and references there in to the original literature). A good guess then is that the map that we are looking for should be a "dressed" version of the former. Let us pause for a moment to discuss the phase space distribution idea [**35**]. A particle (or statistical system) in a quantum theory is described by giving its density matrix ρ. The result of any measurement can be seen as an expectation value of an appropriate operator which is given explicitly by

$$\langle \mathcal{O} \rangle_\rho = \mathrm{Tr}(\rho\, \mathcal{O}) \tag{2.1}$$

[12]Though it would be interesting to study in some detail the connection between the two.

[13]We recall that the general philosophy is that for BPS black objects we hopefully can take a large N limit which preserves the microstates (because they are BPS) while suppressing stringy corrections.

This is reminiscent of classical statistical mechanics where the measurements are averages of appropriate quantities using some statistical distribution,

$$\langle \mathcal{O} \rangle_w = \int dpdq\, w(p,q)\, \mathcal{O}(p,q) \tag{2.2}$$

where the integration is over the full phase space. One can wonder at this point if it is possible to construct a density $w(p,q)$ so that one can rewrite equation (2.1) as equation (2.2). The answer is affirmative: for every density matrix ρ there is an associated phase space distribution w_ρ such that for all operators A the following equality holds:

$$\int dpdq\, w_\rho(p,q)\, A(p,q) = \mathrm{Tr}\left(\rho\, A(\hat{p},\hat{q})\right) \tag{2.3}$$

What about the uniqueness of w_ρ? Recall that in a quantum theory we have to face the question of operator ordering. This comes about because the operators $\hat{q}$ and their dual momenta $\hat{p}$ don't commute with each other. This means that the distribution w_ρ should somehow include information about the chosen order of $\hat{p}$ and $\hat{q}$. As a result there does not exist a unique phase space distribution. For example, the distribution corresponding to Weyl ordering is the Wigner distribution, which is given by

$$w(p,q) \sim \int dy\, \langle q-y|\rho|q+y\rangle\, e^{2ipy}$$

This distribution suffers from the fact that it is not positive definite in general. It is also quite sensitive to the physics at the quantum scale [**35, 16**], as it usually has fluctuations of order $\hbar$. Another drawback of this distribution is that it is difficult to work with from a computational standpoint. There is another commonly used distribution which is positive definite: the Husimi distribution. It is roughly the convolution of the Wigner distribution with a Gaussian. This eliminates most of the fluctuations of order $\hbar$. The price that one pays for this is that the resulting operators must be anti-normal ordered. However, for semiclassical states, which by definition are states for which the classical limit is unambiguous, $w_\rho(p,q)$ should be independent of the choice of ordering prescription in the classical limit as well, so this is not actually much of a problem.

2.4. Typical states versus coarse grained geometry. Let us recapitulate what we have discussed so far and what are the missing steps to be taken to achieve our goal. On the gravity side we have geometries with certain asymptotics that in principle yield the one-point functions of the dual operators in the CFT. On top of that we have in principle a way to quantize the reduced phase space of solutions by using the induced symplectic form.

On the CFT side we can consider density matrices ρ and find the corresponding expectation values of operators. The only missing ingredient is to construct the dual effective geometry. At this stage we have two options: (i) we first select a typical representative from the CFT ensemble of states and then map this typical state directly to geometry, or (ii) we somehow average over all the geometries dual to the states in the ensemble.

2.4.1. *Typical states/geometries.* A typical state in an ensemble is one for which the expectation values of macroscopic observables agree to within the observable accuracy with the average of the observable in the entire ensemble. Obviously, this notion depends on the appropriate notions of macroscopic observables and observable accuracy, but in the examples we describe we will usually have a reasonably educated guess as to what the typical states are. Given a typical state, we can try to map it directly to a solution of supergravity (this may still be a formidable task), after which one still needs to verify that the resulting geometry has no pathologies. This approach was followed for example by Mathur et al (see the references in [**36**]).

2.4.2. *Average/coarse graining.* Instead, we can try to average over states and geometries directly. On the CFT side this is trivial, but the coarse graining procedure on the gravity side is hard to implement. The reason for this is of course that gravity is a non-linear theory. However, in all examples that we will study, the equations of motion of supergravity in the BPS sector will effectively be linearized, which allows us to solve the equations in terms of harmonic functions with sources. In addition, the space of solution will be in one-to-one correspondence with distributions of the sources. This immediately suggests a suitable coarse-graining procedure: we simply smear the harmonic functions against the phase space density which describes the density matrix in question. This will be the basic idea in the three cases we discuss in sections 3–5, but the details will be quite different in each case. It would be interesting to explore in more detail whether this method gives rise to the appropriate averaging of the one-point functions, and to what extent it agrees with the approach based on typical states that we described in the previous paragraph.

3. $\mathbf{AdS_5 \times S^5}$

We start with the best understood case $AdS_5 \times S^5$, whose AdS/CFT dictionary is well developed. The dual CFT is $\mathcal{N} = 4$ $U(N)$ super Yang-Mills where N is the number of $D3$ branes that generate the geometry. Many supergravity solutions that asymptote to $AdS_5 \times S^5$ are known, including ones with 1/2, 1/4, 1/8 and 1/16 of the original supersymmetries preserved. Black holes with a macroscopic horizon only exist either in the 1/16 BPS case [**37**] or without any supersymmetry. The latter include AdS-Schwarzschild black holes and some of their qualitative properties can be reproduced from the dual CFT [**16**]. However, we are going to restrict ourselves to the 1/2

BPS case, where completely explicit descriptions of both the supergravity solutions as well as the CFT states are known. Therefore, this provides an excellent testing ground to test the general philosophy that we have been advocating. Our exposition will be necessarily brief; we refer the reader to [**16**] for further details.

3.1. The 1/2 BPS sector in field theory. The Hilbert space of 1/2 BPS states in $\mathcal{N}=4$ $U(N)$ super Yang-Mills is isomorphic to the Hilbert space of N fermions in a harmonic oscillator potential as shown in [**38, 39**]. The latter can conveniently be enumerated in terms of Young diagrams with N rows as follows. The ground state is composed of fermions (labelled by $i=1,\dots,N$) with energies $E_i^g=[(i-1)+1/2]\hbar$; this is the Fermi sea of the system. When we excite these fermions, the energies become $E_i=(e_i+1/2)\hbar$ for some positive disjoint integers $e_i\geq i-1$. Because we are dealing with fermions wave functions are completely antisymmetrized, and we can always order $\{e_i\}$ in ascending order $e_1<e_2<\cdots<e_N$. As a result, the numbers r_i defined by

$$r_i=e_i-i+1$$

form a non-decreasing set of integers which can be encoded in a Young diagram, where r_i describes the length of the i-th row. It is convenient to also introduce variables c_j which count the number of columns of length j. They are related to the r_i via

$$c_N=r_1, \qquad c_{N-i}=r_{i+1}-r_i, \qquad i=1,2,...,(N-1),$$

and clearly

$$r_{i+1}=e_{i+1}-i=c_{N-i}+\cdots+c_N.$$

A property of Young diagrams that will be useful later is that a single Young diagram corresponds to a geometry with U(1) symmetry in the bulk. This comes about because a single Young diagram is associated to a density matrix of a pure state, i.e. of the form

$$\rho=|\psi\rangle\langle\psi|,$$

where $|\psi\rangle$ has a fixed energy eigenvalue (simply given by the total number of boxes). The density matrix therefore commutes with the Hamiltonian which generates rotations in phase space, and according to the table in section 2.2 the corresponding supergravity solution should also possess this rotational invariance.

3.1.1. *"Average" state as a typical state.* Highly excited states of the large N free fermion system can reliably be discussed in a canonical ensemble in which temperature rather than energy is held fixed[14]. Using the canonical

[14]Often, it is also interesting to study more general canonical ensembles where additional "chemical potentials" are included such as the angular velocity coupling to angular momentum, or electrostatic potentials coupling to charges. We will see some examples in the case of $\mathrm{AdS}_3\times \mathrm{S}^3$.

ensemble one can easily calculate the average energy as well as the expectation values of c_i.

It turns out that in the thermodynamical limit $N \gg 1$ where one rescales the Young diagram by a factor of $\sqrt{N}$, the Young diagram approaches a "*limiting shape*" with probability 1 [**40**]. In other words, in the large N limit almost all states/operators belonging to the canonical ensemble under study will have associated Young diagrams that have vanishingly small fluctuations around this limit curve. One can check this claim by calculating the variance; see [**16**] for further details. The limiting shape Young diagram describes a "*typical*" Young diagram in this ensemble (see section 2.4).

To describe this limiting shape in some more detail, let us introduce two coordinates x and y along the rows and columns of the Young diagram. We adopt the convention where the origin $(0,0)$ is the bottom left corner of the diagram, and x increases going up while y increases to the right. In fermion language, x labels the particle number and y its excitation above the vacuum. One then has

$$y(x) = \langle \mathbf{y}(x) \rangle = \sum_{i=N-x}^{N} \langle c_i \rangle$$

In the large N limit, x and y can be treated as continuous variables and the summation above becomes an integral. Since the c_i are independent random variables in the canonical ensemble, it is straightforward to evaluate $y(x)$ explicitly, and one obtains an equation for the limit shape of the form

$$(1-q^N)q^y + q^{N-x} = 1 \tag{3.1}$$

where q is related to the temperature of the system.

3.2. The 1/2 BPS sector in supergravity. All 1/2 BPS solutions in supergravity are given by the LLM geometries [**4**]

$$ds^2 = -h^{-2}(dt + V_i dx^i)^2 + h^2(d\eta^2 + dx^i dx^i) = \eta e^G d\Omega_3^2 = \eta e^{-G} d\tilde{\Omega}_3^2 \tag{3.2}$$

$$h^{-2} = 2\eta \cosh G, \quad \eta \partial_\eta V_i = \epsilon_{ij}\partial_j z, \quad \eta(\partial_i V_j - \partial_j V_i) = \epsilon_{ij}\partial_\eta z \tag{3.3}$$

$$z = \frac{1}{2}\tanh G, \quad z(\eta, x_1, x_2) = \frac{\eta^2}{\pi}\int dy_1\, dy_2\, \frac{\frac{1}{2} - u(0; y_1, y_1)}{[(\vec{x}-\vec{y})^2 - \eta^2]} \tag{3.4}$$

where $i = 1, 2$. In addition there is a self-dual 5-form field strength that depends on the function z. It is clear from the above equations that the full geometry is specified by choosing a boundary function $u(0; y_1, y_2)$. The requirement of smoothness of the geometry forces $u \in \{0, 1\}$. So one can see the function u as defining a droplet in the (x_1, x_2)-plane whose boundary separates the region where $u = 1$ from the region where $u = 0$. This means smooth 1/2 BPS geometries are in one to one correspondence with droplets on the (x_1, x_2)-plane.

Following [**4**], in order to match these solutions with states in the field theory, consider geometries for which the regions in the (x_1, x_2)-plane where $u = 1$ are compact. Quantization of the flux in the geometry leads to the following identifications:

$$\hbar \leftrightarrow 2\pi l_p^4, \qquad N = \int \frac{d^2x}{2\pi\hbar}\, u(0; x_1, x_2) \tag{3.5}$$

The conformal dimension[15] Δ of a given configuration is

$$\Delta = \frac{1}{2}\int \frac{d^2x}{2\pi\hbar}\,\frac{x_1^2 + x_2^2}{\hbar}\, u(0; x_1, x_2) - \frac{1}{2}\left(\int \frac{d^2x}{2\pi\hbar}\, u(0; x_1, x_2)\right)^2 \tag{3.6}$$

The formulas above suggest to interpret $u(0; x_1, x_2)$ as a density. They indeed have a remarkably simple interpretation [**4**] in terms of the hydrodynamic limit of the phase space of the dual fermionic system, once we identify the (x_1, x_2)-plane with the single particle phase space of the fermions. This has been confirmed by directly quantizing the phase space of smooth gravitational solutions [**41, 14, 42**]. According to the general strategy, we should therefore try to identify $u(0; x_1, x_2)$ directly with the one-particle phase space density for any density matrix in the quantum-mechanical fermion system [**16**].

3.3. Geometry versus field theory states. Let us explore the map between states and geometries in some detail. In particular, we are interested in states and ensembles with a well-defined classical limit. We claim that a sufficient condition for having a well-defined semiclassical limit is that the Young diagrams approach a fixed limiting shape with probability one in the large N limit. For the canonical ensemble, this limiting shape was given in (3.1) but for other states and ensembles different limit curves may arise in the large N limit. We will continue to denote those curves by $y(x)$. They will describe the effective, coarse grained geometry corresponding to the states/ensembles. To extract the geometry, we use the fact that it should be rotationally invariant, and by matching energy $\leftrightarrow$ conformal dimension, flux $\leftrightarrow$ rank of the gauge group (=number of fermions), we get

(3.7)

$$N = \int dx = \int \frac{u(0; r^2)}{2\hbar}\, dr^2, \qquad E = \int (x + y(x))dx = \int \frac{r^2\, u(0; r^2)}{4\hbar^2}\, dr^2 = \Delta.$$

The above equations should not just hold at the level of integrals but also at the level of integrands, so that

$$\frac{u(0; r^2)}{2\hbar}\, dr^2 = dx, \qquad \frac{r^2\, u(0; r^2)}{4\hbar^2}\, dr^2 = (y(x) + x)dx.$$

[15] The bulk interpretation of the conformal dimension is energy.

Combining these yields

$$y(x) + x = \frac{r^2}{2\hbar}$$

and taking derivatives with respect to x, we obtain the identification [**16**]

$$u(0; r^2) = \frac{1}{1 + y'}. \tag{3.8}$$

So given a Young diagram with a limit shape $y(x)$, one can associate to it a geometry generated by $u(0; r^2)$ according to (3.8). For the limit curve of the canonical ensemble (3.1) the resulting phase space distribution is just that of a finite temperature Fermi gas system which was considered in [**43**]. Unfortunately, no one has succeeded in writing the corresponding metric explicitly in closed form. Since it would describe our best guess for a "1/2 BPS black hole," it would be interesting to know what it looks like.

Notice that in our conventions $y' \geq 0$ and therefore the associated geometry generically has null singularities with $0 < u(0; r^2) < 1$; the unphysical cases with $u < 0$ or $u > 1$, which give rise to naked timelike singularities [**44**], do not appear.

We conclude this section with some remarks:

- Most states yield ambiguous "quantum foam" geometries with string scale signature.
- Semi-classical states yield well-defined but still mildly singular spacetimes.
- Geometries could be coarse grained thanks to the linear description of the solutions of the field equations.
- Non-rotationally invariant configurations can also be studied but do not correspond to a single limiting Young diagram.
- Given a limiting shape, we can associate an entropy to it by counting all the states that approach the limiting shape in the large N limit (in other words, that are macroscopically indistinguishable from it). It would be interesting to provide an explicit expression for this entropy and to match it to the bulk geometry[16].

4. $\mathbf{AdS_3 \times S^3}$

In this section we are going to discuss the bound states of $D1$ and $D5$ branes in type IIB string theory compactified on[17] $T^4 \times S^1$. These are 1/2 BPS states (preserving 8 supercharges) that describe a black hole without a classical horizon in 5 dimensions. However, we are going to work in 6 dimensions, keeping track explicitly of the S^1. One of the reasons behind this decision is that in this way one gets solutions that are asymptotically

[16]By dividing the phase space into Planck-size cells, the obvious guess for the information-theory inspired bulk entropy would be $S \sim -\int d^2r\, u \log u$.

[17]The same story carries over to the case of $K3 \times S^1$.

$AdS_3\times S^3$ after taking a suitable decoupling limit. Thus we can employ the AdS/CFT machinery and benefit from the known properties of the dual two-dimensional conformal field theory.

As is well-known, the 1/2 BPS states in the CFT dual to the $D1$-$D5$ system can be identified with the states at level $L_0 = N_1N_5$ in a system with $b_1 + b_3$ chiral fermions and $b_0 + b_2 + b_4$ chiral bosons, where $b_i = \dim H^i(M_4)$. Here N_1 and N_5 are the quantized number of $D1$ and $D5$ branes. Notice that this identification of 1/2 BPS states with a system of free bosons and fermions is only valid at the level of the Hilbert space, not at the level of correlation functions. Thus, we would ideally like to be able to find a detailed map between states/ensembles in this auxiliary theory of free bosons and fermions and half-BPS solutions of six-dimensional supergravity. In what follows we will describe such a map. We will first review the known supergravity solutions and their quantization, and then propose a map which is again based on the notion of phase space densities. We conclude this section by discussing various relevant examples.

4.1. The supergravity solution and its quantization. Starting with a fundamental string with transversal profile $\mathbf{F}(s) \subset \mathbb{R}^4$ then dualizing, one gets the following solutions [**6, 5, 33**], written in the string frame[18]

$$
\begin{aligned}
ds^2 &= \frac{1}{\sqrt{f_1 f_5}}\left[-(dt+A)^2 + (dy+B)^2\right] + \sqrt{f_1 f_5}\,d\mathbf{x}^2 + \sqrt{f_1/f_5}\,d\mathbf{z}^2 \\
e^{2\Phi} &= \frac{f_1}{f_5}, \qquad C = \frac{1}{f_1}(dt+A)\wedge(dy+B) + \mathcal{C}
\end{aligned}
\tag{4.1}
$$

where y parametrizes a circle with coordinate radius R, z^i are coordinates on T^4 with coordinate volume V_4, the Hodge star $*_4$ is defined with respect to the 4-dimensional non-compact space spanned by x^i and

$$
\begin{aligned}
&dB = *_4 dA, \quad d\mathcal{C} = -*_4\, df_5, \quad A = \frac{Q_5}{L}\int_0^L \frac{F_i'(s)ds}{|\mathbf{x}-\mathbf{F}(s)|^2} \\
&f_5 = 1 + \frac{Q_5}{L}\int_0^L \frac{ds}{|\mathbf{x}-\mathbf{F}(s)|^2}, \quad f_1 = 1 + \frac{Q_5}{L}\int_0^L \frac{|\mathbf{F}'(s)|^2 ds}{|\mathbf{x}-\mathbf{F}(s)|^2}.
\end{aligned}
\tag{4.2}
$$

The solutions are asymptotically $\mathbb{R}^{1,4}\times S^1\times T^4$. We can take a decoupling limit, which simply amounts to erasing the 1 from the harmonic functions. The resulting metric will then be asymptotically $AdS_3\times S^3\times T^4$.

As mentioned above, the solutions are parametrized in terms of a closed curve

$$
x_i = F_i(s), \qquad 0 < s < L,\ i = 1,\ldots,4.
\tag{4.3}
$$

[18] We are going to follow the conventions of reference [**33**].

In the sequel we are going to ignore oscillations in the T^4 direction as well as fermionic excitations, for a further discussion of these degrees of freedom see [**45, 46**]. The $D1$ ($D5$) charge Q_1 (Q_5) satisfies

$$L = \frac{2\pi\, Q_5}{R}, \qquad Q_1 = \frac{Q_5}{L}\int_0^L |\mathbf{F}'(s)|^2 ds. \tag{4.4}$$

It turns out that the space of classical solutions has finite volume and therefore will yield a finite number of quantum states. To see this, one first starts by expanding $\mathbf{F}$ in oscillators:

$$\mathbf{F}(s) = \mu \sum_{k=1}^{\infty} \frac{1}{\sqrt{2k}} \left(\mathbf{c}_k e^{i\frac{2\pi k}{L}s} + \mathbf{c}_k^{\dagger} e^{-i\frac{2\pi k}{L}s} \right) \tag{4.5}$$

where $\mu = \frac{g_s}{R\sqrt{V_4}}$. Then one computes the restriction of the Poisson bracket to the space of solutions (4.1) which turns out to be [**32, 33**]

$$[c_k^i, c_{k'}^{j\dagger}] = \delta^{ij}\delta_{kk'}. \tag{4.6}$$

After quantization, the relation between Q_1 and Q_5 reads

$$\left\langle \int_0^L : |\mathbf{F}'(s)|^2 : ds \right\rangle = \frac{(2\pi)^2}{L}\,\mu^2 N \tag{4.7}$$

where

$$N_1 = \frac{g_s}{V_4}\,Q_1, \quad Q_5 = g_s N_5, \quad N \equiv N_1 N_5 = \sum_{k=1}^{\infty} k \left\langle \mathbf{c}_k^{\dagger}\mathbf{c}_k \right\rangle. \tag{4.8}$$

N_1, N_5 count the number of $D1$ and $D5$ branes respectively. The modes c_k^i become the creation and annihilation modes of four of the total of $b_0 + b_2 + b_4$ bosons; one can check that the four that appear are precisely the ones associated to the $H^{(0,0)}(M)$, $H^{(2,0)}(M)$, $H^{(0,2)}(M)$, $H^{(2,2)}(M)$ cohomology. Finally, notice that the number of states and hence the entropy can easily be extracted from the known partition functions of chiral bosons and fermions.

4.2. Geometries from states. The Hilbert space is spanned by

$$|\psi\rangle = \prod_{i=1}^{4}\prod_{k=1}^{\infty} (c_k^{i\dagger})^{N_{k_i}}|0\rangle, \qquad \sum k N_{k_i} = N \tag{4.9}$$

Given a state, or more generically a density matrix in the CFT

$$\rho = \sum_{ij} c_{ij} |\psi_i\rangle\langle\psi_j| \tag{4.10}$$

we wish to associate to it a density on phase space. The phase space is given by classical curves which we will parametrize as (note that d and $\bar{d}$ are now complex numbers, not operators)

$$\mathbf{F}(s) = \mu \sum_{k=1}^{\infty} \frac{1}{\sqrt{2k}} \left(\mathbf{d}_k e^{i\frac{2\pi k}{L} s} + \bar{\mathbf{d}}_k e^{-i\frac{2\pi k}{L} s} \right) \tag{4.11}$$

and which obey the classical constraint (4.4).

We now propose to associate to a density matrix of the form (4.10) a phase space density (compare to the general discussion in section 2.1) of the form [**18**]

$$f(d, \bar{d}) = \sum_i \frac{\langle 0| e^{\mathbf{d}_k \mathbf{c}_k} |\psi_i\rangle \langle \psi_i | e^{\bar{\mathbf{d}}_k \mathbf{c}_k^\dagger} |0\rangle}{\langle 0| e^{\mathbf{d}_k \mathbf{c}_k} e^{\bar{\mathbf{d}}_k \mathbf{c}_k^\dagger} |0\rangle}. \tag{4.12}$$

The distribution corresponding to a generic state $|\psi\rangle = \prod_{k=1}^{\infty} (c_k^{i\dagger})^{N_{k_i}} |0\rangle$ can be easily computed to be

$$f(d, \bar{d}) = \prod_{k,i} (d_k^i \bar{d}_k^i)^{N_{k_i}} e^{-d_k^i \bar{d}_k^i}. \tag{4.13}$$

Notice that our phase space density (4.12), as written, is a function on a somewhat larger phase space as $d, \bar{d}$ do not have to obey (4.4). To cure this discrepancy we are going to include an extra factor $\exp(-\beta \hat{N})$ in the calculations, where we choose β such that the expectation value of $\hat{N}$ is precisely N. This is just like passing from a microcanonical ensemble to a canonical one, and for many purposes this is probably a very good approximation. For a thorough discussion of this point see [**18**].

To further motivate (4.12) we notice that it associates to a coherent state a density which is a Gaussian centered around a classical curve, in perfect agreement with the usual philosophy that coherent states are the most classical states. It is then also clear that given a classical curve (4.11) we wish to associate to it the density matrix

$$\rho = P_N e^{\mathbf{d}_k \mathbf{c}_k} |0\rangle \langle 0| e^{\bar{\mathbf{d}}_k \mathbf{c}_k^\dagger} P_N \tag{4.14}$$

where P_N is the projector onto the actual Hilbert space of states of energy N as defined in (4.9). Because of this projector, the phase space density associated to a classical curve is not exactly a Gaussian centered around the classical curve but there are some corrections due to the finite N projections. Obviously, these corrections will vanish in the $N \to \infty$ limit.

The density (4.12) has the property that for any function $g(d, \bar{d})$

$$\int\int_{d,\bar{d}} f(d, \bar{d}) g(d, \bar{d}) = \sum_i \langle \psi_i | : g(c, c^\dagger) :_A |\psi_i\rangle \tag{4.15}$$

where $:g(c,c^\dagger):_A$ is the anti-normal ordered operator associated to $g(c,c^\dagger)$, and $\int_{d,\bar{d}}$ is an integral over all variables d_i. Since the theory behaves like a $(1+1)$-dimensional field theory the natural thing to do is to calculate expectation values of normal ordered operators in order to avoid infinite normal ordering contributions. Besides, everything we do is limited by the fact that our analysis is in classical gravity and therefore can at best be valid up to quantum corrections. As a result a further modification to our proposal will be to redefine $g(d,\bar{d})$ by subtracting the anti-normal ordering effects.

Since the harmonic functions appearing in (4.2) can be arbitrarily superposed, we finally propose to associate to (4.10) the geometry

$$
\begin{aligned}
f_5 &= 1+\frac{Q_5}{L}\mathcal{N}\int_0^L\int_{d,\bar{d}}\frac{f(d,\bar{d})ds}{|\mathbf{x}-\mathbf{F}(s)|^2}\\
f_1 &= 1+\frac{Q_5}{L}\mathcal{N}\int_0^L\int_{d,\bar{d}}\frac{f(d,\bar{d})|\mathbf{F}'(s)|^2ds}{|\mathbf{x}-\mathbf{F}(s)|^2}\\
A^i &= \frac{Q_5}{L}\mathcal{N}\int_0^L\int_{d,\bar{d}}\frac{f(d,\bar{d})\mathbf{F}_i'(s)ds}{|\mathbf{x}-\mathbf{F}(s)|^2}
\end{aligned}
\tag{4.16}
$$

with the normalization constant

$$
\mathcal{N}^{-1}=\int_{d,\bar{d}}f(d,\bar{d})
\tag{4.17}
$$

In [**5**] it was shown that the geometries corresponding to a classical curve are regular provided $|\mathbf{F}'(s)|$ is different from 0 and the curve is not self-intersecting. In our setup we sum over continuous families of curves with some weighting factor which can introduce singularities. We expect these singularities to be rather mild, certainly for semiclassical density matrices, and in addition in various examples the averages will turn out to be completely smooth anyway (see section 4.3). Another point worth mentioning is that the average will no longer solve the vacuum type IIB equations of motion, instead a small source will appear on the right hand side of the equations. Since these sources are subleading in the $1/N$ expansion and vanish in the classical limit, they are in a regime where classical gravity is not valid and they may well be cancelled by higher order contributions to the equations of motion. To have an idea about these sources let us study the circular profile.

We consider the following profile

$$
F^1(s)=a\cos\frac{2\pi k}{L}s,\qquad F^2(s)=a\sin\frac{2\pi k}{L}s,\qquad F^3(s)=F^4(s)=0
\tag{4.18}
$$

which describes a circular curve winding k times around the origin in the $(1,2)$-plane. In order to simplify our discussion, we focus on the simplest harmonic function f_5. In order to evaluate the various integrals it will be convenient to Fourier transform the x-dependence. Using

$$\frac{1}{|\mathbf{x}|^2} = \frac{1}{4\pi^2}\int d^4u \frac{e^{i\mathbf{u}\cdot\mathbf{x}}}{|\mathbf{u}|^2} \tag{4.19}$$

Classically $\Box f_5$ is a delta function with a source at the location of the classical curve; to be precise,

$$\Box f_5 = -4\pi Q_5 \delta(x_1^2 + x_2^2 - a^2)\delta(x_3)\delta(x_4). \tag{4.20}$$

Now in the quantum theory, we associate to the classical circular curve (4.18) the density matrix (4.14) and subsequently the phase space density (4.12). Working this out we find out that

$$f(d,\bar{d}) = ((d_k^1 + id_k^2)(\bar{d}_k^1 - i\bar{d}_k^2))^{N/k} e^{-\sum_{l,i} d_l^i \bar{d}_l^i}. \tag{4.21}$$

We have ignored the delta function coming from the projection here and expect (4.21) to be valid for large values of N/k. It is therefore better thought of as a semiclassical profile rather than the full quantum profile.

According to (4.16) the harmonic function f_5 is now given by

$$f_5 = 1 + \frac{Q_5}{4\pi^2}\mathcal{N}\int_0^L ds \int_{d,\bar{d}} f(d,\bar{d}) \int d^4u \frac{1}{|\mathbf{u}|^2} e^{i\mathbf{u}\cdot(\mathbf{x}-\mathbf{F}(s)) + \sum_l \frac{u^2\mu^2}{2l}} \tag{4.22}$$

where we have used (4.19) and the constant $\sum_l \frac{u^2\mu^2}{2l}$ appears due to the fact that we want to compute a normal ordered quantity instead of an anti-normal ordered one. The function $\mathbf{F}(s)$ depends on an infinite set of complex oscillators $d_{\bar{l}}^i$. It can be easily seen that the contribution for the oscillators different from d_k^1 and d_k^2 cancels exactly against the normal ordering constant $u^2\mu^2/2l$ mentioned above.

So $\Box f_5$ for this case reads

$$\Box f_5 = -4\pi Q_5 \delta(x_3)\delta(x_4) A(x_1,x_2) \tag{4.23}$$

$$A(x_1,x_2) = \int_0^\infty d\rho \rho J_0\left(\sqrt{x_1^2 + x_2^2}\rho\right) L_{N/k}\left(\frac{a^2\rho^2}{4N/k}\right) \tag{4.24}$$

Until here we have not used any approximation. Using the identity

$$L_N(x) = \frac{e^x}{N!}\int_0^\infty e^{-t} t^N J_0(2\sqrt{tx})dt$$

and approximating $\exp\left(\frac{a^2\rho^2}{4N/k}\right) \approx 1$ one obtains

$$A(x_1,x_2) = \frac{e^{-N/k\; r^2/a^2}\left((N/k)\; r^2/a^2\right)^{N/k}}{((N/k)-1)!a^2} \tag{4.25}$$

with $r^2 = x_1^2 + x_2^2$. In the limit $N/k \to \infty$, $A(x_1, x_2)$ approaches $\frac{\delta(r^2/a^2-1)}{a^2}$ and the classical and quantum results agree. For large N/k, $A(x_1, x_2)$ is approximately a Gaussian around $r^2 \approx a^2$ and width $1/\sqrt{N/k}$; indeed, using Stirling's formula

$$A(x_1, x_2) \approx \frac{\sqrt{N/k}}{\sqrt{2\pi}} e^{-N/k(r^2/a^2-1)} (r^2/a^2)^{N/k} \tag{4.26}$$

So the quantum geometry corresponds to a solution of the equations of motion in the presence of smeared sources. The width of the smeared source goes to zero in the limit $N/k \to \infty$, as expected.

4.3. Thermal ensembles. In the following we consider the geometry of some thermal ensembles of interest.

4.3.1. *$M = 0$ BTZ.* The corresponding thermal ensemble is characterized by the following density matrix[19]:

$$\rho = \sum_{N_k, \tilde{N}_k} \frac{|N_k\rangle\langle N_k|e^{-\beta\hat{N}}|\tilde{N}_k\rangle\langle\tilde{N}_k|}{\mathrm{Tr}\, e^{-\beta\hat{N}}} \tag{4.27}$$

where $|N_k\rangle$ is a generic state labelled by collective indices N_k,

$$|N_k\rangle = \prod_k \frac{1}{\sqrt{N_k!}} (\mathbf{c}_k^\dagger)^{N_k} |0\rangle$$

and we have chosen a normalization so that $\langle N_k|\tilde{N}_k\rangle = \delta_{N_k, \tilde{N}_k}$. The value of the potential β has to be adjusted such that $\langle\hat{N}\rangle = N$. It is clear that

$$\rho = \prod_n \rho_k, \qquad \rho_k = (1 - e^{-k\beta}) \sum_{n=0}^{\infty} e^{-nk\beta} |k, n\rangle\langle k, n| \tag{4.28}$$

with $|k, n\rangle = \frac{1}{\sqrt{n!}} (\mathbf{c}_k^\dagger)^n |0\rangle$. Then the full distribution will simply be the product $f(d, \bar{d}) = \prod_k f^{(k)}_{\mathbf{d}_k, \bar{\mathbf{d}}_k}$ with

$$\begin{aligned} f^{(k)}_{\mathbf{d}_k, \bar{\mathbf{d}}_k} &= (1 - e^{-k\beta}) e^{-\mathbf{d}_k \bar{\mathbf{d}}_k} \sum_{n=0}^{\infty} \frac{e^{-nk\beta}}{n!} (\mathbf{d}_k \bar{\mathbf{d}}_k)^n \\ &= (1 - e^{-k\beta}) \exp\left(-(1 - e^{-k\beta}) d_k \bar{d}_k\right). \end{aligned} \tag{4.29}$$

[19]We are going to ignore the i-index in some equations where it does not play any role. We hope that this will not create any confusion.

The needed harmonic functions (4.2) are deduced from the following generating function

$$f_v = \frac{Q_5}{4\pi^2 L}\mathcal{N}\int d^4u \int_0^L dr \int_{d,\bar{d}} f(d,\bar{d}) \times \frac{e^{\sum_k \left(\frac{|\mathbf{u}|^2\mu^2}{2k} - \frac{2\pi^2 k\mu^2|\mathbf{v}|^2}{L^2}\right)} e^{i\mathbf{u}\cdot(\mathbf{x}-\mathbf{F}(r)) + i\mathbf{v}\cdot\mathbf{F}'(r)}}{|\mathbf{u}|^2} \tag{4.30}$$

which gives

$$f_5 = Q_5\,\frac{1-e^{-\frac{3\beta}{\pi^2\mu^2}x^2}}{x^2}, \qquad f_1 = Q_1\,\frac{1-e^{-\frac{3\beta}{\pi^2\mu^2}x^2}}{x^2}$$
$$A_i = 0, \quad \beta \approx \pi\sqrt{\frac{2}{3N}}. \tag{4.31}$$

A final comment is in order. The geometry obtained differs from the *classical* $M = 0$ BTZ black hole by an exponential piece. Following [**47, 36**] we could put a stretched horizon at the point where this exponential factor becomes of order one, so that the metric deviates significantly from the classical $M = 0$ BTZ solution. Thus, using this criterion we find for the radius of the stretched horizon[20]

$$r_0 \approx \frac{\mu}{\beta^{1/2}} \tag{4.32}$$

with corresponding entropy proportional to $N^{3/4}$. This exceeds the entropy of the mixed state from which the geometry was obtained, the latter grows as $N^{1/2}$. This does not contradict any known laws of physics, and in addition we should remember that the notion of stretched horizon depends on the choice of observer. It is quite likely that for a suitable choice of observer the entropy of the stretched horizon agrees with the entropy obtained from the dual CFT. For a further discussion of this point see [**18, 48**].

4.3.2. *The small black ring.* In this section we consider a slightly more complicated example, namely an ensemble consisting of a condensate of J oscillators of level q plus a thermal ensemble of effective level $N - qJ$. As argued in [**49–51, 17**] such an ensemble should describe (in a certain region of parameter space) a small black ring of angular momentum J and dipole (or Kaluza-Klein) charge q.

[20]The same value is obtained if we compute the average size of the curve in $\mathbb{R}^4$, $r_0^2 \approx \langle |F|^2 \rangle$.

Using the techniques developed in the previous sections we can compute the generating harmonic function for this case as well and we find

$$f_v = Q_5 L_J\left(\frac{\mu^2}{4q}\left[\left(\frac{2\pi q}{L}v_2 + i\partial_1\right)^2 + \left(\frac{2\pi q}{L}v_1 - i\partial_2\right)^2\right]\right)$$

$$\times\, e^{-\frac{\mu^2\pi^2|v|^2}{2L^2}(N-qJ)}\frac{1-e^{-\frac{2|\mathbf{x}|^2}{\mu^2 D}}}{|\mathbf{x}|^2} \tag{4.33}$$

where $D = \pi\sqrt{2/3}(N-qJ)^{1/2}$ so that the geometry is purely expressed in terms of the macroscopic quantities N, J and q.

We would like to make contact between this geometry and the geometry corresponding to small black rings studied in [**17**]. As we will see, in the limit of large quantum numbers both geometries reproduce the same one-point functions.

In order to see this, first note that the exponential factor $e^{-\frac{2|\mathbf{x}|^2}{\mu^2 D}}$ will not contribute (as it vanishes faster than any power at asymptotic infinity). Secondly one has the formal expansion

$$L_J\left(\frac{\mu^2}{4q}\mathcal{O}\right) = J_0\left(\mu\sqrt{\frac{J}{q}}\mathcal{O}^{1/2}\right) + \cdots \tag{4.34}$$

In order to estimate the validity of this approximation we can think of $\mathcal{O}$ as being proportional to $1/|\mathbf{x}|^2$. On the other hand $\mu\sqrt{J/q}$ can be roughly interpreted as the radius of the black ring (see [**52, 17**], where this parameter is called R). Hence the approximation is valid for large values of J at a fixed distance compared to the radius of the ring.

Using the above approximations it is straightforward to compute the harmonic functions

$$f_5 = \frac{Q_5}{r^2 + \mu^2\frac{J}{q}\cos\theta}, \qquad f_1 = \frac{Q_1}{r^2 + \mu^2\frac{J}{q}\cos\theta} \tag{4.35}$$

where we have used the following coordinate system:

$$\begin{aligned} x_1 &= (r^2+a^2)^{1/2}\sin\theta\cos\varphi\,, \qquad x_2 = (r^2+a^2)^{1/2}\sin\theta\sin\varphi \\ x_3 &= r\cos\theta\cos\psi\,, \qquad x_4 = r\cos\theta\sin\psi. \end{aligned} \tag{4.36}$$

Hence in this approximation the geometry reduces exactly to that of the small black ring studied in [**17**].

4.3.3. *Generic thermal ensemble.* In the following we consider a generic thermal ensemble, where each oscillator c_{ki} is occupied thermally with a temperature β_{ki}. We further will assume that $\beta_{k\pm}$ for the directions $1, 2$ is equal to $\beta_{k\pm}$ for the directions $3, 4$. Restricting to, say, directions $1, 2$ we are led to consider the following distribution:

$$f(d,\bar{d}) = \exp\left(-(1-e^{-\beta_{k+}})d_k^+\bar{d}_k^+ - (1-e^{-\beta_{k-}})d_k^-\bar{d}_k^-\right). \tag{4.37}$$

Following the same steps as for the case of the small black ring we obtain

$$\begin{aligned}
(4.38)\quad f_5 &= Q_5 \frac{1-e^{-\frac{2|\mathbf{x}|^2}{\mu^2 D}}}{|\mathbf{x}|^2} \\
(4.39)\quad f_1 &= Q_1 \left(\frac{1-e^{-\frac{2|\mathbf{x}|^2}{\mu^2 D}}}{|\mathbf{x}|^2} - \frac{J^2}{4N\mu^4 D^2} e^{-\frac{2|\mathbf{x}|^2}{\mu^2 D}} \right) \\
(4.40)\quad A &= \frac{\mu^2 J}{2} \left(2\frac{e^{-\frac{2|\mathbf{x}|^2}{\mu^2 D}}}{\mu^2 D} - \frac{1-e^{-\frac{2|\mathbf{x}|^2}{\mu^2 D}}}{|\mathbf{x}|^2} \right) (\cos^2\theta d\phi + \sin^2\theta d\psi)
\end{aligned}$$

where $(|\mathbf{x}|, \theta, \phi, \psi)$ are standard spherical coordinates on $\mathbb{R}^4$.

We see that, rather surprisingly, the geometry depends only on a few quantum numbers N, J and D, which are given in terms of the temperatures by

$$N = 2\sum_k k \left(\frac{e^{-\beta_{k+}}}{1-e^{-\beta_{k+}}} + \frac{e^{-\beta_{k-}}}{1-e^{-\beta_{k-}}} \right) \tag{4.41}$$

$$J = 2\sum_k \left(\frac{e^{-\beta_{k+}}}{1-e^{-\beta_{k+}}} - \frac{e^{-\beta_{k-}}}{1-e^{-\beta_{k-}}} \right) \tag{4.42}$$

$$D = 2\sum_k \frac{1}{k} \left(\frac{e^{-\beta_{k+}}}{1-e^{-\beta_{k+}}} + \frac{e^{-\beta_{k-}}}{1-e^{-\beta_{k-}}} \right). \tag{4.43}$$

As a result, the information carried by the geometry is much less than that carried by the ensemble of microstates. In fact, only N and J are visible at infinity while D sets the size of the "core" of the geometry. We also find that D is precisely the expectation value of the dipole operator introduced in [**17**]. Its presence in the density matrix is supported by an analysis of the first law of thermodynamics [**53**]. It is a non-conserved charge which makes its extension to interacting theories an interesting open problem.

We interpret the above remark as a manifestation of the no-hair theorem for black holes. The derivation in this section assumes that the temperatures are all sufficiently large. By tuning the temperatures, it is possible to condense one (like in the small black ring case) or more oscillators. If this happens, we should perform a more elaborate analysis, and we expect that the dual geometrical description[21] corresponds to concentric small black rings. In this case the configuration will depend on more quantum numbers than just N, J, D, in particular we will find solutions where the small black rings carry arbitrary dipole charge. Thus, once we try to put hair on the small

[21] It is not difficult to see that the harmonic functions now will take the form of multiple Laguerre polynomials with differential operator arguments acting on the generating harmonic function of the $M = 0$ BTZ solution.

black hole by tuning chemical potentials appropriately, we instead find a phase transition to a configuration of concentric small black rings, each of which is still characterized by just a few quantum numbers.

5. $AdS_3 \times S^2$

Although the $D1$-$D5$ system has proved a fertile example in which to test various ideas of the "fuzzball" proposal it is susceptible to the criticism that the solution is not a black hole but rather a naked singularity with a horizon possibly being generated by higher derivative corrections. It would be desirable to be able to study a similar class of solutions for a system where the total charge corresponds to a black hole with a macroscopic horizon scale (i.e. a three-charge black hole). Such black holes are 1/8 BPS solutions in the full string theory or can emerge as 1/2 BPS solutions of $\mathcal{N} = 2$ supergravity in four or five dimensions (i.e. string or M-theory reduced on a Calabi-Yau).

5.1. Solution spaces. In four dimensions Bates and Denef [**11**] have constructed general multicentered BPS solutions of generic $\mathcal{N} = 2$ supergravity theories, and in [**8**] Bena and Warner classify the full set of BPS solutions for the special case of of the five-dimensional $\mathcal{N} = 2$ supergravity theory which is the truncation of the $\mathcal{N} = 8$ theory (i.e. the theory is invariant under 8 instead of 32 supersymmetries). The latter require specifying a four-dimensional base metric that is restricted to be hyperkähler. A particularly appealing class of hyperkähler manifolds are Gibbons-Hawking or multi-Taub-NUT geometries, which are asymptotically $\mathbb{R}^3 \times S^1$ and for which we have explicit metrics. Moreover, it has been shown that the five-dimensional solutions constructed using a Gibbons-Hawking base manifold [**7**] correspond to the four-dimensional ones via the 4d/5d connection [**54**] [**12**], making them an interesting class of solutions to study.

The five-dimensional solutions, although relatively complicated, are determined entirely in terms of $2b_2 + 2$ harmonic functions, where b_2 is the second Betti number of the compactification Calabi-Yau, X,

$$
\begin{aligned}
\mathsf{H}^0 &= \sum_a \frac{p_a^0}{|\mathsf{x} - \mathsf{x}_a|} + h^0\,, & \mathsf{H}^A &= \sum_a \frac{p_a^A}{|\mathsf{x} - \mathsf{x}_a|} + h^A, \\
\mathsf{H}_A &= \sum_a \frac{q_A^a}{|\mathsf{x} - \mathsf{x}_a|} + h_A\,, & \mathsf{H}_0 &= \sum_a \frac{q_0^a}{|\mathsf{x} - \mathsf{x}_a|} + h_0\,.
\end{aligned}
\tag{5.1}
$$

Here the coordinate vector x_a gives the position in the spatial $\mathbb{R}^3$ of the a-th center with charge $\Gamma_a = (p_a^0, p_a^A, q_A^a, q_0^a)$ (note here that A runs from $1, \ldots, b_2$). The IIA interpretation of these charges is ($D6$, $D4$, $D2$, $D0$) wrapping cycles of X while in M-theory the charge vector is (KK, $M5$, $M2$, P). Note that the harmonics have $2b_2 + 2$ constants $h = (h^0, h^A, h_A, h_0)$ that together determine the asymptotic behaviour of the harmonics and hence the solutions. We will also have frequent occasion to use the notation $\Gamma = (p^0, p^A, q_A, q_0)$ to refer to the total charge $\Gamma = \sum_a \Gamma_a$.

The position vectors have to satisfy the integrability constraints

$$\sum_b \frac{\langle \Gamma_a, \Gamma_b \rangle}{|\mathsf{x}_a - \mathsf{x}_b|} = \langle h, \Gamma_a \rangle \,, \tag{5.2}$$

where we define the symplectic intersection product

$$\langle \Gamma_1, \Gamma_2 \rangle := -p_1^0 q_0^2 + p_1^A q_A^2 - q_A^1 p_2^A + q_0^1 p_2^0. \tag{5.3}$$

By summing (5.2) over a we find that the constants h have to obey $\langle h, \Gamma \rangle = 0$. Note that even once the charges of each center have been fixed there is a large space of solutions that may even have several disconnected components. In particular, the constraint (5.2) implies that the positions of the centers are generally restricted, defining a complicated moduli space of (generically) bound solutions.

The metric, gauge field and Kähler scalars of the solution are now given in terms of the harmonics by

$$\begin{aligned} ds_{5d}^2 &= 2^{-2/3}\, Q^{-2} \left[-(\mathsf{H}^0)^2 (dt+\omega)^2 - 2L(dt+\omega)(d\psi + \omega_0) \right. \\ &\quad \left. + \, \Sigma^2 (d\psi + \omega_0)^2 \right] + 2^{-2/3}\, Q \, d\mathsf{x}^i d\mathsf{x}^i, \\ A_{5d}^A &= \frac{-\mathsf{H}^0}{Q^{3/2}} (dt + \omega) + \frac{1}{\mathsf{H}^0} \left(\mathsf{H}^A - \frac{L y^A}{Q^{3/2}} \right) (d\psi + \omega_0) + \mathcal{A}_d^A, \\ Y^A &= \frac{2^{1/3} y^A}{\sqrt{Q}}, \end{aligned} \tag{5.4}$$

where $\mathsf{x}^i \in \mathbb{R}^3$ and ψ is an angular coordinate with period 4π, and the functions appearing satisfy the relations

$$\begin{aligned} d\omega_0 &= \star d\mathsf{H}^0, \\ d\mathcal{A}_d^A &= \star d\mathsf{H}^A, \\ \star d\omega &= \langle d\mathsf{H}, \mathsf{H} \rangle \\ \Sigma &= \sqrt{\frac{Q^3 - L^2}{(\mathsf{H}^0)^2}}, \\ L &= \mathsf{H}_0 (\mathsf{H}^0)^2 + \frac{1}{3} D_{ABC} \mathsf{H}^A \mathsf{H}^B \mathsf{H}^C - \mathsf{H}^A \mathsf{H}_A H^0, \\ Q &= (\frac{1}{3} D_{ABC} y^A y^B y^C)^{2/3}, \\ D_{ABC} y^A y^B &= -2 \mathsf{H}_C \mathsf{H}^0 + D_{ABC}\, \mathsf{H}^A \mathsf{H}^B. \end{aligned} \tag{5.5}$$

Here the Hodge star is with respect to the flat $\mathbb{R}^3$ spanned by the coordinates x^i and D_{ABC} are the triple intersection numbers of the chosen basis of $H^2(X)$. Note that the only equation in (5.5) for which there is no general solution in closed form is the last one. In some cases, e.g. when $b_2 = 1$, it is even possible to obtain a solution, in closed form, to this equation.

From (5.4) and (5.5) it may seem that the solutions are singular if H^0 vanishes but this is not the case as various terms in Q and L cancel any possible divergences due to negative powers of H^0 (in fact, the BTZ black hole can, in the decoupling limit introduced in the next section, be mapped to such a solution with H^0 vanishing everywhere).

An additional complication is the fact that even solutions satisfying the constraint equation (5.2) may still suffer from various pathologies, most notably closed time-like curves (CTCs). For instance, the prefactor to the $d\psi^2$ term in the metric may become negative if Σ becomes imaginary[22]. Unfortunately there is no simple criterion which can be used to determine if a given solution is pathology-free. To fill in this gap [**22**] and [**24**] conjectured the *attractor flow conjecture*, a putative criterion for the existence of (well-behaved) solutions which we will relate in section 5.4.

An essential feature of these solutions is that they are stationary but not static. In particular they carry quantized intrinsic angular momentum associated with the crossed electric and magnetic fields of the dyonic centers [**22**]

$$\vec{J} = \frac{1}{2}\sum_{a<b} \frac{\langle \Gamma_a, \Gamma_b\rangle\, \vec{x}_{ab}}{r_{ab}} . \tag{5.6}$$

This will be important when quantizing the solution space as it is a necessary (but not sufficient) condition for the latter to be a proper phase space with a non-degenerate symplectic form. A solution space with vanishing angular momentum does not enjoy this property and must be completed to a phase space by the addition of velocities (see e.g. [**55**]).

5.2. Decoupling. Some specific five-dimensional solutions obtained from an $\mathcal{N} = 2$ truncation of $\mathcal{N} = 8$ (i.e. compactification on T^6) have been studied using AdS/CFT [**49, 52**] by taking a decoupling limit of a dualized form of the solutions. This cannot be generalized to an arbitrary compact CY because the duality group of the latter is not known. To study the full class of solutions using AdS/CFT it is desirable to have a more general decoupling limit that can embed a larger class of these solutions in an AdS_3 throat. The limit we will describe can be taken in a proper $\mathcal{N} = 2$ theory and will yield an $\mathrm{AdS}_3 \times \mathrm{S}^2$ near horizon geometry [**20**]. This decoupling limit only works for microstate solutions whose total charge does not contain any overall $D6$/KK-monopole charge so the relevant CFT is essentially

[22] As described in [**12**] this would also imply that the 4-dimensional metric associated with this 5d solution (via the 4d/5d connection of [**54**]) becomes imaginary as Σ appears directly in the former.

the MSW CFT[23]. Although the latter is not under very good control it is nonetheless possible to determine, from the asymptotics of a given geometry, the CFT quantum numbers of the corresponding state. It is also possible to use general CFT properties to determine various quantities such as the number of states in a given charge sector. It would, of course, be desirable to make progress in understanding this CFT as this may yield significant insight into the microstate geometries.

In **[20]** the decoupling limit of the solutions described above is defined by taking $\ell_5 \to 0$ (ℓ_5 is the 5d Planck length) while keeping fixed the mass of $M2$ branes stretched between the various centers and wrapping the M-theory circle. In doing so we also fix the volume of the Calabi-Yau and the length of the M-theory circle, R, as measured in 5-d Planck units. Since the mass of such membranes is given by $m_{M2} \sim R\Delta\mathsf{x}/\ell_5^3$, the coordinate distances between the centers, $\Delta\mathsf{x}$, must be rescaled as ℓ_5^3. Alternatively, we can see this limit as a rescaling of the 5d metric under which the Einstein part of the action is invariant.

We now define new rescaled coordinates, x^i, and harmonic functions, H, as

$$x^i = \ell_5^{-3}\mathsf{x}^i \qquad H = \ell_5^{3/2}\mathsf{H} \tag{5.7}$$

By restricting to the region of finite x^i we are essentially keeping the mass of transverse, open membranes fixed while rescaling the original coordinates, x^i. One can see that, in these new variables, the structure of the solution (in terms of the harmonics) does not change in the decoupling limit except for an overall scaling of the metric by a factor of ℓ_5^{-2}. The rescaled harmonics do take a new form, however,

$$\begin{aligned} H^0 &= \sum_a \frac{p_a^0}{|x - x_a|}, & H^A &= \sum_a \frac{p_a^A}{|x - x_a|}, \\ H_A &= \sum_a \frac{q_A^a}{|x - x_a|}, & H_0 &= \sum_a \frac{q_0^a}{|x - x_a|} - \frac{R^{3/2}}{4}. \end{aligned} \tag{5.8}$$

In particular note that all the constants have disappeared except the $D0$ brane constant which now takes a fixed value. Related to this is the fact that the asymptotic value of the moduli are forced to the attractor point, $Y^A \sim p^A$ (this corresponds to sending the 4d Kähler moduli to $J^A = \infty\, p^A$).

Recall that the coordinate locations of the centers must satisfy the integrability constraint (5.2) and that this constraint depends on the value of the constants in the harmonic functions. As a consequence it is possible that, in taking the decoupling limit, some solutions cease to exist. For instance if

[23]The "MSW" CFT is the theory that arises as the low-energy description of $M5$ branes wrapping an ample divisor in the Calabi-Yau. It is an $N = (0, 4)$ superconformal field theory and it owes its name to the three authors of **[2]**.

we consider two $D4$-$D2$-$D0$ centers, Γ_a and Γ_b, then the constraint equation in the decoupling limit is

$$\frac{\langle\Gamma_a, \Gamma_b\rangle}{x_{ab}} = h_0 p^0 = 0 \tag{5.9}$$

Implying that x_{ab}, the inter-center distance, must be infinite (unless the charges are parallel, $\langle\Gamma_a, \Gamma_b\rangle = 0$) so one of the centers is forced out of the finite region of the rescaled coordinates, x^i, as we take the limit. We interpret this as implying that such centers cannot sit in the same decoupled AdS_3 throat.

Because only the $D0$ constant survives, centers without $D6$ charge are no longer bound together. A related fact, which will emerge presently, is that the solution space of such centers does not have a non-degenerate symplectic form (because, in the decoupling limit, intrinsic angular momentum is proportional to $D6$ dipole-moment) on it and hence cannot be quantized without the addition of additional degrees of freedom.

Even if a set of charges admits a solutions that satisfies the constraint equations in the decoupling limit the solution may develop other pathologies such as CTCs. To study these we may resort once more to the attractor flow conjecture of [**22**] which is argued, in [**19**], to extend to the decoupled solutions. We will not discuss this in any depth here except to note that the fact that the asymptotic moduli, Y^A, are forced to the attractor point for the total charge implies that only attractor flow trees which can be extended to trees starting from this value of the asymptotic moduli will survive the decoupling limit. Further discussion can be found in Section 5.4 and [**19**].

The decoupled solutions are asymptotically $AdS_3 \times S^2$ and their asymptotic form is given by

$$\begin{aligned} ds^2_{5D} &= -\frac{\rho^2}{4U^2} dv d\psi + \frac{U^{-4}}{4}\left[-R^2(d^0)^2 dv^2 + \mathcal{D} d\psi^2\right] \\ &\quad +4U^2\frac{d\rho^2}{\rho^2} + U^2\left(d\theta^2 + \sin^2\theta d\alpha^2\right) + \mathcal{O}\left(\frac{1}{\rho^2}\right), \end{aligned} \tag{5.10}$$

$$A^A_{5D} = -p^A \cos\theta d\alpha - D^{AB} q_B d\psi + \mathcal{O}\left(\frac{1}{\rho^2}\right), \tag{5.11}$$

$$Y^A = \frac{p^A}{U} + \mathcal{O}\left(\frac{1}{r^2}\right). \tag{5.12}$$

where we have introduced some coordinate redefinitions. We first introduce standard spherical coordinates, (r, θ, ϕ), on the base spatial $\mathbb{R}^3$ of the solution with the axis of the sphere (the z-axis) aligned with the $D6$ dipole moment of the solution

$$\vec{d}^0 := \sum_a p^0_a \vec{x}_a \tag{5.13}$$

For brevity we introduce the notation $d^0 = |\vec{d}^0|$ and $\vec{e} = \vec{r}/r$ so d^0 is the norm of the dipole moment and $\vec{e}$ is a unit vector in the radial direction. We then make a further coordinate redefinition

$$v = t - \frac{R}{4}\psi\,, \qquad \alpha = \phi + Rd^0\left(\frac{p^3}{3}\right)^{-1} v\,, \qquad U^3 = \frac{p^3}{6} \quad \text{and}$$

$$\mathcal{D} = \frac{p^3}{3}\left(D_{q_A}^{AB} q_B - 2q_0\right), \tag{5.14}$$

and define a new radial coordinate ρ via

$$\frac{\rho^2}{4U^2} = -\frac{U^{-4}}{2}R\left(\frac{e\cdot d^A D_{ABC}p^B p^C}{3} - \frac{p^A q_A d^0 \cos\theta}{3}\right) + \frac{R}{U}r\,, \tag{5.15}$$

Here $\vec{d}^A$ is the $D4$ dipole moment, defined analogously to $\vec{d}^0$. To make a connection to the dual CFT the solution needs to be reduced along the S^2 to give a theory defined purely on AdS_3 with KK modes in representations of $SU(2)$. This procedure is reviewed in some detail in [**56**] with particular attention to the subtleties involved in reducing the 5d Chern-Simons terms in the supergravity action.

The resultant metric is asymptotically AdS_3 and from this, and the asymptotic form of the gauge field, we can determine the charges in dual field theory (see [**20**] for details). In particular we find that

$$\begin{aligned} L_0 &= \frac{(p^A q_A)^2}{2p^3} - q_0 + \frac{p^3}{24}\,, \\ \tilde{L}_0 &= \frac{(p^A q_A)^2}{2p^3} + \frac{p^3}{24}\,. \end{aligned} \tag{5.16}$$

and that the $SU(2)$ R-symmetry charge associated with a solution is determined entirely in terms of its $D6$ dipole moment

$$J_0^3 = -\frac{R^2 d^0}{8} \tag{5.17}$$

While (5.16) gives the charges expected from general considerations of the MSW CFT associated with the total charge Γ, the $SU(2)$ charge, J_0^3, depends on a dipole moment and, as such, is absent in the single center solution. In fact, it is nothing more than the intrinsic angular momentum, J, defined in (5.6) which, in the decoupling limit, can be shown to be proportional to the $D6$ dipole moment.

In the next section we will see that this charge plays a distinguished role in the quantization of the system as its presence is necessary in order to have a non-trivial symplectic form on the phase space.

5.3. Quantization. As anticipated in [**15**], since the dual (0, 4) CFT of $\mathcal{N} = 2$ black holes (lifted to 5-d) is less well understood than its (4, 4) cousin, in this case one can hope to make progress by quantizing the phase space of the supergravity solutions directly. The quantization we will perform will be quite general in that it will cover the original 4-d multi-center black hole configurations [**11**], their 5-d uplift discussed in section 5.1, and the decoupled version of the latter (which can be related to the (0, 4) CFT). For $\mathcal{N} = 2$ black holes coming from Calabi-Yau compactifications it is likely that a large portion of the entropy will come from degrees of freedom in the Calabi-Yau so it may not be possible to account for a finite fraction of the entropy using supergravity alone. Whether this is the case or not is an important question, though the answer may be difficult to determine as the solutions are rather complicated and even the classical phase space is quite non-trivial.

To a given total charge, Γ, a fuzzball-like prescription might associate all possible decompositions of Γ into N charges, Γ_a, positioned at different centers, such that the solutions are smooth. The associated phase space for each decomposition would then be the (possibly disconnected) solution space for the charges. Quantization of this phase space should, in principle, yield certain microstates of the black hole and the set of (supergravity) microstates[24] should come from summing over all possible decompositions. The notion of smoothness is not necessarily (duality frame) invariant so a more precise criterion might be that the constituent charges, Γ_a, should have no entropy (even microscopic) associated with them.

For a given decomposition into N centers the phase space will be the $(2N-2)$-dimensional submanifold of $\mathbb{R}^{3N-3}$ given by solving the constraint equations (5.2) for the positions $\vec{x}_a$. Note that in arriving at this counting we have subtracted the 3 center of mass degrees of freedom; while these are present they generally decouple. As mentioned, this manifold may be disconnected and may possess a rather complicated topology. Moreover, for a given total charge, Γ, there will be many possible decompositions into different numbers of centers, implying that the total phase space will be a disconnected sum of many manifolds of different dimension. Determining the symplectic structure of even the lowest dimensional solution spaces is already challenging [**20**].

5.4. Split attractors and state counting. In [**22**] a conjecture is proposed whereby pathology-free solutions are those with a corresponding *attractor flow tree* in the moduli space. This conjecture was first posed for multicentered four-dimensional solutions so we will introduce some four-dimensional terminology here. The four-dimensional moduli, $t^A(\vec{x}) = B^A(\vec{x}) + iJ^A(\vec{x})$, are the complexified Kähler moduli of the Calabi-Yau.

[24]Actually, the full set of supergravity microstates probably requires considering more general solutions than those with Gibbons-Hawking base but one can at least hope that the latter set contributes a finite fraction of the entropy.

The relation between these moduli and their five-dimensional counterparts can be found in [**54, 19**]. To each charge vector, Γ_i, we can associate a complex number, the central charge, by

$$Z(\Gamma_i; t) := \langle \Gamma_i, \Omega(t) \rangle \qquad \Omega(t) := -\frac{e^t}{\sqrt{\frac{4}{3}J^3}} \tag{5.18}$$

Note that, since t^A is a two-form, Ω is a sum of even degree forms. The phase of the central charge, $\alpha(\Gamma_i) := \arg[Z(\Gamma_i; t)]$, encodes the supersymmetry preserved by that charge at the given value of the moduli. The even form Ω is related, asymptotically, to the constants in the harmonics (5.1) (which define both the 4-d and 5-d solutions) as

$$h = -2\,\mathrm{Im}(e^{-i\alpha(\Gamma)}\,\Omega)|_\infty \tag{5.19}$$

For more details of the 4-d solution the reader should consult [**11**].

An *attractor flow tree* is a graph in the Calabi-Yau moduli space beginning at the moduli at infinity, $t^A|_\infty$, and ending at the attractor points for each center. The edges correspond to single center flows towards the attractor point for the sum of charges further down the tree. Vertices can occur where single center flows (for a charge $\Gamma = \Gamma_1 + \Gamma_2$) cross walls of marginal stability where the central charges are all aligned ($|Z(\Gamma)| = |Z(\Gamma_1)| + |Z(\Gamma_2)|$). The actual flow of the moduli $t^A(\vec{x})$ for a multicentered solution will then be a thickening of this graph (see [**22, 24**] for more details). According to the conjecture a given attractor flow tree will correspond to a single connected set of solutions to the equations (5.2), all of which will be well-behaved. An example of such a flow is given in Figure 2.

The intuition behind this proposal is based on studying the two-center solution for charges Γ_1 and Γ_2. The constraint equations (5.2) imply that when the moduli at infinity are moved near a wall of marginal stability (where Z_1 and Z_2 are parallel) the centers are forced infinitely far apart

$$r_{12} = \frac{\langle \Gamma_1, \Gamma_2 \rangle}{\langle h, \Gamma_1 \rangle} = \left.\frac{\langle \Gamma_1, \Gamma_2 \rangle\, |Z_1 + Z_2|}{2\,\mathrm{Im}(\bar{Z}_2 Z_1)}\right|_\infty \tag{5.20}$$

In this regime the actual flows in moduli space are well approximated by the split attractor trees since the centers are so far apart that the moduli will assume single-center behaviour in a large region of spacetime around each center. Thus in this regime the conjecture is well motivated. Varying the moduli at infinity continuously should not alter the BPS state count, which corresponds to the quantization of the two-center moduli space, so unless the moduli cross a wall of marginal stability we expect solutions smoothly connected to these to also be well defined. Extending this logic to the general N-center case requires an assumption that it is always possible to tune the moduli such that the N centers can be forced to decay into two clusters that effectively mimic the two-center case. There is no general argument that this

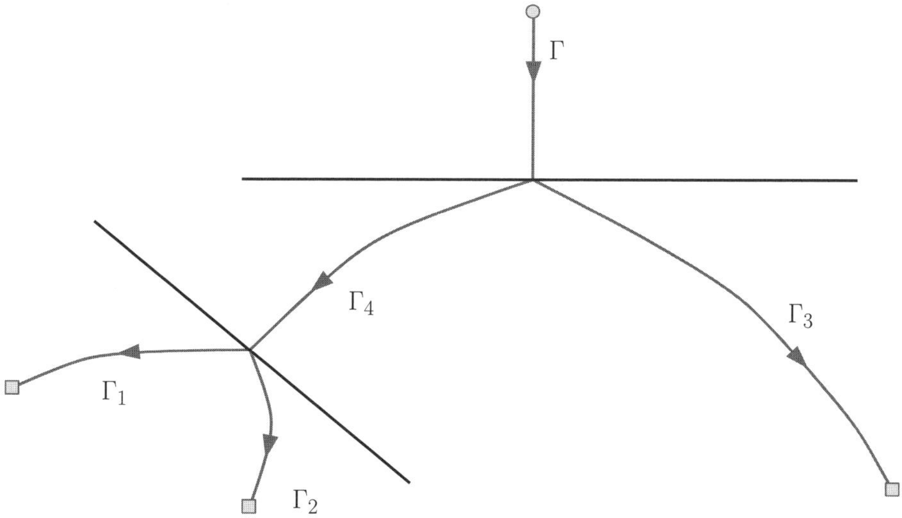

FIGURE 2. Sketch of a 3-center attractor flow tree from [**24, 19**]. The straight lines without arrows are lines of marginal stability, the curved lines with arrows are single center attractor flows. The tree starts at the small circle at top (the moduli at infinity) and flows towards the attractor points indicated by the three small boxes along the bottom. Note here that $\Gamma_4 = \Gamma_1 + \Gamma_2$ and $\Gamma = \Gamma_4 + \Gamma_3$. On the walls of marginal stability the moduli are such that $|Z(\Gamma;t)| = |Z(\Gamma_3;t)| + |Z(\Gamma_4;t)|$ (horizontal wall on top) and $|Z(\Gamma_4;t)| = |Z(\Gamma_1;t)| + |Z(\Gamma_2;t)|$ (diagonal wall on bottom left).

should be the case but one can run the logic in reverse, building certain large classes of solutions by bringing in charges pairwise from infinity, and this can be understood in terms of attractor flow trees. What is not clear is that all solutions can be constructed in this way. For more discussion the reader should consult [**23**].

Although this conjecture was initially proposed for asymptotically flat solutions, in [**20**] it was argued that the essential features of the attractor flow conjecture continue to hold in the decoupling limit.

Before proceeding to count the number of states associated to an attractor flow tree we should mention an important subtlety in using the attractor flow conjecture to classify and validate solutions. Certain classes of charges will admit so-called *scaling solutions* [**57, 24**] which are not amenable to study via attractor flows. These solutions are characterized by the fact that the constraint equations (5.2) have solutions that continue to exist at any value of the asymptotic moduli. For instance, for the three-center case, with charges Γ_1, Γ_2, and Γ_3, if the intersection products $\langle\Gamma_1, \Gamma_2\rangle$, $\langle\Gamma_2, \Gamma_3\rangle$ and $\langle\Gamma_3, \Gamma_1\rangle$ all have the same sign and their absolute values satisfy the triangle inequalities then we can always solve (5.2) by setting $r_{ij} =$

$\lambda |\langle\Gamma_i, \Gamma_j\rangle| + \mathcal{O}(\lambda^2)$ and taking $\lambda \to 0$. Thus, not only do we have a solution for any value of the moduli but the solution admits a scaling in the small separation regime so that we can bring the centers arbitrarily close together in coordinate distance. Even though the coordinate separation goes to zero, the actual metric distance between the centers remains finite because the warp factor in the metric also diverges as the centers approach each other. Thus, as the centers collide, they generate an infinitely deep throat. The only constraint (for three centers) for such solutions to exist is that the intersection products of the charges must satisfy triangle inequalities; this can be generalized to more centers but the constraint becomes more complicated. As such they cannot be decomposed (into clusters, as described above) by moving towards walls of marginal stability (since no such walls exist) and so the standard argument for the attractor flow conjecture does not apply to such solutions.

Because an infinitely deep throat develops for these scaling solutions, [**57**] and [**24**] have suggested that it is, in fact, exactly these configurations which correspond to single-centered black hole microstates. In [**24**] a calculation in the Higgs-branch of the associated gauge theory is used to argue that a macroscopic entropy emerges precisely when the charges are of scaling type. These scaling solutions were further studied in [**58–60**]. Understanding the exact role of scaling solutions and determining if a modified version of the attractor flow conjecture may still apply to them are outstanding problems whose resolution is quite important to the $\mathcal{N} = 2$ microstate counting program. These solutions will not, however, play a very significant role in the rest of these proceedings.

Interestingly, the attractor flow conjecture also provides a way to determine the entropy of a given solution space. The idea is that the entropy of a given total charge is the sum of the entropy of each possible attractor flow tree associated with it. Thus the partition function receives contributions from all possible trees associated with a given total charge and specific *moduli at infinity*. An immediate corollary of this is that, as emphasized in [**24**], the partition function depends on the asymptotic moduli. As the latter are varied certain attractor trees will cease to exist; specifically, a tree ceases to contribute when the moduli at infinity cross a wall of marginal stability (MS) for its first vertex, $\Gamma \to \Gamma_1 + \Gamma_2$, as is evident from (5.20).

For two-center solutions one can determine the entropy most easily near marginal stability where the centers are infinitely far apart. In this regime locality suggests that the Hilbert state contains a product of three factors[25] [**24**]

$$\mathcal{H}(\Gamma_1 + \Gamma_2; t_{ms}) \supset \mathcal{H}_{\text{int}}(\Gamma_1, \Gamma_2; t_{ms}) \otimes \mathcal{H}(\Gamma_1; t_{ms}) \otimes \mathcal{H}(\Gamma_2; t_{ms}) \tag{5.21}$$

[25]Since attractor flow trees do not *have* to split at walls of marginal stability, there will in general be other contributions to $\mathcal{H}(\Gamma_1 + \Gamma_2; t_{ms})$ as well.

Since the centers move infinitely far apart as t_{ms} is approached we do not expect them to interact in general. There is, however, conserved angular momentum carried in the electromagnetic fields sourced by the centers and this also yields a non-trivial multiplet of quantum states. Thus the claim is that $\mathcal{H}_{\text{int}}$ is the Hilbert space of a single spin J multiplet where $J = \frac{1}{2}(|\langle\Gamma_1, \Gamma_2\rangle| - 1)$[26]. $\mathcal{H}(\Gamma_1)$ and $\mathcal{H}(\Gamma_2)$ are the Hilbert spaces associated with BPS brane excitations in the Calabi-Yau and their dimensions are given in terms of a suitable entropy formula for the charges Γ_1 and Γ_2 valid at t_{ms}.

Thus, if the moduli at infinity were to cross a wall of marginal stability for the two-center system above the associated Hilbert space would cease to contribute to the entropy (or the index). A similar analysis can be applied to a more general multicentered configuration like that in Figure 2 by working iteratively down the tree and treating subtrees as though they correspond to single centers with the combined total charge of all their nodes. The idea is, once more, that we can cluster charges into two clusters by tuning the moduli and then treat the clusters effectively like individual charges. We can then iterate these arguments within each cluster. This counting argument mimics the constructive argument for building the solutions by bringing in charges from infinity and is hence subject to the same caveats, discussed above.

Altogether the above ideas allow us to determine the entropy associated with a particular attractor tree, which, by the split attractor flow conjecture corresponds to a single connected component of the solutions space. The entropy of a tree is the product of the angular momentum contribution from each vertex (i.e. $|\langle\Gamma_1, \Gamma_2\rangle|$, the dimension of $\mathcal{H}_{\text{int}}$) times the entropy associated to each node. When we want to compare against the number of states derived from quantizing the classical phase space, however, the latter factor (from the nodes) will not be included as it is not visible in the supergravity solutions.

In the following sections we will show that it is also possible, in the two- and three-center cases, to quantize the solution space directly and to match the entropy so derived with the entropy calculated using the split attractor tree. This provides a non-trivial check of both calculations.

5.5. Symplectic form. In order to quantize the phase spaces described in section 5.3 we will need to determine the symplectic structure on these spaces. This can, in principle, be derived from the supergravity action as was done, for instance, in [**14**]. In this case, however, it is far more tractable to take a different approach [**20**]. As discussed in [**61**], the four-dimensional multicentered solutions can also be analyzed in the probe approximation by studying the quiver quantum mechanics of D-branes in a multicentered supergravity background. Moreover, a non-renormalization theorem [**61**] implies that the terms in the quiver quantum mechanics

[26]The unusual -1 in the definition of J comes from quantizing additional fermionic degrees of freedom [**61, 20**].

Lagrangian linear in the velocities do not receive corrections, either perturbatively or non-perturbatively. We can use this fact to calculate the symplectic form in the probe regime and extend it to the fully back-reacted solution; this is because, for time-independent solutions, the symplectic form depends only on the terms in the action linear in the velocity.

For this approach to be consistent it is necessary that the BPS solution space, which we interpret as a phase space, of the four- and five-dimensional supergravity theories, as well as the probe theory, all match. This follows from the fact that they are all governed by the same equation, (5.2) [**61**]. For instance, one can see that a probe brane of charge Γ_a in the background generated by a charge Γ_b is forced off to infinity as a wall of marginal stability is approached [**61**] analogous to what was described below equation (5.9) for the corresponding supergravity solutions.

In [**20**] the symplectic form on the solution space is determined. We will not review the derivation in detail but simply note that it arises from the term coupling the probe brane to the background gauge field, $\dot{x}^i A_i$, giving

$$\tilde{\Omega} = \frac{1}{2}\sum_p \delta x_p^i \wedge \langle \Gamma_p, \delta \mathcal{A}_d^i(x_p)\rangle. \tag{5.22}$$

where $\mathcal{A}_d$ is the "spatial" part of the gauge field defined in (5.4) (this descends naturally to the spatial part of the 4-d gauge field). Using the definition of $\mathcal{A}_d$ we can further manipulate this expression [**20**] and put it in the form

$$\tilde{\Omega} = \frac{1}{4}\sum_{p\neq q}\langle \Gamma_p, \Gamma_q\rangle \frac{\epsilon_{ijk}(\delta(x_p - x_q)^i \wedge \delta(x_p - x_q)^j)\,(x_p - x_q)^k}{|\mathbf{x}_p - \mathbf{x}_q|^3}. \tag{5.23}$$

This is a two-form on the $(2N-2)$-dimensional solution space which is a submanifold of $\mathbb{R}^{3N-3}$ defined by (5.2). Moreover, one can show that, on this submanifold, this form is closed and, in the cases we will investigate below, non-degenerate. Thus it imbues the solution space with the structure of a phase space. Note that, as anticipated, the center of mass degrees of freedom do not appear in the symplectic form above and hence decouple in the quantization of the system. They will, in principle, yield an overall pre-factor in the partition function which we will not take into account.

Although the constraint equations (5.2) are invariant under global SO(3) rotations these are nonetheless (generically) degrees of freedom of the system and this is reflected in the symplectic form. If we contract (5.23) with the vector field that generates rotations around the 3-vector n^i (i.e. we take $\delta x_{pq}^i = \epsilon^{ijk} n^j x_{pq}^k$) then the symplectic form reduces to

$$\tilde{\Omega} \;\rightarrow\; n^i \delta J^i \tag{5.24}$$

where J^i are the components of the angular momentum vector defined in (5.6).

This is nothing more than the statement that the components J^i are the conjugate momenta associated to global SO(3) rotations. In general the symplectic form on any of our phase spaces[27] will have terms like the above coming from the global SO(3) rotations, in addition to terms depending on other degrees of freedom. As advertised (5.24) implies that solution spaces with any $J^i = 0$ will have a degenerate symplectic form and will therefore not constitute a proper phase space[28].

5.6. Quantizing the two-center phase space. The inter-center position of a two-center configuration is fixed in terms of the charges and the moduli at infinity but the axis of the centers can still be rotated so, neglecting the center of mass degree of freedom, we are left with a solution space that is topologically a two-sphere with diameter

$$x_{12} = \frac{\langle h, \Gamma_1 \rangle}{\langle \Gamma_1, \Gamma_2 \rangle}. \tag{5.25}$$

The symplectic form (5.23) is proportional to the standard volume form on the two-sphere and is entirely of the form (5.24) (note here that, as mentioned in footnote 27, collinearity of the solution implies that one $U(1) \subset SO(3)$ decouples). In terms of standard spherical coordinates it is given by

$$\tilde{\Omega} = \frac{1}{2}\langle \Gamma_1, \Gamma_2 \rangle \sin\theta \, d\theta \wedge d\phi = |J| \sin\theta \, d\theta \wedge d\phi. \tag{5.26}$$

We can now quantize the moduli space using the standard rules of geometric quantization. We introduce a complex variable z by

$$z^2 = \frac{1 + \cos\theta}{1 - \cos\theta} e^{2i\phi} \tag{5.27}$$

and find that the Kähler potential corresponding to $\tilde{\Omega}$ is given by

$$K = -2|J| \log(\sin\theta) = -|J| \log\left(\frac{z\bar{z}}{(1 + z\bar{z})^2} \right). \tag{5.28}$$

The holomorphic coordinate z represents a section of the line bundle L whose first Chern class equals $\tilde{\Omega}/2\pi$. The Hilbert space consists of global holomorphic sections of this line bundle and a basis of these is given by

[27] This does not hold for solution spaces with unbroken rotational symmetries, such as solution spaces containing only collinear centers or only a single center. In these cases some SO(3) rotations act trivially, do not correspond to genuine degrees of freedom and do not appear in the symplectic form.

[28] As mentioned in footnote 27 this does not hold in the two-center case where some SO(3) directions decouple. There are also potential subtleties with solution spaces where $J = 0$ at a single point but we neglect these for now.

$\psi_m(z) = z^m$. However, not all of these functions are globally well-behaved. By examining the norm of ψ_m given by

(5.29)

$$|\psi_m|^2 \sim \int \mathrm{dvol} e^{-K} |\psi_m(z)|^2 \sim \int d\cos\theta\theta \, d\phi \, (1+\cos\theta)^{|J|+m} (1-\cos\theta)^{|J|-m}$$

one finds that ψ_m only has a finite norm if $m \geq -|J|$ and $m \leq |J|$. The total number of states equals $2|J|+1$. This is in agreement with the wall-crossing formula up to a shift by 1. It can be shown that the inclusion of fermionic degrees of freedom will get rid of this unwanted shift [**20**].

The integrand in (5.29) is a useful quantity as it is also the phase space density associated to the state ψ_m. According to the logic in [**16, 18**] (reviewed in section 2) the right bulk description of one of the microstates ψ_m should be given by smearing the gravitational solution against the appropriate phase space density, which here is naturally given by the integrand in (5.29). Since there are only $2|J|+1$ microstates, we cannot localize the angular momentum arbitrarily sharply on the S^2; rather, it will be spread out over an area of approximately $\pi/|J|$ on the unit two-sphere. It is therefore only in the limit of large angular momentum that we can trust the description of the two-centered solution (with two centers at fixed positions) in supergravity.

5.7. Quantizing the three-center phase space. The phase space of the three-center case is four-dimensional. Placing one center at the origin (fixing the translational degrees of freedom) leaves six coordinate degrees of freedom but these are constrained by two equations. This leaves four degrees of freedom, of which three correspond to rotations in SO(3) and one of which is related to the separation of the centers.

This space is most easily visualized in the decoupling limit[29] for the case when one of the centers has no $D6$ brane charge. In that case the solution has an angular momentum vector J^i directed between the two centers with $D6$ charge and the orientation of the direction of this vector defines an S^2 in the phase space. The third center is free to rotate around the axis defined by this vector providing an additional $U(1)$, which we will coordinatize by an angle σ, fibred non-trivially over the S^2. Finally the angular momentum has a length which may be bounded from both below and above and this provides the final coordinate in the phase space. This construction is perhaps not the most obvious one from a coordinate space perspective but in these coordinates the symplectic form takes a simple and convenient form. This can also be used when all the centers have $D6$ charge but then the

[29]We will work in the decoupling limit simply because in this limit $\vec{J}$ is directly proportional to the $D6$ dipole moment so it is easier to visualize it. The general characteristics of the solution space described hold for all incarnations of the solutions we have considered – the 4-d solutions, their 5-d uplift and the decoupling limit of the latter. We will use the language of 4-d charges simply for brevity.

relation between these coordinates and the locations of the centers is less straightforward.

The symplectic form in these coordinates is (see [**20**] for a derivation):

$$\tilde{\Omega} = j \sin\theta \, d\theta \wedge d\phi - dj \wedge D\sigma \tag{5.30}$$

with $D\sigma = d\sigma - A$, $j = |\vec{J}|$, and $dA = \sin\theta \, d\theta \wedge d\phi$, so that A is a standard monopole one-form on S^2. The gauge field A implements the non-trivial fibration of σ over the S^2. A convenient choice for A is $A = -\cos\theta \, d\phi$, so that finally the symplectic form can be written as a manifestly closed two-form

$$\tilde{\Omega} = -d(j\cos\theta) \wedge d\phi - dj \wedge d\sigma. \tag{5.31}$$

Let us consider the shape of the solution space spanned by coordinates $[\theta, \phi, j, \sigma]$. θ and ϕ are standard spherical coordinates with the latter defining a $U(1)$ that degenerates at $\theta = 0, \pi$. As has already been mentioned the angular momentum usually spans some range $j \in [j_-, j_+]$ though there are cases where, for fixed charges, the angular momentum spans two separate ranges resulting in a solution space with two disconnected components (which must be quantized separately). On the boundaries of these regions, at $j = j_-$ or $j = j_+$, the centers are collinear so the coordinate σ degenerates. Thus the phase space is a symplectic manifold with two $U(1)$ actions (corresponding to rotations in σ and ϕ) which degenerate at special points. In fact the manifold is a toric Kähler manifold and can thus be quantized using the technology of [**62, 63**].

This is done in detail in [**20**] and here we will report only the results. Using the technology of geometric quantization we can determine the basis of states spanning the Hilbert space (defined as the space of normalizable holomorphic sections of an appropriate line bundle over the phase space). It turns out that the number of such states is given by

$$\mathcal{N} = (j_+ - j_- + 1)(j_+ + j_- + 1). \tag{5.32}$$

In fact this is not quite correct as we have neglected fermionic degrees of freedom in defining our phase space. It is possible to include these degrees of freedom and quantize the resulting system [**20**] and this slightly changes the set of states. The final result becomes

$$\mathcal{N} = (j_+ - j_-)(j_+ + j_-) \tag{5.33}$$

This is the number we now wish to compare to the entropy determined by the wall crossing formula.

Let us consider the attractor flow tree depicted in Figure 2. For the given charges, Γ_1, Γ_2, and Γ_3, there are, in fact, many different possible trees but, in terms of determining the relevant number of states, the only thing that matters is the branching order. In Figure 2 the first branching is into charges

Γ_3 and $\Gamma_4 = \Gamma_1 + \Gamma_2$ so the degeneracy associated with this split is $|\langle\Gamma_4, \Gamma_3\rangle|$ and the degeneracy of the second split is $|\langle\Gamma_1, \Gamma_2\rangle|$, giving a total number of states

$$\mathcal{N}_{\text{tree}} = |\Gamma_{12}|\,|(\Gamma_{13} + \Gamma_{23})| \tag{5.34}$$

where we have adopted an abbreviated notation $\Gamma_{ij} = \langle\Gamma_i, \Gamma_j\rangle$.

To compare this with the number of states arising from geometric quantization of the solution space, (5.33), we need to determine j_+ and j_-. Recall that the latter correspond to two different collinear arrangements of the centers and, in a connected solution space, there can be only two such configurations [**20**]. To relate this to a given attractor flow tree we will *assume part of the attractor flow conjecture*, namely that we can tune the moduli to force the centers into two clusters as dictated by the tree. For the configuration in Figure 2, for instance, this implies we can move the moduli at infinity close to the first wall of marginal stability (the horizontal straight line) which will force Γ_3 very far apart from Γ_1 and Γ_2. In this regime it is clear that the only collinear configurations are Γ_1-Γ_2-Γ_3 and Γ_2-Γ_1-Γ_3; it is not possible to have Γ_3 in between the other two charges. Since j_+ and j_- always correspond to collinear configurations they must, up to signs, each be one of

$$j_1 = \frac{1}{2}(\Gamma_{12} + \Gamma_{13} + \Gamma_{23}) \tag{5.35}$$

$$j_2 = \frac{1}{2}(-\Gamma_{12} + \Gamma_{13} + \Gamma_{23}) \tag{5.36}$$

j_+ will correspond to the larger of j_1 and j_2 and j_- to the smaller but, from the form of (5.33), we see that this will only effect $\mathcal{N}$ by an overall sign (which we are not tracking carefully in any case). Thus

$$\mathcal{N} = \pm(j_1 - j_2)(j_1 + j_2) = \pm\Gamma_{12}\,(\Gamma_{13} + \Gamma_{23}) \tag{5.37}$$

which nicely matches (5.34).

Of course to obtain this matching we have had to assume the attractor flow conjecture itself (in part) so it does not serve as an entirely independent verification. Demonstrating this matching more carefully would help validate both methods of state counting [**20**].

6. Conclusions

In these conclusions we would briefly like to mention some open problems and possible generalizations of the results that we have been describing. It is not intended as a complete overview of recent developments in the subject, nor does it pretend to contain accurate historical references for the various issues we mention; rather, it should be viewed as hopefully providing some food for thought, inspiration for further work as well as some pointers to the literature.

One obvious limitation of our discussion has been the restriction to extremal supersymmetric black holes. Although these are the most tractable, eventually we would like to deal with non-supersymmetric, realistic black holes that Hawking radiate. In interesting recent work a connection between non-supersymmetric black holes and fluid dynamics was found [**64, 65**] which suggests that the type of dual descriptions in terms of free gases of particles that we used in the supersymmetric cases may not be terribly appropriate. This does not mean that such black holes cannot be studied using the microstate approach, but it does indicate that the nature of microstates as well as the extrapolation from weak to strong coupling is much more complicated. For some recent progress in obtaining Hawking radiation from a microstate point of view, see e.g. [**66**]. See also [**16**] for a naive and qualitative description of the microstates of an AdS-Schwarzschild in the weakly coupled gauge theory.

We have neither mentioned, nor studied in detail, an important feature that microstates should have. Microstates that describe black holes should roughly have an extent that agrees with that of the black hole. In particular, they should grow as the string coupling is increased in the same way as the black hole horizon grows, a property emphasized and explicitly shown for the three-charge supertube in [**67**]. This extended nature of microstates is also what gives black holes their "fuzzball" character and plays a crucial role in resolving the information paradox [**68**].

One may rightfully ask how well-motivated the assumption is that black holes can be completely resolved in terms of supergravity microstates alone. For large black holes, there is the following argument which suggests that this may not be possible. In AdS/CFT, the states that are dual to single particle supergravity modes in the bulk are BPS states and their descendants. Therefore, one might expect that supergravity solutions only contain information about products of BPS operators and their descendants which are the duals of general multiparticle supergraviton states. Now in general, the phase in which AdS contains a thermal gas of supergravitons is separated from the phase in which the AdS space contains a black hole via a phase transition. This seems to indicate that supergravitons alone do not have enough degrees of freedom to account for the black hole entropy. For example, for $\mathrm{AdS}_3 \times \mathrm{S}^3$ one can show explicitly that this is the case [**69**]. It would be interesting to explore this further and to try to consider stringy microstates.

Along similar lines, we would like to explore in more detail to what extend the multicentered solutions we considered in section 5 can possibly account for a finite fraction of the entropy of the corresponding macroscopic black hole. They are not the most general 5d supersymmetric solutions, as we took special hyperkähler manifolds as our base, and in addition we ignored Calabi-Yau excitations and stringy excitations. Even if they do not contribute a finite fraction of the entropy, one may wonder whether one may at least find some typical black hole microstates in this class. This question is very difficult to answer, because it is not clear what set of macroscopic observables we should precisely include in our definition of typicality. Finally, we should be careful to not view all smooth multicentered solutions as microstates of a single black hole. Many are more properly thought of as microstates of multicentered black holes. One may expect that only the solutions which are described by a single-centered attractor flow (this will include in particular many scaling solutions) are honest microstates of a single black hole. Obviously, much more work is required in order to apply the philosophy outlined in these proceedings in greater detail to macroscopic supersymmetric black holes.

Another main open problem is to provide a more detailed map between black objects in the bulk and ensembles in the boundary CFT. What distinguishes semiclassical ensembles from non-classical ones? What possible first laws of thermodynamics can these black object have? How many chemical potentials do we typically need to include in their dual description? Is there a natural way to describe multicentered bound states in the dual field theory? We have only begun to see some glimpses of answers to these questions in the examples we have described, but clearly a lot of work remains to be done.

While these proceedings have focused on the results of [**16**–**20**] there is, of course, a huge body of work attempting to build a coherent picture of black hole microstates. Although related, the various approaches focus on different questions and often explore them in different settings (sometimes related by dualities). To assist the interested reader in accessing the literature we provide, below, a short survey of a limited number of other works in the field, attempting to organize them into several general research branches and to describe how they relate to each other and to the material described in these proceedings. We hope this will be of value to the reader.

7. Literature survey

The "*microstate geometry*" program has, by now, generated a large volume of literature which would be difficult to survey in any great depth.

Due to time and space constraints we will not attempt such an in-depth or comprehensive review but will rather merely point the reader in the direction of many relevant works and attempt to give a sense of the status of various branches of the field. In so doing we will no doubt miss many important developments but hopefully references to these are contained in the works cited below.

7.1. The $D1$-$D5$(-P) and Related Systems.

i- The D1-D5 system

This configuration arises in type-IIB string compactified on $\mathcal{M}\times$ S^1, where $\mathcal{M}$ is a four-dimensional Ricci-flat manifold. The supergravity solution describes N_5 D5-branes wrapping the full compact space and N_1 $D1$-branes wrapping the S^1; the most general metric corresponding to this configuration was constructed in [**6**]. Originally there were neither internal nor fermionic excitations. In [**5**], the internal excitations were included, and [**45**] included the fermionic excitations. The near horizon geometry of this system is $AdS_3 \times S^3 \times \mathcal{M}$, so the dual theory is a CFT_2. It was in the context of this system that the fuzzball proposal was first proposed [**21**].

A nice, but slightly outdated, review of the $D1$-$D5$ system and the fuzzball proposal is contained in [**36**]. Significant progress has occurred since the appearance of this review: the supergravity phase space has been directly quantized [**32, 33**], and the fuzzball proposal has been studied in even greater detail in the context of this system [**50, 70, 71, 18, 72, 48, 46**]. Supporting evidence for the stretched horizon idea was given in [**70, 48**]. In [**70**] it was argued that quantum gravity effects become important at scales larger than the Planck/string scales. In [**48**] a sub-ensemble of the original thermal ensemble was considered. It was shown that, even in this case, the area of stretched horizon matches the microscopic entropy of the subsystem to leading order. This continues to hold even after the inclusion of 1-loop string corrections. In [**50, 18, 72, 46**] various quantities in the CFT and supergravity were compared.

ii- Beyond the D1-D5 system

Further generalizations of the $D1$-$D5$ system followed. They can be collected under two general themes: adding more charges and breaking supersymmetry. In the first case the aim was to add some charges (momentum most of the time) to reduce the amount of preserved supersymmetry without breaking all the supersymmetry. An incomplete list of references in this direction includes [**10, 73–79**].

In [**73**] a first attempt to construct dual geometries for $D1$-$D5$-P microstates was undertaken; here the momentum was added as a small perturbation of the $D1$-$D5$ system. The first success was achieved in [**10**] which was followed by other works [**74, 75, 79**]. In [**78**] some arguments were given to support the claim that higher order correction to these 3-charge geometries would not generate a horizon or a singularity if they were not present at tree level.

There are also studies of microstates using a 1/4 BPS probe approach; see *e.g.* [**80, 28**].

A class of five-dimensional smooth solutions, closely related to the $D1$-$D5$-P system, have also been focus of an extended research program [**49, 8, 9, 7, 81, 57, 82, 83, 58–60, 19, 84, 20**]. These solutions are discussed in section 5 of these proceedings.

On the other hand, less work has been done on non-BPS solutions [**85–88, 66**]. In [**85**] smooth non-supersymmetric geometries were constructed. They were then proven to be classically unstable in [**86**]. This classical instability was shown to give rise to Hawking radiation in [**66**]. Other properties of these solutions were studied in [**87**]. Another set of non-supersymmetric solutions in 4 dimensions appeared in [**88**].

iii- **Bound versus unbound systems**

The fuzzball proposal was put forward to explain the entropy of a black hole, among other issues. To be sure that we are not over-counting one needs to know that the solution one is dealing with describes a bound system. For the $D1$-$D5$ system, this can be achieved by adding $D3$ charge. This system has been studied in [**89**] in the context of the fuzzball proposal.

Other works in this direction are [**71, 76, 84**]. In [**76**] known $D1$-$D5$ ($D1$-$D5$-P) geometries were rewritten in a *fiber* $\times$ *base* form in order to gain some insight into their structure. In [**71**] another route was followed. Using the $D1$-$D5$ system as a prototype, this paper studied the dynamical behavior of such systems and put forth a conjecture to distinguish bound systems. In [**19**] three-charge systems were studied in an $\mathrm{AdS}_3 \times \mathrm{S}^2$ decoupling limit and a criterion for bound configurations was proposed in terms of split attractor trees [**22**]. In [**20**] some of these solutions were quantized, allowing a direct analysis of whether the relevant state is bound or not. In [**84**] another criterion was given using spectral flow in $\mathrm{AdS}_3 \times \mathrm{S}^3$.

7.2. Asymptotically $\mathbf{AdS_5 \times S^5}$ and related solutions. In $\mathrm{AdS}_5 \times \mathrm{S}^5$ one can, in principle, consider 1/2, 1/4, 1/8 and 1/16 excitations of the full string theory. The only known example of a supersymmetric black hole in AdS_5 is the Gutowski-Reall black hole [**90, 37**] and this solution is only 1/16 BPS. There are 1/2 BPS black hole solutions [**91–93**], known as superstars, but these have no horizon in two-derivative supergravity. Nonetheless, significant effort has gone into studying the possible smooth, asymptotically AdS_5 geometries with varying amounts of supersymmetry. For the 1/2 BPS case, as we discussed above, the solutions were completely classified in [**4**].

i- **The 1/2 BPS case**

Various generalizations of the LLM [**4**] geometries have been considered. The notion of typical states in an ensemble was explored in [**16**], while in [**25**] a "metric" operator was defined in the CFT and used to establish a criterion to determine which states will have a well-defined classical dual geometry. Another direction of research, closely related to topics discussed in these proceedings, involves quantization of the original LLM geometries using phase space techniques [**14–15**].

Thermal properties and Wilson loops in the CFT dual to these geometries were studied in [**94, 95**] while in [**96–98**] 1/2 BPS geometries corresponding to defects in the CFT were considered. In [**99**] the relation between phase space densities in the fermion formulation of the theory and generalized Young tableaux was studied. In [**100**] the on-shell action for the LLM geometries was derived.

There is, in fact, a large volume of literature on the 1/2 BPS case and our short survey is by no means intended to be comprehensive. For more references the reader should consult the works cited above.

ii- Less SUSY

Backgrounds which preserve only 1/4, 1/8 and 1/16 supersymmetry have also been considered and explicit supergravity solutions have been constructed. For instance smooth 1/4 BPS solutions were constructed in [**101–104**] and an LLM-like prescription to derive them from droplets on the plane is related to constraints on brane webs in [**105, 106**].

Probe solutions (giant gravitons) preserving 1/8 of the supersymmetry were studied in [**107–111**] and the backreaction of such probes was worked out in [**112, 113**]. The 1/8 BPS sector of the dual CFT was explored in [**114**].

There has also been recent progress in generalizing the notion of typical states from $\mathcal{N} = 4$ SYM theory to more general four-dimensional SCFT's [**115**].

As mentioned above, the 1/16 BPS sector is distinguished by having black holes with macroscopic horizons [**37, 90, 116–118**]. The 1/16 sector of the CFT has also been studied; operators potentially related to the black hole have been identified [**119**] and the entropy has been (qualitatively) reproduced [**120**]. Giant gravitons preserving 1/16 of the supersymmetry were found in [**121**].

Attempts to treat the full set of 1/2, 1/4, and 1/8 solutions in a common framework can be found in [**122, 123**]. Other related work includes [**124–126**]. Once more, this list is only intended to serve as an introduction to the literature and no doubt has failed to include many important works.

iii- No supersymmetry

Another direction of investigation involves breaking all supersymmetries. Some research in this direction includes [**127–129**].

Other backgrounds (an M-theory solution) are also discussed in [**130**]. In [**131**] quite general considerations are applied to study the variances between correlators of an operator in a typical state and a thermal state.

Acknowledgements

VB and JdB would like to thank the organizers of the Sowers workshop, and in particular Mark Sowers, for the opportunity to present this work at a very stimulating meeting.

The work of VB is supported in part by DOE grant DE-FG02-95ER40893. The work of JdB, SES and IM is supported financially by the Foundation of Fundamental Research on Matter (FOM).

References

[1] A. Strominger and C. Vafa, *Microscopic origin of the Bekenstein-Hawking entropy*, *Phys. Lett.* **B379** (1996) 99–104, [hep-th/9601029].
[2] J. M. Maldacena, A. Strominger, and E. Witten, *Black hole entropy in M-theory*, *JHEP* **12** (1997) 002, [hep-th/9711053].
[3] J. M. Maldacena and A. Strominger, *AdS_3 black holes and a stringy exclusion principle*, *JHEP* **12** (1998) 005, [hep-th/9804085].
[4] H. Lin, O. Lunin, and J. M. Maldacena, *Bubbling AdS space and 1/2 BPS geometries*, *JHEP* **10** (2004) 025, [hep-th/0409174].
[5] O. Lunin, J. M. Maldacena, and L. Maoz, *Gravity solutions for the D1-D5 system with angular momentum*, hep-th/0212210.
[6] O. Lunin and S. D. Mathur, *Metric of the multiply wound rotating string*, *Nucl. Phys.* **B610** (2001) 49–76, [hep-th/0105136].
[7] I. Bena and N. P. Warner, *Bubbling supertubes and foaming black holes*, *Phys. Rev.* **D74** (2006) 066001, [hep-th/0505166].
[8] I. Bena and N. P. Warner, *One ring to rule them all ... and in the darkness bind them?*, *Adv. Theor. Math. Phys.* **9** (2005) 667–701, [hep-th/0408106].
[9] P. Berglund, E. G. Gimon, and T. S. Levi, *Supergravity microstates for BPS black holes and black rings*, *JHEP* **06** (2006) 007, [hep-th/0505167].
[10] O. Lunin, *Adding momentum to D1-D5 system*, *JHEP* **04** (2004) 054, [hep-th/0404006].
[11] B. Bates and F. Denef, *Exact solutions for supersymmetric stationary black hole composites*, hep-th/0304094.
[12] M. C. N. Cheng, *More bubbling solutions*, *JHEP* **03** (2007) 070, [hep-th/0611156].
[13] R. Emparan, D. Mateos, and P. K. Townsend, *Supergravity supertubes*, *JHEP* **07** (2001) 011, [hep-th/0106012].
[14] L. Grant, L. Maoz, J. Marsano, K. Papadodimas, and V. S. Rychkov, *Minisuperspace quantization of 'bubbling AdS' and free fermion droplets*, *JHEP* **08** (2005) 025, [hep-th/0505079].
[15] L. Maoz and V. S. Rychkov, *Geometry quantization from supergravity: The case of 'bubbling AdS'*, *JHEP* **08** (2005) 096, [hep-th/0508059].
[16] V. Balasubramanian, J. de Boer, V. Jejjala, and J. Simon, *The library of Babel: On the origin of gravitational thermodynamics*, *JHEP* **12** (2005) 006, [hep-th/0508023].
[17] L. F. Alday, J. de Boer, and I. Messamah, *What is the dual of a dipole?*, *Nucl. Phys.* **B746** (2006) 29–57, [hep-th/0511246].
[18] L. F. Alday, J. de Boer, and I. Messamah, *The gravitational description of coarse grained microstates*, *JHEP* **12** (2006) 063, [hep-th/0607222].
[19] J. de Boer, F. Denef, S. El-Showk, I. Messamah, and D. Van den Bleeken, *Black hole bound states in* $AdS_3 \times S^2$, 0802.2257.
[20] J. de Boer, S. El-Showk, I. Messamah, and D. Van den Bleeken, *Quantizing the moduli space of $\mathcal{N} = 2$ solutions*, . (to appear).
[21] O. Lunin and S. D. Mathur, *AdS/CFT duality and the black hole information paradox*, *Nucl. Phys.* **B623** (2002) 342–394, [hep-th/0109154].
[22] F. Denef, *Supergravity flows and D-brane stability*, *JHEP* **08** (2000) 050, [hep-th/0005049].
[23] F. Denef, *On the correspondence between D-branes and stationary supergravity solutions of type II Calabi-Yau compactifications*, hep-th/0010222.

[24] F. Denef and G. W. Moore, *Split states, entropy enigmas, holes and halos*, hep-th/0702146.

[25] V. Balasubramanian, B. Czech, K. Larjo, D. Marolf, and J. Simón, *Quantum geometry and gravitational entropy*, *JHEP* **12** (2007) 067, [0705.4431].

[26] K. Larjo, *On the existence of supergravity duals to D1–D5 CFT states*, *JHEP* **07** (2007) 041, [0705.4433].

[27] P. Dedecker, *Calcul des variations, formes différentielles et champs geodesiques*, Geometrie différentielle, Colloq. Intern. du CNRS LII, Strasbourg (1953) 17–34.

[28] S. Raju, *Counting giant gravitons in AdS_3*, 0709.1171.

[29] C. Crnkovic and E. Witten, *Covariant description of canonical formalism in geometrical theories*, . Print-86-1309 (PRINCETON).

[30] G. J. Zuckerman, *Action principles and global geometry*, Print-89-0321 (YALE).

[31] J. Lee and R. M. Wald, *Local symmetries and constraints*, *J. Math. Phys.* **31** (1990) 725–743.

[32] A. Donos and A. Jevicki, *Dynamics of chiral primaries in* $\mathrm{AdS}_3 \times \mathrm{S}^3 \times \mathrm{T}^4$, *Phys. Rev.* **D73** (2006) 085010, [hep-th/0512017].

[33] V. S. Rychkov, *D1-D5 black hole microstate counting from supergravity*, *JHEP* **01** (2006) 063, [hep-th/0512053].

[34] J. M. Maldacena, *The large N limit of superconformal field theories and supergravity*, *Adv. Theor. Math. Phys.* **2** (1998) 231–252, [hep-th/9711200].

[35] M. Hillery, R. F. O'Connell, M. O. Scully, and E. P. Wigner, *Distribution functions in physics: Fundamentals*, *Phys. Rept.* **106** (1984) 121–167.

[36] S. D. Mathur, *The fuzzball proposal for black holes: An elementary review*, *Fortsch. Phys.* **53** (2005) 793–827, [hep-th/0502050].

[37] J. B. Gutowski and H. S. Reall, *General supersymmetric AdS_5 black holes*, *JHEP* **04** (2004) 048, [hep-th/0401129].

[38] S. Corley, A. Jevicki, and S. Ramgoolam, *Exact correlators of giant gravitons from dual $\mathcal{N} = 4$ SYM theory*, *Adv. Theor. Math. Phys.* **5** (2002) 809–839, [hep-th/0111222].

[39] D. Berenstein, *A toy model for the AdS/CFT correspondence*, *JHEP* **07** (2004) 018, [hep-th/0403110].

[40] A. M. Vershik, *Statistical mechanics of combinatorial partitions, and their limit configurations*, *Funkts. Anal. Prilozh* **02** (1996), (English translation: *Funct. Anal. Appl*, 30, 02:90-105, 1996) 19–30.

[41] G. Mandal, *Fermions from half BPS supergravity*, *JHEP* **08** (2005) 052, [hep-th/0502104].

[42] Y. Takayama and A. Tsuchiya, *Complex matrix model and fermion phase space for bubbling AdS geometries*, *JHEP* **10** (2005) 004, [hep-th/0507070].

[43] A. Buchel, *Coarse-graining 1/2 BPS geometries of type IIB supergravity*, *Int. J. Mod. Phys.* **A21** (2006) 3495–3502, [hep-th/0409271].

[44] G. Milanesi and M. O'Loughlin, *Singularities and closed time-like curves in type IIB 1/2 BPS geometries*, *JHEP* **09** (2005) 008, [hep-th/0507056].

[45] M. Taylor, *General two-charge geometries*, *JHEP* **03** (2006) 009, [hep-th/0507223].

[46] I. Kanitscheider, K. Skenderis, and M. Taylor, *Fuzzballs with internal excitations*, *JHEP* **06** (2007) 056, [0704.0690].

[47] O. Lunin and S. D. Mathur, *Statistical interpretation of Bekenstein entropy for systems with a stretched horizon*, *Phys. Rev. Lett.* **88** (2002) 211303, [hep-th/0202072].

[48] S. D. Mathur, *Black hole size and phase space volumes*, 0706.3884.

[49] I. Bena and P. Kraus, *Microscopic description of black rings in AdS/CFT*, *JHEP* **12** (2004) 070, [hep-th/0408186].

[50] V. Balasubramanian, P. Kraus, and M. Shigemori, *Massless black holes and black rings as effective geometries of the D1-D5 system*, *Class. Quant. Grav.* **22** (2005) 4803–4838, [hep-th/0508110].

[51] N. Iizuka and M. Shigemori, *A note on D1-D5-J system and 5D small black ring*, *JHEP* **08** (2005) 100, [hep-th/0506215].

[52] H. Elvang, R. Emparan, D. Mateos, and H. S. Reall, *Supersymmetric black rings and three-charge supertubes*, *Phys. Rev.* **D71** (2005) 024033, [hep-th/0408120].

[53] K. Copsey and G. T. Horowitz, *The role of dipole charges in black hole thermodynamics*, *Phys. Rev.* **D73** (2006) 024015, [hep-th/0505278].

[54] D. Gaiotto, A. Strominger, and X. Yin, *New connections between 4D and 5D black holes*, *JHEP* **02** (2006) 024, [hep-th/0503217].

[55] J. Michelson and A. Strominger, *Superconformal multi-black hole quantum mechanics*, *JHEP* **09** (1999) 005, [hep-th/9908044].

[56] P. Kraus, *Lectures on black holes and the AdS_3/CFT_2 correspondence*, hep-th/0609074.

[57] I. Bena, C.-W. Wang, and N. P. Warner, *Mergers and typical black hole microstates*, *JHEP* **11** (2006) 042, [hep-th/0608217].

[58] F. Denef, D. Gaiotto, A. Strominger, D. Van den Bleeken, and X. Yin, *Black hole deconstruction*, hep-th/0703252.

[59] E. G. Gimon and T. S. Levi, *Black ring deconstruction*, 0706.3394.

[60] I. Bena, C.-W. Wang, and N. P. Warner, *Plumbing the abyss: Black ring microstates*, 0706.3786.

[61] F. Denef, *Quantum quivers and Hall/hole halos*, *JHEP* **10** (2002) 023, [hep-th/0206072].

[62] M. Abreu, *Kahler geometry of toric manifolds in symplectic coordinates*, 2000.

[63] V. Guillemin, *Kaehler structures on toric varieties*, *J. Differential Geometry* **40** (1994) 285–309.

[64] S. Bhattacharyya, V. E. Hubeny, S. Minwalla, and M. Rangamani, *Nonlinear fluid dynamics from gravity*, *JHEP* **02** (2008) 045, [0712.2456].

[65] S. Bhattacharyya, S. Lahiri, R. Loganayagam, and S. Minwalla, *Large rotating AdS black holes from fluid mechanics*, 0708.1770.

[66] B. D. Chowdhury and S. D. Mathur, *Radiation from the non-extremal fuzzball*, 0711.4817.

[67] I. Bena and P. Kraus, *Three-charge supertubes and black hole hair*, *Phys. Rev.* **D70** (2004) 046003, [hep-th/0402144].

[68] S. D. Mathur, *What exactly is the information paradox?*, 0803.2030.

[69] J. de Boer, *Large N elliptic genus and AdS/CFT correspondence*, *JHEP* **05** (1999) 017, [hep-th/9812240].

[70] S. D. Mathur, *The quantum structure of black holes*, *Class. Quant. Grav.* **23** (2006) R115, [hep-th/0510180].

[71] S. Giusto, S. D. Mathur, and Y. K. Srivastava, *Dynamics of supertubes*, *Nucl. Phys.* **B754** (2006) 233–281, [hep-th/0510235].

[72] I. Kanitscheider, K. Skenderis, and M. Taylor, *Holographic anatomy of fuzzballs*, *JHEP* **04** (2007) 023, [hep-th/0611171].

[73] S. D. Mathur, A. Saxena, and Y. K. Srivastava, *Constructing 'hair' for the three-charge hole*, *Nucl. Phys.* **B680** (2004) 415–449, [hep-th/0311092].

[74] S. Giusto, S. D. Mathur, and A. Saxena, *Dual geometries for a set of three-charge microstates*, *Nucl. Phys.* **B701** (2004) 357–379, [hep-th/0405017].

[75] S. Giusto, S. D. Mathur, and A. Saxena, *Three-charge geometries and their CFT duals*, *Nucl. Phys.* **B710** (2005) 425–463, [hep-th/0406103].

[76] S. Giusto and S. D. Mathur, *Geometry of D1-D5-P bound states*, *Nucl. Phys.* **B729** (2005) 203–220, [hep-th/0409067].

[77] A. Saxena, *Geometry of the D1-D5-P system*, UMI-31-41729.

[78] S. Giusto and S. D. Mathur, *Fuzzball geometries and higher derivative corrections for extremal holes*, *Nucl. Phys.* **B738** (2006) 48–75, [hep-th/0412133].

[79] J. Ford, S. Giusto, and A. Saxena, *A class of BPS time-dependent three-charge microstates from spectral flow*, *Nucl. Phys.* **B790** (2008) 258–280, [hep-th/0612227].

[80] G. Mandal, S. Raju, and M. Smedback, *Supersymmetric giant graviton solutions in AdS_3*, *Phys. Rev.* **D77** (2008) 046011, [0709.1168].

[81] I. Bena, *The foaming three-charge black hole*, hep-th/0604110.

[82] V. Balasubramanian, E. G. Gimon, and T. S. Levi, *Four-dimensional black hole microstates: From D-branes to spacetime foam*, *JHEP* **01** (2008) 056, [hep-th/0606118].

[83] I. Bena and N. P. Warner, *Black holes, black rings and their microstates*, hep-th/0701216.

[84] I. Bena, N. Bobev, and N. P. Warner, *Spectral flow, and the spectrum of multi-center solutions*, 0803.1203.

[85] V. Jejjala, O. Madden, S. F. Ross, and G. Titchener, *Non-supersymmetric smooth geometries and D1-D5-P bound states*, *Phys. Rev.* **D71** (2005) 124030, [hep-th/0504181].

[86] V. Cardoso, O. J. C. Dias, J. L. Hovdebo, and R. C. Myers, *Instability of non-supersymmetric smooth geometries*, *Phys. Rev.* **D73** (2006) 064031, [hep-th/0512277].

[87] E. G. Gimon, T. S. Levi, and S. F. Ross, *Geometry of non-supersymmetric three-charge bound states*, *JHEP* **08** (2007) 055, [0705.1238].

[88] S. Giusto, S. F. Ross, and A. Saxena, *Non-supersymmetric microstates of the D1-D5-KK system*, *JHEP* **12** (2007) 065, [0708.3845].

[89] O. Lunin, S. D. Mathur, I. Y. Park, and A. Saxena, *Tachyon condensation and 'bounce' in the D1-D5 system*, *Nucl. Phys.* **B679** (2004) 299–328, [hep-th/0304007].

[90] J. B. Gutowski and H. S. Reall, *Supersymmetric AdS_5 black holes*, *JHEP* **02** (2004) 006, [hep-th/0401042].

[91] K. Behrndt, A. H. Chamseddine, and W. A. Sabra, *BPS black holes in $\mathcal{N} = 2$ five dimensional AdS supergravity*, *Phys. Lett.* **B442** (1998) 97–101, [hep-th/9807187].

[92] K. Behrndt, M. Cvetič, and W. A. Sabra, *Non-extreme black holes of five dimensional $\mathcal{N} = 2$ AdS supergravity*, *Nucl. Phys.* **B553** (1999) 317–332, [hep-th/9810227].

[93] M. Cvetič *et. al.*, *Embedding AdS black holes in ten and eleven dimensions*, *Nucl. Phys.* **B558** (1999) 96–126, [hep-th/9903214].

[94] T. Okuda, *A prediction for bubbling geometries*, *JHEP* **01** (2008) 003, [0708.3393].

[95] C.-Y. Liu, *Thermal phase in bubbling geometries*, 0707.1382.

[96] E. D'Hoker, J. Estes, and M. Gutperle, *Exact half-BPS type IIB interface solutions. II: Flux solutions and multi-janus*, *JHEP* **06** (2007) 022, [0705.0024].

[97] E. D'Hoker, J. Estes, and M. Gutperle, *Exact half-BPS type IIB interface solutions I: Local solution and supersymmetric janus*, *JHEP* **06** (2007) 021, [0705.0022].

[98] J. Gomis and S. Matsuura, *Bubbling surface operators and S-duality*, *JHEP* **06** (2007) 025, [0704.1657].

[99] L. D'Errico, W. Mueck, and R. Pettorino, *Stretched horizon and entropy of superstars*, *JHEP* **05** (2007) 063, [hep-th/0703223].

[100] M. Sato, *On-shell action for bubbling geometries*, hep-th/0703063.

[101] J. P. Gauntlett, D. Martelli, J. Sparks, and D. Waldram, *Supersymmetric AdS_5 solutions of M-theory*, *Class. Quant. Grav.* **21** (2004) 4335–4366, [hep-th/0402153].

[102] A. Donos, *A description of 1/4 BPS configurations in minimal type IIB SUGRA*, *Phys. Rev.* **D75** (2007) 025010, [hep-th/0606199].

[103] A. Donos, *BPS states in type IIB SUGRA with $SO(4) \times SO(2)$ (gauged) symmetry*, *JHEP* **05** (2007) 072, [hep-th/0610259].

[104] Z. W. Chong, H. Lu, and C. N. Pope, *BPS geometries and AdS bubbles*, *Phys. Lett.* **B614** (2005) 96–103, [hep-th/0412221].

[105] O. Lunin, *Strings ending on branes from supergravity*, *JHEP* **09** (2007) 093, [0706.3396].

[106] O. Lunin, *Brane webs and 1/4 BPS geometries*, 0802.0735.

[107] A. Mikhailov, *Giant gravitons from holomorphic surfaces*, *JHEP* **11** (2000) 027, [hep-th/0010206].

[108] G. Mandal and N. V. Suryanarayana, *Counting 1/8 BPS dual-giants*, *JHEP* **03** (2007) 031, [hep-th/0606088].

[109] I. Biswas, D. Gaiotto, S. Lahiri, and S. Minwalla, *Supersymmetric states of $\mathcal{N} = 4$ Yang-Mills from giant gravitons*, *JHEP* **12** (2007) 006, [hep-th/0606087].

[110] S. Kim and K.-M. Lee, *BPS electromagnetic waves on giant gravitons*, *JHEP* **10** (2005) 111, [hep-th/0502007].

[111] D. Berenstein, *Large N BPS states and emergent quantum gravity*, *JHEP* **01** (2006) 125, [hep-th/0507203].

[112] E. Gava, G. Milanesi, K. S. Narain, and M. O'Loughlin, *1/8 BPS states in AdS/CFT*, *JHEP* **05** (2007) 030, [hep-th/0611065].

[113] E. Gava, G. Milanesi, K. S. Narain, and M. O'Loughlin, *BPS solutions in AdS/CFT*, *Fortsch. Phys.* **55** (2007) 781–786, [hep-th/0701119].

[114] J. Lucietti and M. Rangamani, *Asymptotic counting of BPS operators in superconformal field theories*, 0802.3015.

[115] V. Balasubramanian, B. Czech, Y.-H. He, K. Larjo, and J. Simon, *Typicality, black hole microstates and superconformal field theories*, 0712.2434.

[116] Z. W. Chong, M. Cvetič, H. Lu, and C. N. Pope, *Five-dimensional gauged supergravity black holes with independent rotation parameters*, *Phys. Rev.* **D72** (2005) 041901, [hep-th/0505112].

[117] Z. W. Chong, M. Cvetič, H. Lu, and C. N. Pope, *General non-extremal rotating black holes in minimal five-dimensional gauged supergravity*, *Phys. Rev. Lett.* **95** (2005) 161301, [hep-th/0506029].

[118] H. K. Kunduri, J. Lucietti, and H. S. Reall, *Supersymmetric multi-charge AdS_5 black holes*, *JHEP* **04** (2006) 036, [hep-th/0601156].

[119] M. Berkooz, D. Reichmann, and J. Simon, *A fermi surface model for large supersymmetric AdS_5 black holes*, *JHEP* **01** (2007) 048, [hep-th/0604023].

[120] J. Kinney, J. M. Maldacena, S. Minwalla, and S. Raju, *An index for four-dimensional superconformal theories*, *Commun. Math. Phys.* **275** (2007) 209–254, [hep-th/0510251].

[121] S. Kim and K.-M. Lee, *1/16 BPS black holes and giant gravitons in the* $\mathrm{AdS}_5 \times \mathrm{S}^5$ *space*, *JHEP* **12** (2006) 077, [hep-th/0607085].

[122] J. T. Liu, H. Lu, C. N. Pope, and J. F. Vazquez-Poritz, *New supersymmetric solutions of $\mathcal{N} = 2$, $D = 5$ gauged supergravity with hyperscalars*, *JHEP* **10** (2007) 093, [0705.2234].

[123] B. Chen *et. al.*, *Bubbling AdS and droplet descriptions of BPS geometries in IIB supergravity*, *JHEP* **10** (2007) 003, [0704.2233].

[124] J. Hoppe and K.-M. Lee, *New BPS configurations of BMN matrix theory*, 0712.3616.

[125] S. Dutta and R. Gopakumar, *Free fermions and thermal AdS/CFT*, 0711.0133.

[126] E. Gava, G. Milanesi, K. S. Narain, and M. O'Loughlin, *Half BPS states in* $\mathrm{AdS}_5 \times \mathrm{Y}^{p,q}$, *JHEP* **02** (2008) 050, [0709.2856].

[127] V. Balasubramanian, J. de Boer, V. Jejjala, and J. Simon, *Entropy of near-extremal black holes in AdS_5*, 0707.3601.

[128] J. T. Liu, H. Lu, C. N. Pope, and J. F. Vazquez-Poritz, *Bubbling AdS black holes*, *JHEP* **10** (2007) 030, [hep-th/0703184].

[129] H.-Y. Chen, D. H. Correa, and G. A. Silva, *Geometry and topology of bubble solutions from gauge theory*, *Phys. Rev.* **D76** (2007) 026003, [hep-th/0703068].

[130] H.-H. Shieh, G. van Anders, and M. Van Raamsdonk, *Coarse-graining the Lin-Maldacena geometries*, *JHEP* **09** (2007) 059, [0705.4308].

[131] V. Balasubramanian *et. al.*, *Typicality versus thermality: An analytic distinction*, hep-th/0701122.

David Rittenhouse Laboratories, University of Pennsylvania, Philadelphia, PA, 19104, USA
E-mail address: vijay@physics.upenn.edu

Instituut voor Theoretische Fysica, Valckenierstraat 65, 1018XE Amsterdam, The Netherlands
E-mail address: jdeboer@science.uva.nl

Instituut voor Theoretische Fysica, Valckenierstraat 65, 1018XE Amsterdam, The Netherlands
E-mail address: sheer@science.uva.nl

Instituut voor Theoretische Fysica, Valckenierstraat 65, 1018XE Amsterdam, The Netherlands
E-mail address: imessama@science.uva.nl

AMS/IP Studies in Advanced Mathematics
Volume 44, 2008

Geometry of Supersymmetric Type II Solutions

Alessandro Tomasiello

ABSTRACT. If supersymmetry is realized in nature, it is necessarily spontaneously broken. Nevertheless, for various phenomenological reasons, one would like the scale of supersymmetry breaking to be much smaller than the scale of quantum gravity. For this reason, it is often useful to study supersymmetric solutions and then break the symmetry in a second stage. In this talk, we will review a general characterization of supersymmetric solutions of type II supergravity of the form Minkowski$_4 \times M_6$ and AdS$_4 \times M_6$, for M_6 any internal manifold. This characterization has been obtained in the papers [**1, 2, 3**], and it uses mathematical concepts introduced in [**4, 5**].

1. The Calabi-Yau case

For the case in which the various field-strengths of the theory (the RR fluxes $F_{\rm RR}$ and the NSNS fluxes H) are set to zero,

$$F_{\rm RR} = H = 0, \tag{1.1}$$

such a characterization was already obtained in [**6**]. In that case, the internal manifold M_6 should have holonomy contained in SU(3). Such an M_6 is also called a Calabi-Yau space.

For the case with $F_{\rm RR}$ and H different from zero, holonomy can be as large as SO(6), and it hence ceases to be a useful classification tool. There is another way of defining a Calabi-Yau, however, which remains useful in the case with fluxes. It involves the so-called *calibrations*: these are differential forms on M_6, and can be thought of as "geometrical fluxes" that measure volumes of subspaces of M_6. This means that they contain some of the information of a metric on M_6; as it turns out, they actually contain all of it.

2000 *Mathematics Subject Classification.* Primary 81T60, 83E50; Secondary 53C10, 53C80, 81T30, 83E30.

DEFINITION 1.1. An SU(3) structure is a pair of forms (J, Ω) such that

- J is a real two-form, Ω is a complex three-form and *decomposable* (*i.e.* locally the wedge product of three complex one-forms); Ω then determines an almost complex structure I;
- $\frac{4}{3}J^3 = i\Omega \wedge \bar{\Omega} \neq 0$ everywhere;
- $J \wedge \Omega = 0$;
- the tensor $g = JI$ (which is symmetric because of the conditions above) is positive definite.

Notice that with respect to I, it is easy to see from the above conditions that J is $(1,1)$ and Ω is $(3,0)$.

A Calabi-Yau manifold can be defined then as a manifold on which

$$dJ = 0 = d\Omega \qquad \text{(Calabi-Yau)}. \tag{1.2}$$

In the case with fluxes, however, we will see that the condition (1.2) is no longer relevant.

2. SU(3) × SU(3) structures

In fact, we will need to generalize even the definition of SU(3) structure. Let us stress that until theorem 3.2, we will not assume any *differential* (as opposed to algebraic) property for the forms involved.

DEFINITION 2.1. Given two differential forms ω_1, ω_2, their pairing (ω_1, ω_2) is a function given by

$$(\omega_1, \omega_2)\,\text{vol} \equiv \Big(\omega_1 \wedge \lambda(\omega_2)\Big)_{\text{top}}, \tag{2.1}$$

where for any form C_k of degree k we denote

$$\lambda(C_k) = (-)^{\lfloor \frac{k}{2} \rfloor} C_k. \tag{2.2}$$

DEFINITION 2.2. A *pure spinor* Φ on M_6 is a complex differential form (of mixed degree; thus a section of $\oplus_k \Lambda^k T^* \otimes \mathbb{C}$)

- whose annihilator in $(T \oplus T^*) \otimes \mathbb{C}$ has complex dimension 6.
- such that $(\bar{\Phi}, \Phi) \neq 0$ everywhere on the manifold.

One can associate to any pure spinor Φ a generalized almost complex structure $\mathcal{J}$ (see [**4, 5**]), namely an almost complex structure on $T \oplus T^*$ (the sum of the tangent bundle and the cotangent bundle) which is Hermitian with respect to the natural pairing between vectors and one-forms.

DEFINITION 2.3. A *compatible* pair of pure spinors $\Phi_{1,2}$ is then a pair of pure spinors such that

- the condition

$$(\Phi_1, X\Phi_2) = 0 = (\bar{\Phi}_1, X\Phi_2) \qquad \forall X \in T \oplus T^* \tag{2.3}$$

  is satisfied; this implies that the pair $\Phi_{1,2}$ defines a metric g and an antisymmetric two-form b (see [**5, 2**] for details);
- g should be positive definite.

A compatible pair is also called an $\mathrm{SU}(3) \times \mathrm{SU}(3)$ structure. This means that the structure group on $T \oplus T^*$ can be reduced to $\mathrm{SU}(3) \times \mathrm{SU}(3)$. For an example of a compatible pair, we turn to the "SU(3) structure" we defined in the previous section. That name refers to the structure group on the tangent bundle T. The relation between the two is as follows: An $\mathrm{SU}(3) \times \mathrm{SU}(3)$ structure on $T \oplus T^*$ can be projected to the tangent bundle T to give two SU(3) structures. If these two coincide, one can write

$$\Phi_+ = e^{-iJ}, \qquad \Phi_- = \Omega \tag{2.4}$$

with (J, Ω) the forms defining an SU(3) structure as defined in the previous section.[1] There exist other types of compatible pairs.

In the next subsection, we will apply these $\mathrm{SU}(3) \times \mathrm{SU}(3)$ structures to type II theories. One can see that in a pure compatible pair $\Phi_{1,2}$, one of the two pure spinors is an even differential form Φ_+ and the other one is an odd form Φ_-. From now on we will take

$$\mathrm{IIA}: \begin{array}{c} \Phi_1 = \Phi_+ \\ \Phi_2 = \Phi_- \end{array}, \qquad \mathrm{IIB}: \begin{array}{c} \Phi_1 = \Phi_- \\ \Phi_2 = \Phi_+ \end{array}. \tag{2.5}$$

3. Supersymmetric Minkowski vacua

3.1. In supergravity.

DEFINITION 3.1. A manifold M_6 *supports a Minkowski type II solution* if there exists a supersymmetric solution of either type IIA or type IIB supergravity such that none of the fields in the solution breaks four-dimensional Poincaré invariance. Notice that this implies that the metric decomposes as

$$g_{10} = e^{2A} g_{\mathrm{Mink}_4} + g_6 \tag{3.1}$$

for some function A, which is then called the *warping*. It also implies that

$$dF_{\mathrm{RR}} = \delta \tag{3.2}$$

---

[1] e^{-iJ} has an annihilator given by the combinations $\{\iota_{\partial_m} + iJ_{mn}dx^n\wedge, m = 1, \cdots, 6\}$; Ω has an annihilator given by the three complex $(1,0)$-forms (which exist because Ω is decomposable by assumption) and by the three $(0,1)$-vectors. The other conditions for a compatible pair reduce to the ones listed in the definition of SU(3) structure in section 1.

for some distribution δ on M_6; here, $F_{\rm RR}$ is the formal sum $\sum_k F_k$, where k runs over even numbers in IIA and odd numbers in IIB.[2] The solution is called a *RR-flux solution* if there is at least one non-zero Ramond-Ramond field strength.

Let h be a closed three-form on M_6, so that

$$d_h \equiv d + h\wedge \tag{3.3}$$

is a differential. We also have the operator

$$d_h^{\mathcal{J}_1} \equiv e^{-(\pi/2)\mathcal{J}_1} d_h\, e^{(\pi/2)\mathcal{J}_1}, \tag{3.4}$$

with $\mathcal{J}_1$ the generalized almost complex structure associated to Φ_1, and $\mathcal{J}_1\cdot$ its natural action on differential forms.

THEOREM 3.2. *M_6 supports a Minkowski supersymmetric RR-flux solution with warping A and source δ if and only if there exists on M_6 an* $\mathrm{SU}(3)\times\mathrm{SU}(3)$ *structure $\Phi_{1,2}$ and a closed three-form h such that*

$$\boxed{\begin{array}{ll} d_h\Phi_1 = 0, & d_h(e^{-A}\mathrm{Re}\,\Phi_2) = 0, \\ \delta = -8 d_h d_h^{\mathcal{J}_1}(e^{-3A}\mathrm{Im}\,\Phi_2), & ||\Phi_1|| = ||\Phi_2|| \end{array}} \tag{3.5}$$

The proof is scattered in several papers. The main step was obtained in [**1**] and detailed in [**2**, App. A]. Those papers explain how to obtain a slightly different system:

$$\begin{array}{ll} d_h\Phi_1 = 0, & d_h(e^{-A}\mathrm{Re}\,\Phi_2) = 0, \\ \delta = -8 d_h(e^{-4A} * \lambda\, d_h(e^{A}\mathrm{Im}\,\Phi_2)), & ||\Phi_1|| = ||\Phi_2||, \end{array} \tag{3.6}$$

which was then shown to be equivalent to (3.5) in [**3**]. The disadvantage of (3.6) is that it contains the metric explicitly (the one defined by the $\mathrm{SU}(3)\times\mathrm{SU}(3)$ structure $\Phi_{1,2}$) through $*$. A crucial point (namely, that the equation of motion for h is redundant) was shown in [**7**].

Remarks.

- The system (3.5) would be rather easy to solve, were δ to be allowed to be any distribution. We will see in the next subsection that string theory poses some restrictions on δ.
- The first equation in (3.5) says that the internal manifold must be a *generalized Calabi-Yau manifold* in the sense of Hitchin [**4**]. A more appropriate, but less catchy, name would be "generalized complex

[2] One might think one could allow a second term $\mathrm{vol}_4 \wedge \tilde\delta$ in the right hand side of (3.2), but one finds, while showing theorem 3.2, that $\tilde\delta = 0$.

manifold with holomorphically trivial canonical bundle." One can also show that the first equation in (3.5) also implies that

$$d_h^{\mathcal{J}_1} = [d_h, \mathcal{J}_1]. \tag{3.7}$$

- When the conditions in theorem 3.2 are satisfied, the values of the fields are

(3.8)

$$g_{10} = e^{2A} g_{\mathrm{Mink}_4} + g_6, \qquad H = h + db, \qquad \tfrac{1}{\sqrt{8}} e^{3A-\phi} = ||\Phi_1|| = ||\Phi_2||,$$
$$F_{\mathrm{RR}} = f + \mathrm{vol}_4 \wedge *_6 f, \qquad f = -8 d_H^{\mathcal{J}_1} (e^{-3A} \mathrm{Im}\, \Phi_2),$$

where ϕ is the dilaton. As required by the definition of a Minkowski solution, these fields do not break Poincaré invariance in four dimensions, because none of them depends on the coordinates of Mink_4, and the only forms on Mink_4 appearing in (3.8) are 1 and vol_4. Notice that f in (3.8) together with $df = \delta$ yields the third equation in (3.5). An alternative form for f in (3.8) is

$$f = -8 e^{-4A} * \lambda\, d_h (e^A \mathrm{Im}\, \Phi_2) \tag{3.9}$$

that yields the third equation in (3.6).

3.2. In string theory. String theory is more demanding than supergravity in two respects:

- the flux F_{RR} has to obey flux quantization
- the source δ has come from an object present in the theory (an orientifold or a brane).

For the first point, we recall that, for a supergravity solution, F_{RR} is given by the expression in (3.8). The most reasonable way of interpreting flux quantization seems to impose the conditions of [**8**] on the internal fluxes f.

We can say a bit more about the second point.

DEFINITION 3.3. A *supersymmetric cycle* B with respect to the $\mathrm{SU}(3) \times \mathrm{SU}(3)$ structure $\Phi_{1,2}$ is a submanifold of M_6 such that $\iota^* \mathrm{Im}\, \Phi_2 = \mathrm{vol}_B$, where $\iota : B \to M_6$. (This also implies that $\iota^* \mathrm{Re}\, \Phi_2 = 0$ and that $\iota^*(X \cdot \Phi_1) = 0$ for any $X \in T \oplus T^*$.) Here vol_B is the *Born-Infeld volume form* computed using $\det(g + b)$, with g and b determined by $\Phi_{1,2}$. We will also say that B is *almost calibrated* by $\mathrm{Im}\Phi_2$. We will denote by δ_B the current supported on B, normalized such that $(\delta_B, \mathrm{Im}\Phi_2) = \delta(B)$.

Let now σ be an involution on M_6 with fixed locus O_σ.

DEFINITION 3.4. The $\mathrm{SU}(3) \times \mathrm{SU}(3)$ structure $\Phi_{1,2}$ is *compatible with* σ if

$$\sigma^* \Phi_1 = -(-)^{\lfloor \frac{p}{2} \rfloor} \lambda(\Phi_1), \qquad \sigma^* \Phi_2 = (-)^{\lfloor \frac{p-1}{2} \rfloor} \lambda(\bar{\Phi}_2). \tag{3.10}$$

We also call σ an *orientifold involution* for the $\mathrm{SU}(3) \times \mathrm{SU}(3)$ structure $\Phi_{1,2}$.

If σ is an orientifold involution for $\Phi_{1,2}$, the fixed locus O_σ is a supersymmetric cycle with respect to $\Phi_{1,2}$.

There are now two cases: with or without orientifolds.

a) $\delta = \sum_i \delta_{B_i}$, with B_i supersymmetric cycles;
b) there exists an orientifold involution σ for the $\mathrm{SU}(3) \times \mathrm{SU}(3)$ structure $\Phi_{1,2}$, and $\delta = -\delta_{O_\sigma} + \sum_i \delta_{B_i}$, again with B_i supersymmetric cycles; as we noticed before, O_σ is automatically supersymmetric. (One can prove [**9, 2**] that this case is the only possibility if M_6 is compact).

Going back now to (3.5), it follows from the third equation in (3.5) that δ is trivial in cohomology. However, the same equation also says that $\mathrm{Im}\,\Phi_2$ is not closed, so that O_σ and B_i are only "almost" calibrated by it, and we cannot conclude that they are non-trivial in homology (or rather, not even that their sum is).

There exist many solutions of type a) in the literature, in fact predating theorem 3.2; essentially all the solutions used in the context of $\mathrm{AdS}_5/\mathrm{CFT}_4$ are of this type. This is because AdS_5 can be written as a warped product of Mink_4 and $\mathbb{R}$; the same is true for the generalization to RG flows of the correspondence.

Non-trivial compact examples (thus, of type b)) are harder to find. Some can be found in [**2**]. To make (3.5) look less intimidating, we will now massage the system in some particular cases.

3.3. Particular cases. When we specialize Theorem 3.2 to an SU(3) structure, looking at (2.5) we see that the first equation in (3.5) gives

$$\begin{aligned} &\mathrm{IIA}: \ dJ = 0 \qquad (\Rightarrow M_6 \text{ is symplectic}); \\ &\mathrm{IIB}: \ d\Omega = 0 \qquad (\Rightarrow M_6 \text{ is complex}). \end{aligned} \tag{3.11}$$

We will now have a closer look at both cases.

3.3.1. *Symplectic manifolds (from O6-planes).* In IIA, specializing (3.5) to an SU(3) structure gives

$$\begin{aligned} dJ = H = 0, \qquad & d(e^{-A}\mathrm{Re}\,\Omega) = 0, \\ \delta = -dd^\Lambda(e^{-3A}\mathrm{Im}\,\Omega), \qquad & i\Omega \wedge \bar{\Omega} = \tfrac{3}{4}J^3. \end{aligned} \tag{3.12}$$

Here, d^Λ is the Lefschetz differential $d^\Lambda = [d, J^{-1}\llcorner]$. It follows that δ is a three-form. Thus its components B_i and O_σ have to be three-cycles; one might call them "almost special Lagrangian." The "almost" stresses that $\mathrm{Im}\,\Omega$ is not closed, and that hence there is no reason B_i and O_σ should be non-trivial in H_3, as remarked after Theorem 3.2. It also follows from (3.8) that $\phi = 3A + \text{const}$.

Given that the fixed locus of σ is three-dimensional in M_6, one calls O_σ an O6-plane (adding the three Minkowski space dimensions to the count).

Some compact solutions of this kind were found in **[2]**[3], on some particular solvmanifolds that also admit flat metrics.

3.3.2. *Complex manifolds from O5-planes.* In IIB, choosing the $SU(3) \times SU(3)$ structure to come from an $SU(3)$ structure on T still leaves several distinct possibilities open. Technically, this is because, in $\Phi_2 = \rho e^{-iJ}$, different choices for the phase of ρ lead to dramatically different $\mathrm{Re}\,\Phi_2$ and $\mathrm{Im}\,\Phi_2$. In this section we will take ρ to be real. This leads to

$$\begin{aligned} d\Omega &= 0, & dJ^2 &= 0, \\ \delta &= -2i\partial\bar\partial(e^{-2A}J), & i\Omega\wedge\bar\Omega &= \tfrac{3}{4}e^{2A}J^3. \end{aligned} \tag{3.13}$$

This time ∂ is the usual Dolbeault operator, and hence δ is a four-form. Hence B_i and O_σ are two-cycles; they are almost calibrated by J, but given that it is not closed, B_i and O_σ might be trivial cycles. It also follows from (3.8) that $\phi = 2A + \mathrm{const}$.

Counting again also the three space directions in Mink_4, O_σ is called this time an O5-plane.

Examples of this type have again been found in **[2]**; this time a nilmanifold is also possible, one which does not admit any flat metric. The comment in footnote 3 also applies.

4. Supersymmetric anti-de Sitter vacua

An AdS solution is defined just like Minkowski solutions, but replacing Poincaré invariance in four dimensions with invariance under its analogue for negative cosmological constant, $SO(4,2)$. As a consequence, the four-dimensional metric g_{Mink_4} in (3.1) gets replaced by g_{AdS_4}.

The AdS_4 case turns out to be a bit less elegant mathematically, but more forgiving physically.

In supergravity, theorem 3.2 gets replaced by

THEOREM 4.1. *M_6 supports an AdS supersymmetric RR-flux solution with warping A, cosmological constant Λ and source δ if and only if there exists on M_6 an $SU(3)\times SU(3)$ structure $\Phi_{1,2}$ and a three-form h such that*

$$\boxed{\begin{aligned} d_h\Phi_1 &= -2\mu\, e^{-A}\mathrm{Re}\,\Phi_2, \qquad ||\Phi_1|| = ||\Phi_2|| \\ -\tfrac{1}{8}\delta &= d_h d_h^{\mathcal{J}_1}(e^{-3A}\mathrm{Im}\,\Phi_2) + 3\mu\, d_h(e^{-4A}\mathrm{Re}\,\Phi_1), \end{aligned}} \tag{4.1}$$

where

$$\mu \equiv \sqrt{-\frac{\Lambda}{3}}$$

[3]The orientifold and brane sources were actually *smeared* in that paper, that is, replaced with constant forms; this is not really legitimate for orientifolds, but such an approximation often captures solutions that one can control otherwise.

This result was obtained in [**2**]; as in the Minkowski case, one can show that all the equations of motion are implied by (4.1). The redundancy of the equation of motion for H, in particular, was shown in [**10**]. Notice that one could also write in this case an analogue of the equation for $\mathrm{Re}\,\Phi_2$ that appears in (3.5); that equation would, however, be dependent on the ones we already wrote. In [**11, 12**] it was noted that this is because, for AdS vacua in four-dimensional $\mathcal{N}=1$ supergravity, D-terms are dependent on F-terms.

The analogue of (3.6) also exists:

$$\begin{aligned} &d_h\Phi_1 = -2\sqrt{-\Lambda}e^{-A}\mathrm{Re}\,\Phi_2, \qquad ||\Phi_1|| = ||\Phi_2|| \\ &-\tfrac{1}{8}\delta = d_h(e^{-4A} * \lambda\, d_h(e^{A}\mathrm{Im}\,\Phi_2)) + 3\mu\, d_h(e^{-4A}\mathrm{Im}\,\Phi_1). \end{aligned} \tag{4.2}$$

The fields now have the same value as in (3.8), except that

$$f = -8d_H^{\mathcal{J}_1}(e^{-3A}\mathrm{Im}\,\Phi_2) - 24\mu\, e^{-4A}\mathrm{Re}\,\Phi_1, \tag{4.3}$$

or equivalently

$$f = -8e^{-4A} * \lambda d_h(e^{A}\mathrm{Im}\,\Phi_2) - 24\mu\, e^{-4A}\mathrm{Re}\,\Phi_1. \tag{4.4}$$

These two equations, together with $df = \delta$, give respectively (4.1) and (4.2).

Finally, one still has a choice to include an orientifold or not; however, it is no longer necessary to do so to keep compact M_6. This point is illustrated by the many solutions of Freund-Rubin type (for an early review, see [**13**]), and more recently, for example, in [**14, 15**].

References

[1] M. Graña, R. Minasian, M. Petrini, and A. Tomasiello, "Generalized structures of $\mathcal{N}=1$ vacua," *JHEP* **0511** (2005) 020, `hep-th/0505212`.

[2] M. Graña, R. Minasian, M. Petrini, and A. Tomasiello, "A scan for new $\mathcal{N}=1$ vacua on twisted tori," *JHEP* **0705** (2007) 031, `hep-th/0609124`.

[3] A. Tomasiello, "Reformulating supersymmetry with a generalized Dolbeault operator," `arXiv:0704.2613 [hep-th]`.

[4] N. Hitchin, "Generalized Calabi-Yau manifolds," *Quart. J. Math. Oxford Ser.* **54** (2003) 281–308, `math.dg/0209099`.

[5] M. Gualtieri, "Generalized complex geometry," `math.dg/0401221`.

[6] P. Candelas, G. T. Horowitz, A. Strominger, and E. Witten, "Vacuum configurations for superstrings," *Nucl. Phys.* **B258** (1985) 46–74.

[7] P. Koerber and D. Tsimpis, "Supersymmetric sources, integrability and generalized-structure compactifications," *JHEP* **0708** (2007) 082, `arXiv:0706.1244 [hep-th]`.

[8] G. W. Moore and E. Witten, "Self-duality, Ramond-Ramond fields, and K-theory," *JHEP* **0005** (2000) 032, `hep-th/9912279`.

[9] J. M. Maldacena and C. Nuñez, "Supergravity description of field theories on curved manifolds and a no-go theorem," *Int. J. Mod. Phys.* **A16** (2001) 822–855, `hep-th/0007018`.

[10] P. Koerber and L. Martucci, "D-branes on AdS flux compactifications," *JHEP* **0801** (2008) 047, `arXiv:0710.5530 [hep-th]`.

[11] D. Cassani and A. Bilal, "Effective actions and $\mathcal{N} = 1$ vacuum conditions from SU(3) $\times$ SU(3) compactifications," *JHEP* **0709** (2007) 076, arXiv:0707.3125 [hep-th].
[12] P. Koerber and L. Martucci, "From ten to four and back again: how to generalize the geometry," *JHEP* **0708** (2007) 059, arXiv:0707.1038 [hep-th].
[13] M. J. Duff, B. E. W. Nilsson, and C. N. Pope, "Kaluza-Klein supergravity," *Phys. Rept.* **130** (1986) 1–142.
[14] K. Behrndt and M. Cvetič, "General $\mathcal{N} = 1$ supersymmetric fluxes in massive type IIA string theory," *Nucl. Phys.* **B708** (2005) 45–71, hep-th/0407263.
[15] A. Tomasiello, "New string vacua from twistor spaces," arXiv:0712.1396 [hep-th].

DEPARTMENT OF PHYSICS, HARVARD UNIVERSITY, 17 OXFORD STREET, CAMBRIDGE, MA 02138
E-mail address: tomasiel@physics.harvard.edu

AMS/IP Studies in Advanced Mathematics
Volume 44, 2008

Resolving Gravitational Singularities

Finn Larsen

ABSTRACT. We review recent progress on the resolution of gravitational singularities in string theory. The main example is the fundamental string in five dimensions which is singular in the standard supergravity description but regular after taking into account higher derivative corrections determined by anomalies and supersymmetry. The application of the AdS/CFT correspondence to geometries that require higher derivatives for regularity poses interesting challenges.

1. Introduction

In the last several years there has been significant progress on understanding higher derivative corrections to black holes in string theory [**1**–**6**]. An important aspect of this development is that some interesting examples have been identified where gravitational singularities are resolved by higher derivative corrections to the action [**7**–**13**]. The prototypical example is a regular black hole that would have vanishing horizon area were it not for the higher derivative corrections in the Lagrangian. The resolution of gravitational singularities is an important problem in many different contexts so it is worthwhile discussing this aspect of higher derivative corrections separately from perturbative corrections to solutions that are already regular in the leading approximation. That is the focus of this lecture.

The work discussed here is collected from a series of articles by Castro, Davis, Kraus, and the current author [**14**–**16**]. These articles develop the subject in the context of five dimensional supergravity corrected by the mixed gauge-gravitational Chern-Simons term:

$$\mathcal{L}_{\rm CS} = \frac{1}{24\cdot 16\pi^2}\, c_{2I} A^I \wedge {\rm Tr}\ R\wedge R\ , \tag{1.1}$$

2000 *Mathematics Subject Classification.* Primary 83E30; Secondary 81T30, 81T60.

The author was supported in part by DoE grant # DE-FG02-95ER40899.

and terms related to this by supersymmetry [**17**]. In this lecture we sacrifice generality and technical detail in order to keep the text brief and, hopefully, accessible also for interested non-experts. A much more comprehensive review appeared recently in [**18**].

Before considering any of the specifics, it is useful to recall some of the general challenges intrinsic to any attempt at resolving singularities using higher derivative terms:

- *Exact symmetries* of the theory must be preserved by the higher derivative corrections. In our context, the Chern-Simons term (1.1) is required for anomaly cancellation. Furthermore, we go through much effort to preserve supersymmetry.
- *Effective field theory* demands that *all* terms of a given order should be taken into account. We address this by appealing to the uniqueness of the supersymmetric completion of the mixed gauge-gravitational Chern-Simons term (1.1).
- *Field redefinition ambiguities*, such as $g_{\mu\nu} \rightarrow g_{\mu\nu} + AR_{\mu\nu}$ (and generalizations involving matter fields), mix terms of different derivative order so that the singularity could depend on the choice of coordinates in field space. Our construction employs an off-shell formalism that implements supersymmetry at each order in the derivative approximation independently. This precludes most field redefinitions and defines the metric unambiguously.
- *Singularities* make any systematic expansion inherently problematic. In order to cancel leading order singularities, the higher order terms must be comparable to the leading order terms. Then even higher order terms might be important as well, *i.e.* there is no systematic expansion parameter. We will not be able to address this point completely but we will make it very explicit and concrete in due course.

We now turn to the case of small five-dimensional black strings in $N = 2$ supergravity, the example that is the focus of this talk.

2. Singularities or no singularities?

Consider M-theory on $CY_3 \times \mathbf{R}^{4,1}$ for small CY_3 volume. The theory is effectively five-dimensional and in the supergravity approximation it reduces to $N = 2$ supergravity in $D = 5$ coupled to a number of vector multiplets. The Lagrangian is

$$(2.1)\quad \mathcal{L}_0 = R + G_{IJ}\nabla M^I \nabla M^J + \frac{1}{2} G_{IJ} F^I_{ab} F^{Jab} - \frac{1}{24e} c_{IJK} A^I_a F^J_{bc} F^K_{de} \epsilon^{abcde},$$

in a self-explanatory notation (the theory is introduced in more detail in *e.g.* [**19**]).

We seek string solutions to this theory and so consider the *ansatz*:

$$ds^2 = e^{2U_1(r)}\left(dt^2 - dy^2\right) - e^{-4U_2(r)}(dr^2 + r^2 d\Omega_2^2),$$

for the geometry. The strings are supported by magnetic charges p^I with respect to the vector fields A^I, *i.e.* the two-forms F^I

$$F^I = -\frac{1}{2} p^I e^{\hat{\theta}} \wedge e^{\hat{\phi}}, \tag{2.2}$$

give rise to magnetic flux through the transverse two-sphere.

The warp factors $U_{1,2}(r)$ are going to diverge at the origin so there will at least be a coordinate singularity near the string. However, the singularity may just be a coordinate artifact; the actual geometry could be regular. This happens exactly when $e^{-6U_1} = e^{-6U_2} \sim r^{-3}$ as $r \to 0$. Moreover, the resulting regular solutions have $AdS_3 \times S^2$ near string geometry.

The actual behavior of the string solution depends on the charge configuration. An important special case is when the matter supporting the string solution is the two-form $B_{\mu\nu}$ (represented by the one-form gauge-field that it dualizes to in five dimensions) and the dilaton Φ, as in perturbative string theory. In this case the solution to the supergravity equations of motion has $e^{-6U_1} = e^{-6U_2} \sim r^{-1}$ as $r \to 0$. This solution is therefore singular in the supergravity approximation. We want to establish that higher derivative corrections to the theory can modify the warp factors near the string so that in fact $e^{-6U_1} = e^{-6U_2} \sim r^{-3}$ as $r \to 0$, leaving the full geometry regular.

For the general class of string solutions considered here there is a simple criterion determining whether the geometry has a singularity: if the charge configuration is such that $c_{IJK}p^I p^J p^K$ is nonvanishing, then there are magnetic string solutions to standard (two-derivative) supergravity with the regular $AdS_3 \times S^2$ near string geometry with scale set by $c_{IJK}p^I p^J p^K$. However, if the charges are such that the intersection number $c_{IJK}p^I p^J p^K = 0$, then the classical geometry is singular.

The string solutions in five dimensions are interpreted in M-theory as $M5$-branes wrapped on four-cycles P in CY_3. The magnetic charge p^I is the wrapping number around the basis four-cycle P_I. An $M5$-brane wrapping a general four-cycle P, one not coinciding with any basis-cycle P_I, carries several (or many) of the charges p^I. The important combination $c_{IJK}p^I p^J p^K$ can be interpreted as the self-intersection number of the four-cycle P underlying the string solution, because the coupling constants c_{IJK} in the action (2.1) are interpreted microscopically as the intersection numbers of the four-cycles P_I.

We will be particularly interested in the special case where the CY_3 is the product manifold $K3 \times T^2$ and the $M5$-brane wraps the four-cycle $P = K3$. This solitonic string is important because it is the type IIA dual of the heterotic string. Since the four-cycle underlying the dual heterotic string corresponds to one of the basis cycles it has vanishing self-intersection number, $c_{IJK}p^I p^J p^K = 0$. The criterion using the self-intersection number $c_{IJK}p^I p^J p^K = 0$ thus reproduces the result that heterotic strings have singular near string geometry in the supergravity approximation.

3. Indirect resolution of singularities: anomalies

The magnetic string solutions we consider are subject to powerful symmetry principles. These principles suggest when we should expect that gravitational singularities are resolved.

3.1. AdS/CFT correspondence. We must first recall the significance of the $AdS_3 \times S^2$ geometry. The global isometry group of AdS_3 is

$$SO(2,2) \simeq SL(2) \times SL(2) \ .$$

This isometry induces an obvious global $SL(2) \times SL(2)$ symmetry acting on the asymptotic boundary of AdS_3. The important point is that there is much more symmetry[**20**]: bulk diffeomorphisms combine with the global $SL(2) \times SL(2)$ and enhance the full boundary symmetry to $\mathrm{V}ir \times \mathrm{V}ir$. Similarly, the $SO(3) \simeq SU(2)$ isometry group of the S^2 combines with bulk diffeomorphisms and forms the affine current algebra $\widehat{SU(2)}$ acting on the left-movers (the supersymmetric side) in the two-dimensional boundary CFT.

The bulk spacetime supersymmetry complements the bosonic isometries so that there is in fact a superisometry group. Combining this with bulk diffeomorphisms one finds a superconformal algebra on the boundary. In the case of $N = 2$ supersymmetry in bulk the correct superisometry is $SU(1,1|2)$ and the boundary theory becomes a $(4,0)$ superconformal CFT.

Explicit computation from the standard (two-derivative) supergravity action (2.1) determines the spacetime central charges of the Virasoro algebras [**20**]:

$$c_L = c_R = \frac{3\ell_A}{2G_3}, \tag{3.1}$$

where ℓ_A is the length scale of the AdS_3 space. The central charge is a measure of the number of degrees of freedom in the two-dimensional boundary CFT. In bulk it is essentially the size of AdS_3, suggesting that we can use central charge as a convenient proxy for the scale of the geometry.

The affine current algebra $\widehat{SU(2)}$ is similarly found to have level $k = c_L/6$, consistent with the $(4,0)$ superalgebra. Geometrically, this relates the scale of the S^2 to that of AdS_3. We will later see that the precise relation is $\ell_A = 2\ell_S$.

Expressing ℓ_A in terms of the magnetic charges p^I of the string, it can be shown that the central charge (3.1) can be written as:

$$c_L = c_R = c_{IJK}p^I p^J p^K. \tag{3.2}$$

We previously mentioned that the intersection number $c_{IJK}p^I p^J p^K$ sets the scale of the near string solution and now see how this fits very nicely with the relation between central charges and the size of the AdS_3 (and of S^2). For charge configurations such that the central charge (3.2) vanishes in the supergravity approximation, the corresponding geometry is singular. However, it is reasonable to expect that the exact central charge does not vanish. This motivates the expectation that the geometry should be regular when higher derivative corrections are taken into account.

3.2. Local anomalies. The most robust higher derivative corrections are those related to anomaly cancellation. The standard Green-Schwarz terms (or alternatively terms inferred from $M5$-brane anomaly cancellation) give rise in five dimensions to the mixed gauge-gravitational Chern-Simons term

$$\begin{aligned} \mathcal{L}_{\rm CS} &= -\frac{c_{2I}}{96e} A^I \wedge {\rm Tr} R \wedge R \\ &= \frac{c_{2I}}{96e} F^I \wedge \omega_3 + \text{tot.der.} \end{aligned} \tag{3.3}$$

In the first line the interaction was written in a form that violates gauge symmetry, albeit only by a total derivative. The second line introduces the Chern-Simons three-form ω_3 (through ${\rm Tr} R \wedge R = d\omega_3$) to make the term manifestly gauge invariant but then diffeomorphism symmetry is violated, albeit again just by a total derivative.

The AdS/CFT correspondence relates the diffeomorphism violations in bulk to corresponding anomalies in the boundary theory [**12, 13**]. The relation determines the boundary central charges as:

$$c_L = c_{IJK}p^I p^J p^K + \frac{1}{2} c_{2I} p^I, \quad c_R = c_{IJK}p^I p^J p^K + c_{2I} p^I, \tag{3.4}$$

in accordance with other arguments [**21, 22**]. Both central charges are determined because diffeomorphism symmetry in AdS_3 and in S^2 (which is R-symmetry in the dual theory) give two pieces of information. The expressions (3.4) for the central charges are exact because the local symmetries that they enforce are exact. We also note that supersymmetry plays no important role at this point: c_L (the supersymmetric side) and c_R are determined with equal precision from anomalies.

3.3. Resolution of singularities. The exact central charges (3.4) suggest the resolution of gravitational singularities. For this we consider charge vectors such that the intersection number $c_{IJK}p^I p^J p^K = 0$, corresponding to singular geometry in the leading order description. In this case the central charges (3.4) are linear in the magnetic charges

$$c_L = \frac{1}{2} c_{2I} p^I, \quad c_R = c_{2I} p^I, \tag{3.5}$$

and in general they do not vanish. To the extent that central charges can be taken as a measure of the scales of the near horizon $AdS_3 \times S^2$, we see that higher derivative corrections have replaced a singularity (a space with vanishing size) with a regular geometry (with nonvanishing scale). Thus a gravitational singularity has been resolved.

It is of special interest to consider the case of the dual heterotic string. Recall this is the case where the compact Calabi-Yau threefold is $CY_3 = K3 \times T^2$ and the $M5$-brane simply wraps the four-cycle $P = K3$. Since $c_2(K3) = 24$ the central charges (3.5) become

$$c_L = 12p, \ \ c_R = 24p.$$

These are the correct values for p heterotic strings in a physical gauge. (The left movers are 8 bosons and 8 fermions with central charge $c_L = 8 \cdot 1 + 8 \cdot \frac{1}{2} = 12$, the right movers are 24 bosons with central charge $c_R = 24 \cdot 1 = 24$). This gives confidence that both the central charges (3.5) have been correctly determined. It also suggests that the gravitational singularity has been resolved in the specific case of the heterotic string.

4. Explicit singularity resolution

So far we have used symmetries to argue that certain gravitational singularities are resolved. However, there are several ways the argument could fail. For example, we *assume* an $AdS_3 \times S^2$ near string geometry and then anomalies and symmetries determine the central charges. This reasoning would fail if the near string geometry was not $AdS_3 \times S^2$. Another point is that we use central charges as a proxy for the scale of the geometry and this relation is not precise when higher derivative corrections are taken into account. In order to understand these and other issues better we would like to explicitly construct asymptotically flat solutions with the anticipated properties. This is what we turn to next.

4.1. Supersymmetry. As we have already mentioned repeatedly, the essential ingredient is the mixed gauge-gravitational Chern-Simons term (3.3). Since the indirect arguments rely in part on supersymmetry, all terms related to the Chern-Simons term by supersymmetry must also be taken into account.

The direct way to supersymmetrize (3.3) would be to carry out the Noether procedure: act by the standard supersymmetry transformations and identify other terms in the Lagrangian. Then modify the supersymmetry transformations as needed and improve the supersymmetrization iteratively. The problem is that this procedure generates terms of arbitrarily high order so that there is no closed form of the Lagrangian.

This challenge can be addressed by using off-shell supermultiplets identified using the superconformal formalism. In the gauge fixed version of the superconformal formalism that we employ, the only complication relative to the on-shell formalism is that the standard physical fields must be augmented by auxiliary fields. In particular, the Weyl multiplet (*i.e.* gravity) has an auxiliary scalar D, auxiliary two-tensor v_{ab}, and also an auxiliary fermion χ.

The off-shell action is invariant under the supersymmetry transformations:

$$\begin{aligned}
\delta\psi_\mu &= \left(\mathcal{D}_\mu + \frac{1}{2}v^{ab}\gamma_{\mu ab} - \frac{1}{3}\gamma_\mu\gamma\cdot v\right)\epsilon,\\
\delta\Omega^I &= \left(-\frac{1}{4}\gamma\cdot F^I - \frac{1}{2}\gamma^a\partial_a M^I - \frac{1}{3}M^I\gamma\cdot v\right)\epsilon,\\
\delta\chi &= \left(D - 2\gamma^c\gamma^{ab}\mathcal{D}_a v_{bc} - 2\gamma^a\epsilon_{abcde}v^{bc}v^{de} + \frac{4}{3}(\gamma\cdot v)^2\right)\epsilon.
\end{aligned} \tag{4.1}$$

It is important to emphasize that these transformations are symmetries of each order in the action by itself. Results we find by analyzing supersymmetry will therefore apply to the leading order solution, and also to the solution corrected by four-derivative terms. Indeed, they must apply to the exact solution which take into account corrections to all orders.

The supersymmetry variations (4.1) take a form similar to the standard on-shell supersymmetries, except for the unfamiliar appearance of the auxilary fields. We exploit these variations in the familiar way: demanding that the supersymmetry transformations vanish, when evaluated on purely bosonic backgrounds, yields linear differential equations satisfied by BPS-solutions.

As advertised previously, we assume as an *ansatz* that the metric takes the string form:

$$ds^2 = e^{2U_1}(dt^2 - dy^2) - e^{-4U_2}(dr^2 + r^2 d\Omega_2^2).$$

Now the BPS conditions (4.1) impose $U_1 = U_2$ (so we can omit the index $1, 2$ on U) and also determine the auxiliary fields in terms of the metric function U:

$$\begin{aligned}
v &= \frac{3}{4}e^{2U}\partial_r U e^{\hat\theta}\wedge e^{\hat\phi},\\
D &= 6e^{4U}\nabla^2 U,
\end{aligned} \tag{4.2}$$

where ∇^2 is the Laplacian on the three-dimensional space transverse to the string. Finally, the magnetic fields are governed by the scalar fields:

$$F^I = \frac{1}{2}\partial_r\left(M^I e^{-2U}\right)e^{4U}e^{\hat\theta}\wedge e^{\hat\phi}. \tag{4.3}$$

This is the standard attractor flow which represents field strengths as gradient flows, with the scalar fields essentially acting as potentials.

4.2. The Bianchi identity. The scalar fields M^I and the metric function U are not determined by supersymmetry alone. Generally, we must at this point appeal to the equations of motion which depend on the action. However, the present case is special because magnetic fields

$$F^I = -\frac{1}{2}p^I e^{\hat{\theta}} \wedge e^{\hat{\phi}},$$

support the string solution. These fields are topological and so they must be exact. However, the field strengths (4.3) determined from supersymmetry do not automatically take the correct form; they may even fail to satisfy the Bianchi identity. Imposing the Bianchi identity we find the harmonic equation

$$\nabla^2 \left(e^{-2U} M^I\right) = 0.$$

In the spherically symmetric case assumed here the solution is

$$M^I e^{-2U} = H^I = M^I_\infty + \frac{p^I}{2r},$$

where M^I_∞ are integration constants. These are the harmonic functions which underlie many familiar two-derivative solutions. Since we have not yet specified the action, these are in fact *precisely* the standard harmonic functions, with normalizations and other details unchanged from the two-derivative case.

4.3. The equations of motion. At this point we used supersymmetry and the Bianchi identity to determine the magnetic string solution completely, except for the metric factor $U(r)$. It is this function that depends on the detailed action. In order to get some familiarity with the off-shell formalism we first introduce the two-derivative action

$$\begin{aligned} \mathcal{L}_0 = &-\frac{1}{2}D + \frac{3}{4}\mathcal{R} + v^2 + \mathcal{N}\left(\frac{1}{2}D + \frac{1}{4}\mathcal{R} + 3v^2\right) + 2\mathcal{N}_I v^{ab} F^I_{ab} \\ &+ \mathcal{N}_{IJ}\left(\frac{1}{4}F^I_{ab}F^{Jab} + \frac{1}{2}\nabla M^I \nabla M^J\right) + \frac{1}{24e} c_{IJK} A^I_a F^J_{bc} F^K_{de} \epsilon^{abcde}. \end{aligned} \tag{4.4}$$

This action appears much more complicated than the standard supergravity action (2.1). However, the auxiliary fields D and v_{ab} enter algebraically so they can be integrated exactly by imposing their equation of motion. The equations of motion for D gives

$$\mathcal{N} = \frac{1}{6} c_{IJK} M^I M^J M^K = 1. \tag{4.5}$$

This is the familiar special geometry constraint. It ensures that the number of independent scalar fields is one less than the number of vector fields, as must be the case because one of the vector fields is the graviphoton which has no scalar superpartner. Simplifying the Lagrangian (4.4) using the special geometry constraint and the equation of motion for v_{ab} we recover the familiar on-shell $N = 2$ Lagrangian (2.1) with the correct expression

$$G_{IJ} = \frac{1}{2}\left(\mathcal{N}_I\mathcal{N}_J - \mathcal{N}_{IJ}\right),$$

for the metric on moduli space.

We are now ready to introduce the four-derivative action in all its glory [**17**]

(4.6)
$$\begin{aligned}\mathcal{L}_1 = \frac{c_{2I}}{24}\bigg(&\frac{1}{16}\epsilon_{abcde}A^{Ia}R^{bcfg}R^{de}_{\ \ fg} + \frac{1}{8}M^I C^{abcd}C_{abcd} + \frac{1}{12}M^I D^2 \\ &+\frac{1}{6}F^{Iab}v_{ab}D + \frac{1}{3}M^I C_{abcd}v^{ab}v^{cd} + \frac{1}{2}F^{Iab}C_{abcd}v^{cd} \\ &+\frac{8}{3}M^I v_{ab}\mathcal{D}^b\mathcal{D}_c v^{ac} - \frac{16}{9}M^I v^{ac}v_{cb}R_a^{\ b} - \frac{2}{9}M^I v^2 R \\ &+\frac{4}{3}M^I\mathcal{D}^a v^{bc}\mathcal{D}_a v_{bc} + \frac{4}{3}M^I\mathcal{D}^a v^{bc}\mathcal{D}_b v_{ca} - \frac{2}{3}M^I\epsilon_{abcde}v^{ab}v^{cd}\mathcal{D}_f v^{ef} \\ &+\frac{2}{3}F^{Iab}\epsilon_{abcde}v^{cf}\mathcal{D}_f v^{de} + F^{Iab}\epsilon_{abcde}v^c_{\ f}\mathcal{D}^d v^{ef} - \frac{4}{3}F^{Iab}v_{ac}v^{cd}v_{db} \\ &-\frac{1}{3}F^{Iab}v_{ab}v^2 + 4M^I v_{ab}v^{bc}v_{cd}v^{da} - M^I(v^2)^2\bigg),\end{aligned}$$

where the Weyl tensor is

$$C_{abcd} = R_{abcd} - \frac{2}{3}\left(g_{a[c}R_{d]b} - g_{b[c}R_{d]a}\right) + \frac{1}{6}g_{a[c}g_{d]b}R.$$

The first term is the mixed gauge-gravity anomaly (1.1). The second term includes a more conventional *Riemann*-squared term, which was expected to be in the same supermultiplet. The remaining terms are various matter terms and gravity-matter couplings. All of these are of the same order as the first two terms and so must be taken properly into account.

The complete four-derivative Lagrangian is clearly quite complicated and it would be overwhelming to solve the corresponding equations of motion by brute force. However, we have already determined most features of the solution using general principles. All that is needed is a single equation of motion to find the metric factor U. The simplest is to use the equation of motion for the auxiliary D field

$$\mathcal{N} = 1 - \frac{c_{2I}}{72}\left(F^I_{ab}v^{ab} + M^I D\right).$$

Inserting the expressions for the various fields in terms of charges and the metric factor we find

$$e^{-6U} = \frac{1}{6}c_{IJK}H^I H^J H^K + \frac{c_{2I}}{24}\left(\nabla H^I\nabla U + 2H^I\nabla^2 U\right). \tag{4.7}$$

The two-derivative theory corresponds to keeping only the first term on the right hand side and then the equation gives the metric factor U explicitly. In the four-derivative theory the equation has become a second order differential equation for the metric factor U. This describes the deformation of the standard special geometry constraint (4.5) due to higher derivative corrections.

4.4. Attractor solution. In the introduction we noted that, if the string solution is regular, the near string geometry will be $AdS_3 \times S^2$. The near string region is isolated by omitting the constant term in the harmonic functions

$$H^I = p^I/2r,$$

and an $AdS_3 \times S^2$ *ansatz* amounts to

$$e^{-6U} = \ell_S^3/r^3.$$

We have for definiteness written the *ansatz* in terms of ℓ_S, the radius of the S^2, but supersymmetry implies $U_1 = U_2$, which amounts to

$$\ell_A = 2\ell_S,$$

so we could just as well have used ℓ_A, the radius of AdS$_3$.

Now, inserting the *ansatz* into the deformed special geometry constraint (4.7) we see that the attractor *ansatz* is an exact solution with the identification of ℓ_S as

$$\ell_S^3 = \frac{1}{8}\left(\frac{1}{6}c_{IJK}p^Ip^Jp^K + \frac{1}{12}c_{2I}p^I\right). \tag{4.8}$$

The last term in this equation constitutes a correction to the scale ℓ_S. In the case of a string that is classically singular $c_{IJK}p^Ip^Jp^K = 0$ and so this term corrects the scale ℓ_S to a finite value. In this case the higher derivative terms have therefore resolved a singularity.

4.5. c-extremization. In two-derivative gravity the central charge and the scale of the AdS_3 space are related by the Brown-Henneaux formula (3.1). Indeed this was a source of intuition that the singularity should be resolved. At this point have seen how derivative corrections modify the space-time solution but we have yet to compute the corrected central charge.

The total central charge $c = \frac{1}{2}(c_R + c_L)$ is the trace anomaly of the two-dimensional boundary CFT. The dual manifestation in bulk is an anomalous variation of the on-shell action under rescaling of the AdS-geometry, which in turn is essentially the on-shell action itself. The central charge is therefore the on-shell action of the attractor solution, up to a known overall factor [**12**].

Evaluating the full action (4.4), (4.6) on the attractor solution determined in the previous subsection we find

$$c = -12\ell_A^3\ell_S^2(\mathcal{L}_0 + \mathcal{L}_1) = c_{IJK}p^Ip^Jp^K + \frac{3}{4}c_{2I}p^I, \tag{4.9}$$

in agreement with the exact central charges (3.4) determined by anomaly inflow. More generally one can verify that the central charge (4.9) is extremized upon variation of the defining parameters of the attractor geometry. These agreements are sensitive consistency checks because the on-shell action depends on most terms in the actions (4.4), (4.6).

4.6. The interpolating solution. We have focussed so far on the behavior of the string solution in the important near string region. We now determine the complete radial evolution of the metric factor U in the special case of the solitonic heterotic string, represented as p $M5$-branes wrapping the $K3$ of $CY_3 = K3 \times T^2$.

To do this we need to solve the deformed special geometry constraint (4.7). This is a nonlinear second order differential equation which cannot generally be solved analytically. However, we can proceed by solving one region at a time and then patch together partial results to determine the complete solution: first consider the attractor solution applicable at $r \sim p^{1/3}$ and extend it perturbatively to larger distances; and then consider the semi-classical solution around flat space, valid at $r \sim p$, and extend perturbatively towards smaller distances. Alternatively, we can simply find the numerical solution to the full differential equation. The results are presented in figure 1. It is seen that the expansion out from the attractor solution matches up quite well with the expansion in from asymptotically flat space, with the numerical solution shadowing both.

The details of these procedures are given in [**15**] (and the review [**18**]). For the purpose of this talk the main point is simply that all this can be done, providing some confidence that the near horizon region attaches smoothly on to flat space in a rather conventional fashion.

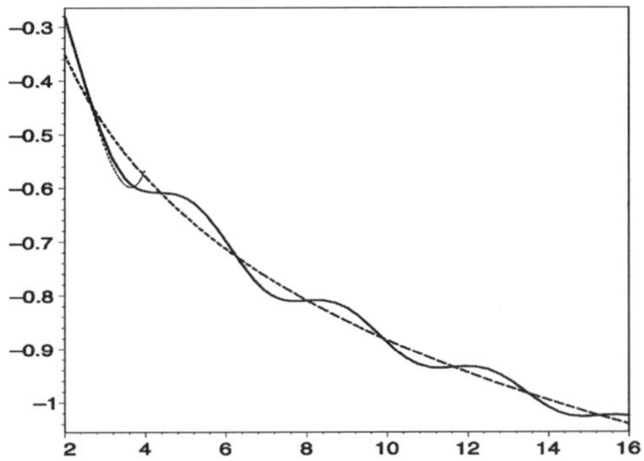

FIGURE 1. Metric factor interpolating between near string region and asymptotic space. *Solid line*: numerical solution. *Dashed line*: expansion around flat space. *Dotted line*: expansion around the near string solution.

4.7. Further corrections? We have included all corrections up to four derivative order in this work but one should ask whether there might further corrections? As mentioned in the introduction, this question is acute when corrections resolve a singularity for then the "corrections" are as important as the leading order terms. In other words, there is generally no systematic expansion parameter and so we must generally keep all orders in the Lagrangian.

In our explicit contruction we have seen that the *form* of the solution is determined completely by supersymmetry. However, the final determination of the *scale* (4.8) requires solving the deformed special geometry constraint, an equation of motion which is expected to receive further corrections. Thus our approach is *not* generally immune to further corrections.

Our result (4.8) for the scale is an expansion with corrections controlled by an effective expansion parameter $1/p^2$. For large magnetic charge p, this structure precludes cancellation between different orders in the derivative expansion. However, for small black holes there is an additional challenge: some of the moduli are inherently small so higher order in inverse moduli can compensate for additional derivatives. For example, the dual heterotic string has strong coupling in the near string region, as one expects for a $M5$-brane. It is not sufficient to consider solutions to the classical equations of motion as we do, since quantum fluctuations around the saddle points are significant.

The issue of quantum fluctuations can be circumvented by dualizing the type IIA solitonic string to the heterotic frame. The metric factor near the string becomes

$$e^{-6U} \to \ell^3/r^3 \ , \ \ell = \sqrt{\alpha'/2} \ .$$

The geometry of the heterotic string is thus $AdS_3 \times S^2$, with the geometries of string scale. Note in particular that the scale of the geometry is independent of the number of fundamental strings. In this frame it is thus manifest that there is no systematic expansion parameter: the full geometry is stringy in nature. On the other hand, in the heterotic frame the dependence on the number of strings enters exclusively through the coupling constant, which is fixed as

$$g_5^{\text{het}} = 2^{-1/4} p^{-1/2},$$

so quantum corrections are under good control for a large number of strings.

In the string geometry the precise definition of scale is ambiguous and one may try to maintain that central charge is a good proxy for scale. In that sense the higher derivative corrections resolve the gravitational singularity, but with other notions of length the resolution is merely qualitative, which are subject to further corrections of the same order as those that have been included.

5. The holographic dual of the heterotic string

The supergravity representation of the heterotic string is of string scale, but it is still reasonable to ask what the holographic dual is. The standard line of reasoning suggests that it should be a $D = 1 + 1$ CFT with $(8, 0)$ supersymmetry and R-symmetry at least SU(2). The classical superisometry group of the solution is $OSp(4*|4)$ [**24**], so one might in fact expect that the dual theory represents an affine extension of that group. The challenge is that there exists no CFT with all these properties. There is simply too much supersymmetry.

A resolution to this conundrum is suggested by the existence of nonlinear superconformal algebras (NSCA's) with the correct symmetries [**26**]. The nonlinearity of these algebras refers to the supercurrent OPE, which takes the form

$$G^i(z)G^j(0) \sim \frac{k}{\chi}\frac{\delta^{ij}}{z^3} + \frac{1}{\chi}\frac{J^a(T^a)^{ij}}{z^2} + \frac{2T\delta^{ij}}{z} + \frac{1}{2\chi}\frac{\partial_z J^a_z T^a}{z} + \frac{1}{2\chi k}\frac{J^a_z J^b_z P^{ij}_{ab}}{z}.$$

The last term is bilinear in the currents and so represents an unfamiliar nonlinearity. The nonlinear term depends on the tensor

$$P^{ij}_{ab} = \frac{1}{2}\{T^a, T^b\}^{ij} - 2\chi\delta_{ab}\delta^{ij},$$

where χ is a constant depending on the group. This tensor is such that the non-linearities vanish for small groups like $SU(2)$ but not for larger R-symmetry groups. The nonlinear superconformal algebras [**27, 28**] are powerful but unfamiliar relatives to W-algebras. In the current setting we consider multi-string states so the suggestion is that NSCA's are important in string field theory.

An intriguing feature of the NSCA's is that the nonlinearities determine the central charge. For example, the conformal extension of $OSp(4*|4)$ has central charge

$$c = -12p\left(1 + \frac{3}{2p}\right). \tag{5.1}$$

The classical (large p) limit gives the Brown-Henneaux formula (except for the sign). However, $p \sim 1/g^2$ so the nonlinear algebra appears to identify nontrivial quantum corrections to the spacetime central charge. Interesting as this prospect may seem it cannot be emphasized enough that there are severe obstacles to this optimistic interpretation. The biggest problem is that the relevant algebras apparently have no unitary representations. This problem is manifest in the expression (5.1) for the central charge, which is in fact negative. Unitarity violation is not only unacceptable on general grounds, it is also the wrong physics for a setting expected to be extremely stable because of the high degree of supersymmetry.

The most probable conclusion is that the symmetry of the heterotic string is unrelated to non-linear superconformal algebras after all. Of course this still leaves the challenge of finding an alternative framework for describing the theory.

6. Discussion

In this talk we discussed the resolution of gravitational singularities with emphasis on the case of 5D string solutions with $AdS_3 \times S^2$ near string geometry. However, the techniques apply in many other cases where one seeks to resolve singularities, or to correct solutions that are already regular in the two-derivative theory. Some of the examples considered hitherto are:

- *Black holes in five dimensions* with $AdS_2 \times S^3$ near horizon geometry. The solutions are supported by electric charges, so the Bianchi identity is trivial. Instead one must work out Gauss' law from the explicit action. Fortunately, it turns out that the end result again reduces to a harmonic equation and the complete solution can be determined explicitly.
- *Rotating black holes* in five dimensions have less symmetry and are correspondingly more complicated. Although all terms in the four-derivative action (4.6) now contribute, Gauss' law remains integrable, and again the complete solution can be determined.
- *Rotating black holes on Taub-NUT base space* are more involved yet but, remarkably, all equations remain integrable in much the same way as the simpler case. The significance of Taub-NUT is that its "cigar"-form interpolates between a central $\mathbf{R}^4$ region and asymptotic $\mathbf{R}^3 \times S^1$. Thus these 5D solutions on Taub-NUT include the 4D black holes that have been considered previously. The detailed comparison reveals discrepancies which can be traced to subtleties in the definition of charge in the presence of higher derivative terms.

These results and others were worked out in [**14**–**16**] and were recently reviewed in detail in [**18**]. The combined body of examples verifies previous indirect arguments by explicit computation, sort out discrepancies in those arguments, and uncovers numerous new results, including many still await interpretation. The explicit construction of supersymmetric solutions to five-dimensional theories with higher derivatives has thus proven a fruitful endeavor.

Among many open problems, the most far-reaching involve the generalizations to other dimensions, including 10 and 11 dimensions, and theories with more general matter content. In such more general contexts there are often very different motivations for considering higher derivative corrections like, for example, the recent reports of the possible finiteness of $N = 8$ supergravity [**29, 30**]. Many of the higher derivative terms identified for different purposes could be employed also in the context of the resolution of gravitational singularities. It is therefore reasonable to expect a fruitful interplay between different perspectives on higher derivative corrections to gravitational theories.

Acknowledgments

I am grateful to A. Castro, J. Davis, P. Kraus, and A. Shah for collaboration on the work reported on here [**14–16, 18, 23**]. I also thank A. Giveon and K. Hanaki for discussions. This work was supported by DOE under grant DE-FG02-95ER40899.

References

[1] K. Behrndt, G. Lopes Cardoso, B. de Wit, D. Lust, T. Mohaupt and W. A. Sabra, "Higher-order black-hole solutions in N = 2 supergravity and Calabi-Yau string backgrounds," Phys. Lett. B **429**, 289 (1998) [arXiv:hep-th/9801081].

[2] G. Lopes Cardoso, B. de Wit and T. Mohaupt, "Corrections to macroscopic supersymmetric black-hole entropy," Phys. Lett. B **451**, 309 (1999) [arXiv:hep-th/9812082].

[3] G. Lopes Cardoso, B. de Wit and T. Mohaupt, "Macroscopic entropy formulae and non-holomorphic corrections for supersymmetric black holes," Nucl. Phys. B **567**, 87 (2000) [arXiv:hep-th/9906094].

[4] G. Lopes Cardoso, B. de Wit, J. Kappeli and T. Mohaupt, "Stationary BPS solutions in N = 2 supergravity with R^2 interactions," JHEP **0012**, 019 (2000) [arXiv:hep-th/0009234].

[5] H. Ooguri, A. Strominger and C. Vafa, "Black hole attractors and the topological string," Phys. Rev. D **70**, 106007 (2004) [arXiv:hep-th/0405146].

[6] E. P. Verlinde, "Attractors and the holomorphic anomaly," arXiv:hep-th/0412139.

[7] A. Dabholkar, "Exact counting of black hole microstates," Phys. Rev. Lett. **94**, 241301 (2005) [arXiv:hep-th/0409148].

[8] A. Dabholkar, R. Kallosh and A. Maloney, "A stringy cloak for a classical singularity," JHEP **0412**, 059 (2004) [arXiv:hep-th/0410076].

[9] V. Hubeny, A. Maloney and M. Rangamani, "String-corrected black holes," JHEP **0505**, 035 (2005) [arXiv:hep-th/0411272].

[10] A. Sen, "How does a fundamental string stretch its horizon?," JHEP **0505**, 059 (2005) [arXiv:hep-th/0411255].

[11] A. Dabholkar, F. Denef, G. W. Moore and B. Pioline, "Exact and asymptotic degeneracies of small black holes," JHEP **0508**, 021 (2005) [arXiv:hep-th/0502157]; "Precision counting of small black holes," JHEP **0510**, 096 (2005) [arXiv:hep-th/0507014].

[12] P. Kraus and F. Larsen, "Microscopic black hole entropy in theories with higher derivatives," JHEP **0509**, 034 (2005) [arXiv:hep-th/0506176].

[13] P. Kraus and F. Larsen, "Holographic gravitational anomalies," JHEP **0601**, 022 (2006) [arXiv:hep-th/0508218].

[14] A. Castro, J. L. Davis, P. Kraus and F. Larsen, "5D attractors with higher derivatives," JHEP **0704**, 091 (2007) [arXiv:hep-th/0702072].

[15] A. Castro, J. L. Davis, P. Kraus and F. Larsen, "5D Black Holes and Strings with Higher Derivatives," JHEP **0706**, 007 (2007) [arXiv:hep-th/0703087].

[16] A. Castro, J. L. Davis, P. Kraus and F. Larsen, "Precision entropy of spinning black holes," JHEP **0709**, 003 (2007) [arXiv:0705.1847 [hep-th]].

[17] K. Hanaki, K. Ohashi and Y. Tachikawa, "Supersymmetric Completion of an R^2 Term in Five-Dimensional Supergravity," Prog. Theor. Phys. **117**, 533 (2007) [arXiv:hep-th/0611329].

[18] A. Castro, J. L. Davis, P. Kraus and F. Larsen, "String Theory Effects on Five-Dimensional Black Hole Physics," arXiv:0801.1863 [hep-th].

[19] F. Larsen, "The attractor mechanism in five dimensions," arXiv:hep-th/0608191.
[20] J. D. Brown and M. Henneaux, "Central Charges in the Canonical Realization of Asymptotic Symmetries: An Example from Three-Dimensional Gravity," Commun. Math. Phys. **104**, 207 (1986).
[21] J. M. Maldacena, A. Strominger and E. Witten, "Black hole entropy in M-theory," JHEP **9712**, 002 (1997) [arXiv:hep-th/9711053].
[22] J. A. Harvey, R. Minasian and G. W. Moore, "Non-abelian tensor-multiplet anomalies," JHEP **9809**, 004 (1998) [arXiv:hep-th/9808060].
[23] P. Kraus, F. Larsen and A. Shah, "Fundamental Strings, Holography, and Nonlinear Superconformal Algebras," arXiv:0708.1001 [hep-th].
[24] J. M. Lapan, A. Simons and A. Strominger, "Nearing the Horizon of a Heterotic String," arXiv:0708.0016 [hep-th].
[25] A. Giveon and D. Kutasov, "Fundamental strings and black holes," JHEP **0701**, 071 (2007) [arXiv:hep-th/0611062].
[26] M. Henneaux, L. Maoz and A. Schwimmer, "Asymptotic dynamics and asymptotic symmetries of three-dimensional extended AdS supergravity," Annals Phys. **282**, 31 (2000) [arXiv:hep-th/9910013].
[27] V. G. Knizhnik, "Superconformal Algebras in Two-Dimensions," Theor. Math. Phys. **66**, 68 (1986) [Teor. Mat. Fiz. **66**, 102 (1986)].
[28] M. A. Bershadsky, "Superconformal Algebras in Two-Dimensions with Arbitrary N," Phys. Lett. B **174**, 285 (1986).
[29] M. B. Green, H. Ooguri and J. H. Schwarz, "Decoupling Supergravity from the Superstring," Phys. Rev. Lett. **99**, 041601 (2007) [arXiv:0704.0777 [hep-th]].
[30] Z. Bern, J. J. Carrasco, D. Forde, H. Ita and H. Johansson, "Unexpected Cancellations in Gravity Theories," Phys. Rev. D **77**, 025010 (2008) [arXiv:0707.1035 [hep-th]].

DEPARTMENT OF PHYSICS, UNIVERSITY OF MICHIGAN, ANN ARBOR MI-48109, USA.
Current address: Theory Division, CERN, CH-1211, Geneva 23, Switzerland.
E-mail address: `larsenf@umich.edu`

AMS/IP Studies in Advanced Mathematics
Volume 44, 2008

Recent Developments in Heterotic Compactifications

Eric Sharpe

ABSTRACT. In this short review, we outline three sets of developments in understanding heterotic string compactifications. First, we outline recent progress in heterotic analogues of quantum cohomology computations. Second, we discuss a potential swampland issue in heterotic strings, and new heterotic string constructions that can be used to fill in the naively missing theories. Third, we discuss recent developments in string compactifications on stacks and their applications, concluding with an outline of work-in-progress on heterotic string compactifications on gerbes.

Contribution to the proceedings of the Virginia Tech Sowers workshop, May 2007.

CONTENTS

1. Introduction

Over the last several years, heterotic strings have been undergoing something of a revival. There has been a lot of interest in heterotic strings on

2000 *Mathematics Subject Classification.* Primary 81T40, 81T45; Secondary 14N35, 81T30.

Key words and phrases. Algebraic geometry.

The author was supported in part by NSF grant DMS-0705381.

non-Kähler manifolds (see *e.g.* [**1**–**8**]), on MSSM derivations from string theory [**9, 10**], as well as many other matters related to heterotic strings, as reflected in Volker Braun's, Savdeep Sethi's, and Li-Sheng Tseng's talks at this workshop.

In this talk, we shall outline three other recent developments in heterotic strings.

First, we shall discuss recent progress in understanding nonperturbative corrections in heterotic strings, the heterotic analogue of curve corrections and quantum cohomology. These will be computed by a heterotic analogue of the two-dimensional topological field theory known as the A model. We will briefly review the A model at the same time as we present its heterotic generalization. This work was motivated by efforts to understand the heterotic generalization of mirror symmetry.

Second, we shall discuss a potential heterotic swampland, referring to the fact that many E_8 bundles with connection cannot be described by the standard heterotic worldsheet construction. We will outline new heterotic worldsheet CFT constructions which will make it possible to describe all the E_8 bundles with connection. Although there will not be a heterotic swampland of this form, this does serve as a warning on the dangers of performing statistical computations within fixed worldsheet constructions.

Finally, we shall discuss the recent understanding of string compactifications on stacks. We shall also outline descriptions of some of those compactifications with gauged linear sigma models. Understanding string compactifications on stacks not only makes predictions for *e.g.* certain Gromov-Witten invariants, but it also yields insight into ordinary-seeming gauged linear sigma models. We shall briefly outline the analysis of the GLSM for the complete intersection Calabi-Yau $\mathbf{P}^7[2,2,2,2]$, which has at its Landau-Ginzburg point another Calabi-Yau given by a branched double cover of $\mathbf{P}^3$. The branched double cover appears physically in a novel way, and furthermore is not birational to the original complete intersection $\mathbf{P}^7[2,2,2,2]$, contradicting some of the lore on gauged linear sigma models. Finally, we shall outline how heterotic string compactifications on special stacks known as gerbes appear to provide new examples of CFT's.

2. Nonperturbative corrections in heterotic strings

In this section, we shall outline the results of [**11, 12**] on nonperturbative corrections to heterotic string compactifications. See [**13**] for another review, and [**14**–**19**] for more recent results.

Roughly, there are two sources of nonperturbative corrections in heterotic strings:

- gauge instantons and five-branes, and
- worldsheet instantons – from strings wrapping minimal area 2-cycles ("holomorphic curves") in spacetime.

In this talk, we shall focus on the latter class, in perturbative worldsheet theories.

Worldsheet instantons generate superpotential terms in the target-space effective field theory. For example, for a heterotic theory with a rank three bundle, breaking an E_8 to E_6, there are

- $\overline{\mathbf{27}}^3$ couplings – on the (2,2) locus, *i.e.* when the gauge bundle is the same as the tangent bundle; these are computed by A model correlation functions.
- $\mathbf{27}^3$ couplings – on the (2,2) locus; these are computed by B model correlation functions.
- Singlet couplings, such as potential terms lifting moduli; these are not computed on the (2,2) locus by any topological field theory, and so in principle are harder to compute directly. See [**20**–**22**] for some (vanishing) results concerning such singlet couplings.

In this talk, we will focus on the first two classes of superpotential terms. Off the (2,2) locus, *i.e.* when the gauge bundle is not the same as the tangent bundle, there exist analogues of the A and B models. These are no longer strictly topological field theories, though they become topological field theories on the (2,2) locus. Nevertheless, although they are not quite the same as topological field theories, they have many of the same properties as topological field theories, and in particular some correlation functions still have a mathematical formulation.

These quasi-topological field theories also have some unusual symmetries; in particular, the (0,2) A model on a space X with gauge bundle $\mathcal{E}$ is isomorphic to the (0,2) B model on X with gauge bundle $\mathcal{E}^\vee$. (For example, this means the (2,2) A model on X is the same as the (0,2) B model on X with gauge bundle T^*X instead of TX.)

In addition to computing superpotential terms, these quasi-topological field theories also have applications to understanding the (0,2) generalization of mirror symmetry. Recall that ordinary mirror symmetry exchanges pairs of topologically distinct spaces, dualizing quantum-corrected computations into classical computations, and exchanging cohomology, in the sense that if X and Y are mirror, then $h^{p,q}(X) = h^{d-p,q}(Y)$, where d is the dimension of X.

In principle, (0,2) mirror symmetry exchanges spaces together with bundles. Instead of swapping ordinary cohomology classes, it exchanges sheaf cohomology groups: if $(X_1, \mathcal{E}_1)$ is exchanged with $(X_2, \mathcal{E}_2)$, then $H^1(X_1, \mathcal{E}_1)$ is exchanged with $H^1(X_2, \mathcal{E}_2^\vee)$. In the special case that $\mathcal{E}_i = TX_i$, (0,2) mirror symmetry reduces to ordinary mirror symmetry.

At the present time, (0,2) mirror symmetry is not well understood. One example of evidence for (0,2) mirror symmetry is numerical computations [**23**], in which sheaf cohomology classes of a large number of examples were computed. When one graphs the set of dimensions of sheaf cohomology groups, the graph is symmetric. Other work on the subject can be found in [**23**–**27**].

In particular, [**27**] proposed that there should exist a (0,2) analogue of 'quantum cohomology' computations, encoding worldsheet instanton corrections, as would be computed ordinarily in the A model topological field theory. The papers [**11**–**13**] and others since (*e.g.* [**14**–**17, 19**]) have begun building the details of those proposed (0,2) quantum cohomology computations, and that is what we shall discuss in this section.

Since ordinary quantum cohomology rings are operator product rings in the A model topological field theory, we shall discuss the (0,2) analogue of the A model. First, let us first review the ordinary A model. This is a two-dimensional quantum field theory with Lagrangian

$$g_{i\bar{\jmath}}\overline{\partial}\phi^i\partial\phi^{\bar{\jmath}} + ig_{i\bar{\jmath}}\psi_-^{\bar{\jmath}}D_z\psi_-^i + ig_{i\bar{\jmath}}\psi_+^{\bar{\jmath}}D_{\bar{z}}\psi_+^i + R_{i\bar{\jmath}k\bar{l}}\psi_+^i\psi_+^{\bar{\jmath}}\psi_-^k\psi_-^{\bar{l}}$$

where the ϕ are maps from the worldsheet Σ into a target space X, and the Grassmann field ψ couples to bundles as follows:

$$\begin{array}{ll} \psi_-^i(\equiv\chi^i)\in\Gamma((\phi^*T^{0,1}X)^\vee) & \psi_+^i(\equiv\psi_z^i)\in\Gamma(K\otimes\phi^*T^{1,0}X) \\ \psi_-^{\bar{\imath}}(\equiv\psi_{\bar{z}}^{\bar{\imath}})\ \in\ \Gamma(\overline{K}\otimes\phi^*T^{0,1}X) & \psi_+^{\bar{\imath}}(\equiv\chi^{\bar{\imath}})\in\Gamma((\phi^*T^{1,0}X)^\vee). \end{array}$$

These fields are not worldsheet fermions, but rather are worldsheet vectors and scalars – a result of the topological twisting. Half of the original supersymmetries form the scalar supercharge or BRST operator, and under the action of that supercharge,

$$\begin{array}{ll} \delta\phi^i\propto\chi^i, & \delta\phi^{\bar{\imath}}\propto\chi^{\bar{\imath}} \\ \delta\chi^i=0, & \delta\chi^{\bar{\imath}}=0 \\ \delta\psi_z^i\neq 0, & \delta\psi_{\bar{z}}^{\bar{\imath}}\neq 0. \end{array}$$

As a result, the BRST-invariant worldsheet scalar states are built from products of χ's, and there is a well-known isomorphism to cohomology given by

$$\begin{array}{rcl} \mathcal{O}\sim b_{i_1\cdots i_p\bar{\imath}_1\cdots\bar{\imath}_q}\chi^{\bar{\imath}_1}\cdots\chi^{\bar{\imath}_q}\chi^{i_1}\cdots\chi^{i_p} & \leftrightarrow & H^{p,q}(X), \\ Q & \leftrightarrow & d. \end{array}$$

Next, let us examine comparable facts about the (0,2) analogue of the A model. This is based on the heterotic Lagrangian

$$g_{i\bar{\jmath}}\overline{\partial}\phi^i\partial\phi^{\bar{\jmath}} + ih_{a\bar{b}}\lambda_-^{\bar{b}}D_z\lambda_-^a + ig_{i\bar{\jmath}}\psi_+^{\bar{\jmath}}D_{\bar{z}}\psi_+^i + F_{i\bar{\jmath}a\bar{b}}\psi_+^i\psi_+^{\bar{\jmath}}\lambda_-^a\lambda_-^{\bar{b}}$$

describing a heterotic string propagating on a space X with left-movers coupling to a holomorphic vector bundle $\mathcal{E}$, in which the Grassmann fields ψ, λ couple to bundles as follows:

$$\begin{array}{ll} \lambda_-^a\in\Gamma(\phi^*\overline{\mathcal{E}}) & \psi_+^i\in\Gamma(K\otimes\phi^*T^{1,0}X) \\ \lambda_-^{\bar{b}}\in\Gamma(\overline{K}\otimes\phi^*\overline{\mathcal{E}}) & \psi_+^{\bar{\imath}}\in\Gamma((\phi^*T^{1,0}X)^\vee). \end{array}$$

As before, the ψ and λ are no longer worldsheet fermions, but rather are scalars and vectors. Because of the asymmetry between left- and right-moving fields, this theory must satisfy the conditions

$$\Lambda^{top}\mathcal{E}^\vee \cong K_X, \quad \mathrm{ch}_2(\mathcal{E}) = \mathrm{ch}_2(TX).$$

The second of these is the well-known anomaly cancellation condition of perturbative heterotic strings, the first is another condition present only in the twisted theory – a close analogue of the anomaly in the closed string B model that makes it well-defined only for complex Kähler manifolds obeying $K_X^{\otimes 2}$ trivial [**12**].

Here, the BRST-invariant worldsheet scalar states that one considers are products of ψ's, λ's, and are in one-to-one correspondence with elements of sheaf cohomology groups

$$\mathcal{O} \sim b_{\bar{\imath}_1 \cdots \bar{\imath}_n a_1 \cdots a_p} \psi_+^{\bar{\imath}_1} \cdots \psi_+^{\bar{\imath}_n} \lambda_-^{a_1} \cdots \lambda_-^{a_p} \;\leftrightarrow\; H^n(X, \Lambda^p \mathcal{E}^\vee).$$

In the special case that $\mathcal{E} = TX$, this pseudo-topological field theory becomes the A model, a true topological field theory. (For example,

$$H^q(X, \Lambda^p(TX)^\vee) = H^{p,q}(X)$$

so the state counting specializes in the desired fashion.)

Next, let us turn to correlation function computations in these theories.

In the A model, the classical contributions to correlation functions are computed as follows. For X compact and of dimension n, there are n χ^i, $\chi^{\bar{\imath}}$ zero modes, plus bosonic zero modes whose moduli space is X itself, so the correlation function reduces to

$$\langle \mathcal{O}_1 \cdots \mathcal{O}_m \rangle = \int_X H^{p_1,q_1}(X) \wedge \cdots \wedge H^{p_m,q_m}(X).$$

(In our notation, we schematically indicate representatives of cohomology classes by writing the cohomology groups themselves.) In addition, there is a selection rule from the left and right $U(1)_R$ symmetries which says the classical contributions to correlation functions are only nonzero when

$$\sum_i p_i = \sum_i q_i = n.$$

Putting this together, we see that the classical contribution has the form

$$\langle \mathcal{O}_1 \cdots \mathcal{O}_m \rangle \sim \int_X (\text{top-form}).$$

Next, let us consider the analogous computation in the (0,2) analogue of the A model. For X compact of dimension n, and $\mathcal{E}$ of rank r, there are

n $\psi_+^{\bar{\imath}}$ zero modes and r λ_-^a zero modes, so the classical contribution to a correlation function is of the form

$$\langle \mathcal{O}_1 \cdots \mathcal{O}_m \rangle = \int_X H^{q_1}(X, \Lambda^{p_1} \mathcal{E}^\vee) \wedge \cdots \wedge H^{q_m}(X, \Lambda^{p_m} \mathcal{E}^\vee).$$

As before, there is a selection rule, which now enforces

$$\sum_i q_i = n, \quad \sum_i p_i = r$$

for classical contributions. Therefore, the classical contributions have the form

$$\langle \mathcal{O}_1 \cdots \mathcal{O}_m \rangle \sim \int_X H^{top}(X, \Lambda^{top} \mathcal{E}^\vee).$$

The constraint $\Lambda^{top}\mathcal{E}^\vee \cong K_X$ makes the integrand a top-form.

Next, let us turn to a non-classical contribution to a correlation function. Again let us first consider the original A model before studying the (0,2) analogue. In the standard A model, the moduli space of bosonic zero modes in some non-classical sector will be a moduli space $\mathcal{M}$ of worldsheet instantons. If there are no $\psi_{\bar{z}}^i$ or $\psi_z^{\bar{\imath}}$ zero modes, then the contribution to a correlation function is of the form

$$\langle \mathcal{O}_1 \cdots \mathcal{O}_m \rangle \sim \int_{\mathcal{M}} H^{p_1,q_1}(\mathcal{M}) \wedge \cdots \wedge H^{p_m,q_m}(\mathcal{M}).$$

More generally, taking into account the possibility of $\psi_{\bar{z}}^i$ or $\psi_z^{\bar{\imath}}$ zero modes, the contribution to a correlation function will be of the form

$$\langle \mathcal{O}_1 \cdots \mathcal{O}_m \rangle \sim \int_{\mathcal{M}} H^{p_1,q_1}(\mathcal{M}) \wedge \cdots \wedge H^{p_m,q_m}(\mathcal{M}) \wedge c_{top}(\mathrm{Obs}).$$

In all cases, the contribution has the form

$$\langle \mathcal{O}_1 \cdots \mathcal{O}_m \rangle \sim \int_{\mathcal{M}} (\text{top form}).$$

Next, let us consider the analogous computations to the (0,2) analogue of the A model. Here, the bundle $\mathcal{E}$ on X induces a holomorphic vector bundle[1] $\mathcal{F}$ of λ zero modes on the moduli space $\mathcal{M}$. Mathematically,

$$\mathcal{F} \equiv R^0\pi_*\alpha^*\mathcal{E}$$

where

$$\pi : \Sigma \times \mathcal{M} \to \mathcal{M}, \quad \alpha : \Sigma \times \mathcal{M} \to X.$$

[1] In general, this will only be a sheaf. However, for the bundle constructions under consideration, over toric varieties, the induced sheaf will always be locally free.

On the (2,2) locus, where $\mathcal{E} = TX$, one has $\mathcal{F} = T\mathcal{M}$. When there are no 'excess' (worldsheet vector) zero modes, the contribution to the correlation function is of the form

$$\langle \mathcal{O}_1 \cdots \mathcal{O}_m \rangle \sim \int_{\mathcal{M}} H^{top}(\mathcal{M}, \Lambda^{top}\mathcal{F}^\vee).$$

When we apply the physical consistency conditions and the Grothendieck-Riemann-Roch theorem, we find

$$\left.\begin{array}{r} \Lambda^{top}\mathcal{E}^\vee \cong K_X \\ \mathrm{ch}_2(\mathcal{E}) = \mathrm{ch}_2(TX) \end{array}\right\} \stackrel{GRR}{\Longrightarrow} \Lambda^{top}\mathcal{F}^\vee \cong K_{\mathcal{M}}$$

so again the integrand is a top-form. In the general case, one has

$$\begin{aligned} \langle \mathcal{O}_1 \cdots \mathcal{O}_m \rangle \sim \int_{\mathcal{M}} & H^{\sum q_i}\left(\mathcal{M}, \Lambda^{\sum p_i}\mathcal{F}^\vee\right) \\ & \wedge H^n\left(\mathcal{M}, \Lambda^n\mathcal{F}^\vee \otimes \Lambda^n\mathcal{F}_1 \otimes \Lambda^n(\mathrm{Obs})^\vee\right) \end{aligned}$$

where

$$\begin{array}{ll} \psi_+^{\bar{\imath}} \sim T\mathcal{M} = R^0\pi_*\alpha^*TX & \lambda_-^a \sim \mathcal{F} = R^0\pi_*\alpha^*\mathcal{E} \\ \psi_+^i \sim \mathrm{Obs} = R^1\pi_*\alpha^*TX & \lambda_-^{\bar{b}} \sim \mathcal{F}_1 \equiv R^1\pi_*\alpha^*\mathcal{E}. \end{array}$$

Applying anomaly constraints as before, we find

$$\left.\begin{array}{r} \Lambda^{top}\mathcal{E}^\vee \cong K_X \\ \mathrm{ch}_2(\mathcal{E}) = \mathrm{ch}_2(TX) \end{array}\right\} \stackrel{GRR}{\Longrightarrow} \Lambda^{top}\mathcal{F}^\vee \otimes \Lambda^{top}\mathcal{F}_1 \otimes \Lambda^{top}(\mathrm{Obs})^\vee \cong K_{\mathcal{M}}$$

so, again, the integrand is a top-form. On the (2,2) locus, this reduces to the previous result via Atiyah classes [**28**].

To do any computations with these general formulas, we need explicit expressions for the space $\mathcal{M}$ and bundle $\mathcal{F}$. Luckily, the gauged linear sigma model naturally provides such expressions. Gauged linear sigma models are two-dimensional gauge theories, generalizations of the $\mathbf{CP}^n$ model, which are important in string compactifications. They renormalization-group flow to nonlinear sigma models and other conformal field theories, and technical questions about the CFT's will (often) become easier questions about the gauged linear sigma model. For example, the worldsheet instantons of the IR nonlinear sigma model are the two-dimensional gauge instantons of the gauge theory.

To be specific, let us consider the example of $\mathbf{CP}^{N-1}$. Physically, this is described by N chiral superfields $x_1, \cdots, x_N$, each of charge 1. For degree d maps, and a genus zero worldsheet, we expand in a basis of zero modes:

$$x_i = x_{i0}u^d + x_{i1}u^{d-1}v + \cdots + x_{id}v^d$$

where u, v are homogeneous coordinates on the worldsheet $\mathbf{P}^1$. Taking the (x_{ij}) to be homogeneous coordinates on $\mathcal{M}$ [**29**], we find $\mathcal{M} = \mathbf{P}^{N(d+1)-1}$.

We can do something very similar to build $\mathcal{F}$. For example, suppose $\mathcal{E}$ is a completely reducible bundle:

$$\mathcal{E} = \oplus_a \mathcal{O}(\vec{n}_a).$$

Expanding the left-moving fermions in a basis of zero modes, on a genus zero worldsheet, we find:

$$\lambda_-^a = \lambda_-^{a0} u^{\vec{n}_a \cdot \vec{d}+1} + \lambda_-^{a1} u^{\vec{n}_a \cdot d} v + \cdots .$$

We then identify each λ_-^{ai} with $\mathcal{O}(\vec{n}_a)$ on $\mathcal{M}$. Thus, in this case,

$$\mathcal{F} = \oplus_a H^0 \left(\mathbf{P}^1, \mathcal{O}\left(\vec{n}_a \cdot \vec{d} \right) \right) \otimes_{\mathbf{C}} \mathcal{O}(\vec{n}_a).$$

There are analogous expressions for more general constructions appearing in (0,2) gauged linear sigma models; see [**11**].

Next, let us quickly review ordinary quantum cohomology, before describing the (0,2) analogue. Quantum cohomology encodes the OPE ring of the A model. Such ideas appear elsewhere in physics; for example there is a close analogy with results on four-dimensional gauge theories of Cachazo-Douglas-Seiberg-Witten [**30**]:

4d $\mathcal{N}=1$ $SU(N)$ SYM	2d SUSY $\mathbf{CP}^{N-1}$
$S^N = \Lambda^{3N}$	$x^N = q$
$W = S\left(1 + \log\left(\Lambda^{3N}/S^N\right)\right)$	$W = \Sigma\left(1 + \log\left(\Lambda^N/\Sigma^N\right)\right)$

In the two-dimensional case, the OPE ring looks like a modification of the classical cohomology ring relation, hence the term "quantum cohomology."

In the case of the $\mathbf{CP}^n$ model, the quantum cohomology ring corresponds to correlation functions of the form

$$\langle x^k \rangle = \begin{cases} q^m & \text{if } k = mN + N - 1 \\ 0 & \text{else.} \end{cases}$$

Now, ordinarily for a well-behaved OPE ring, it is assumed that one needs (2,2) supersymmetry, but it has been found that that condition can be relaxed to (0,2) supersymmetry. This was first conjectured by [**27**], and then [**11**] found strong evidence for this conjecture in our computations of correlation functions. Later [**14**] found a CFT argument explaining why a well-behaved OPE ring should exist with only (0,2) supersymmetry.

In particular, the paper [**27**] studied a (0,2) theory describing $\mathbf{P}^1 \times \mathbf{P}^1$ with gauge bundle given by a deformation of the tangent bundle. Using a duality argument, they argued that the quantum cohomology ring of this theory should be given by

$$\begin{aligned} \tilde{X}^2 &= \exp(it_2) \\ X^2 - (\epsilon_1 - \epsilon_2) X \tilde{X} &= \exp(it_1) \end{aligned}$$

(where $\epsilon_1 - \epsilon_2$ parametrizes the distance from the tangent bundle), which is a deformation of the quantum cohomology ring of $\mathbf{P}^1 \times \mathbf{P}^1$.

In [**11**] we directly computed correlation functions in this theory, using the technology outlined so far, and we found that

$$\begin{aligned}
\langle \tilde{X}^4 \rangle &= \langle 1 \rangle \exp(2it_2) = 0, \\
\langle X\tilde{X}^3 \rangle &= \langle (X\tilde{X})\tilde{X}^2 \rangle, \\
&= \langle X\tilde{X} \rangle \exp(it_2) = \exp(it_2), \\
\langle X^2\tilde{X}^2 \rangle &= \langle X^2 \rangle \exp(it_2) = (\epsilon_1 - \epsilon_2)\exp(it_2), \\
\langle X^3\tilde{X} \rangle &= \exp(it_1) + (\epsilon_1 - \epsilon_2)^2 \exp(it_2), \\
\langle X^4 \rangle &= 2(\epsilon_1 - \epsilon_2)\exp(it_1) + (\epsilon_1 - \epsilon_2)^3 \exp(it_2),
\end{aligned}$$

verifying the prediction of [**27**].

More recently, [**15**] extended the computation of [**11**] to include all the bundle moduli. A comparison of their results to the work of [**27**] showed that to obtain agreement between the Landau-Ginzburg description and the gauged linear sigma model would require finding a set of potentially complicated field redefinitions [**15, 16**]. The work of [**17**] pointed out that these correlators may be efficiently computed on the Coulomb branch of the GLSM. Besides proving an efficient route to the quantum cohomology ring and explicit correlators, these techniques should be helpful in elucidating the details of the field redefinitions relating the GLSM to the dual theory of [**27**]. More recently still, in [**19**] A-twists of heterotic Landau-Ginzburg models in the same universality classes as the heterotic nonlinear sigma models studied in other work were studied. These provide alternative approaches to the same computations, which often make mathematical tricks manifest.

So far we have discussed the (0,2) analogue of the A model. There is also a (0,2) analogue of the B model. The ordinary (2,2) B model gets no nonperturbative corrections at all, whereas the (0,2) B model does get nonperturbative corrections off the (2,2) locus. These two twists are related in a simple way: the (0,2) A twist of the theory with bundle $\mathcal{E}$ on the space X is the same, for trivial field-redefinition reasons, as the (0,2) B twist of the theory with bundle $\mathcal{E}^\vee$ on the space X. This is discussed in more detail in [**12, 14**].

3. A potential heterotic swampland and new heterotic CFT constructions

This section is a summary of the paper [**31**], concerning a potential heterotic swampland and new heterotic worldsheet CFT constructions.

There has recently been a lot of interest in the landscape program, which for those readers not already aware is an attempt to extract phenomenological predictions by doing statistics on the set of string vacua.

One of the technical challenges in the landscape program is that those string vacua are counted by low-energy effective field theories, and it is not clear that all of those have consistent UV completions – not all of them may come from an underlying quantum gravity [**32, 33**].

One potential such problem arises in heterotic $E_8 \times E_8$ strings. The conventional worldsheet construction builds each E_8 using a ($\mathbf{Z}_2$ orbifold of a) set of fermions λ_-.

$$L = g_{\mu\nu}\partial\phi^\mu\partial\phi^\nu + i g_{i\bar{\jmath}}\overline{\psi}_+^{\bar{\jmath}} D_- \psi_+^i + h_{\alpha\beta}\lambda_-^\alpha D_+ \lambda_-^\beta + \cdots .$$

The fermions realize a Spin(16) current algebra at level 1, and the $\mathbf{Z}_2$ orbifold makes the current algebra Spin(16)/$\mathbf{Z}_2$. The group Spin(16)/$\mathbf{Z}_2$ is a subgroup of E_8, and we use it to realize the E_8.

In more detail, the adjoint representation **248** of E_8 decomposes into the adjoint representation **120** of Spin(16)/$\mathbf{Z}_2$ plus a spinor representation **128**:

$$\mathbf{248} = \mathbf{120} + \mathbf{128}.$$

The **120** arises from the left NS sector and the **128** from the left R sector of the $\mathbf{Z}_2$ orbifold. We take currents transforming in the adjoint and spinor representations of Spin(16)/$\mathbf{Z}_2$, and form E_8 via commutation relations. More, in fact: all E_8 degrees of freedom, the entire level 1 Kac-Moody algebra, are realized in this fashion by Spin(16)/$\mathbf{Z}_2$.

This construction has served us well for many years, but, in order to describe an E_8 bundle with connection, that bundle and connection must be reducible to Spin(16)/$\mathbf{Z}_2$. After all, all information is buried in the kinetic term $h_{\alpha\beta}\lambda_-^\alpha D_+ \lambda_-^\beta$, and since the λ_- transform as a vector representation of Spin(16), the connection is necessarily only an adjoint of Spin(16).

So, can an E_8 bundle with connection always be reduced to Spin(16)/$\mathbf{Z}_2$? Very briefly:

- Bundles: in dimension 9 or less, an E_8 bundle can be reduced to a Spin(16)/$\mathbf{Z}_2$ bundle.
- Connections: just because the bundle can be reduced does not mean the connection on the bundle can be reduced. We shall find that connections are not always so reducible – in fact, this is the generic case.

So, is there a heterotic swampland, populated by theories with E_8 bundles with connection which cannot be described in the standard construction?

Let us review the technical issue with connections[2]; readers interested in learning about reducibility of bundles are referred to [**31**].

On a principal G bundle, even a trivial principal G bundle, one can find connections with holonomy that fill out all of G, and so cannot be understood as connections on a principal H bundle for H a subgroup of G: just take a connection whose curvature generates the Lie algebra of G.

[2]The work on reducibility of connections was done in collaboration with R. Thomas.

Now, in heterotic strings, we cannot work with arbitrary connections, but rather we want gauge fields satisfying both the anomaly-cancellation condition $\mathrm{ch}_2(\mathcal{E}) = \mathrm{ch}_2(TX)$ as well as the Donaldson-Uhlenbeck-Yau condition $g^{i\bar{\jmath}}F_{i\bar{\jmath}} = 0$. We shall see that even after imposing these constraints, there are still examples of E_8 bundles with connection that cannot be reduced to $\mathrm{Spin}(16)/\mathbf{Z}_2$.

To build such an example, we shall use the fact that E_8 has an

$$(SU(5) \times SU(5))\,/\mathbf{Z}_5$$

subgroup that does not sit inside $\mathrm{Spin}(16)/\mathbf{Z}_2$ (see figure 1). In particular, $SU(n)$ connections are easy to build (using holomorphic vector bundles), so we shall construct an $(SU(5) \times SU(5))/\mathbf{Z}_5$ connection that is anomaly-free and satisfies Donaldson-Uhlenbeck-Yau.

For simplicity, we shall work on an elliptically-fibered $K3$ surface, and we shall build a stable $SU(5)$ bundle using Friedman-Morgan-Witten [**34**] technology. A rank 5 bundle with $c_1 = 0$, $c_2 = 12$ has a spectral cover in the lienar system $|5\sigma + 12f|$ (σ the class of the base, f the class of the fiber in the elliptically-fibered surface), describing a curve of genus $g = 5c_2 - 5^2 + 1 = 36$, together with a line bundle of degree $-(5 + g - 1) = -40$. The result is a (72-parameter) family of stable $SU(5)$ bundles with $c_2 = 12$ on $K3$, whose

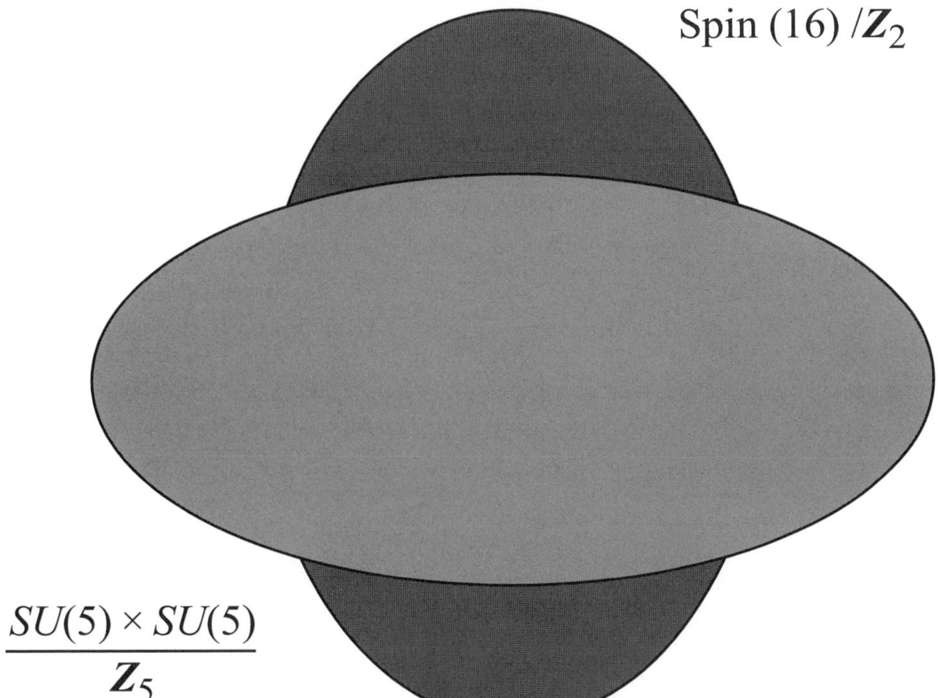

FIGURE 1. The $\mathrm{Spin}(16)/\mathbf{Z}_2$ and $(SU(5) \times SU(5))/\mathbf{Z}_5$ subgroups of E_8.

holonomy generically fills out all of $SU(5)$. If we put two together, and project to the $\mathbf{Z}_5$ quotient, the result is an $(SU(5) \times SU(5))/\mathbf{Z}_5$ bundle with connection that satisfies both anomaly cancellation and Donaldson-Uhlenbeck-Yau.

Moreover, this example is not unique – this describes a 144-dimensional family of examples. The lesson here is that such examples are not exotic special cases, but rather are very common, even generic.

Now that we have set up a problem, we shall see that in fact there is no swampland, by describing alternative worldsheet constructions that can describe more general E_8 bundles with connection. It should be emphasized, however, that even though these more general E_8 gauge fields can be described by some conformal field theory, there is a danger here for anyone doing statistics on standard worldsheet constructions – they do not capture all string vacua, only a nonrepresentative subset, as indicated schematically in figure 2.

We shall begin by describing more general ten-dimensional flat-space constructions. The basic idea is to replace $\text{Spin}(16)/\mathbf{Z}_2$ with other subgroups of E_8, realized as orbifolds of abstract current algebras [since only $U(n)$ and $\text{Spin}(n)$ have level 1 free field representations].

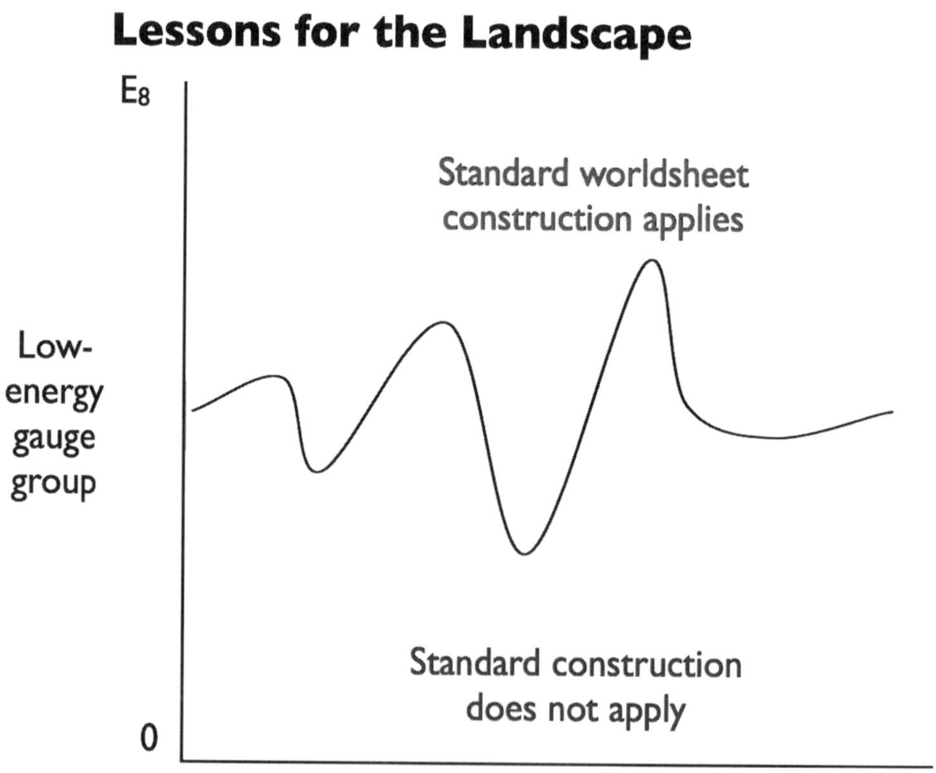

FIGURE 2. Low-energy gauge groups that can be described by the standard heterotic worldsheet construction.

To be specific, we shall describe $(SU(5) \times SU(5))/\mathbf{Z}_5$.

As a first check, let us consider the central charge. The central charge of a current algebra for an ADE group at level 1 is the same as the rank, so the central charge of each $SU(5)$ is 4, hence the central charge of an $SU(5) \times SU(5)$ algebra at level 1 is 8, which will be unchanged by the $\mathbf{Z}_5$ orbifold. This is exactly right to match the central charge of E_8.

A more convincing test is to study the characters used to build a (hopefully modular-invariant) partition function. In the case of Spin$(16)/\mathbf{Z}_2$, corresponding to the decomposition

$$\mathbf{248} = \mathbf{120} + \mathbf{128}$$

of the adjoint representation of E_8, there is a decomposition of characters

$$\chi_{E_8}(\mathbf{1}, q) = \chi_{Spin(16)}(\mathbf{1}, q) + \chi_{Spin(16)}(\mathbf{128}, q)$$

which build the left-moving part of the heterotic partition function. (The adjoint representations are descendants of the identity operator in level 1 current algebras, and so are buried in the characters for $\mathbf{1}$.)

In the case of $SU(5)^2/\mathbf{Z}_5$, from the decomposition

$$\mathbf{248} = (\mathbf{1}, \mathbf{24}) + (\mathbf{24}, \mathbf{1}) + (\mathbf{5}, \overline{\mathbf{10}}) + (\overline{\mathbf{5}}, \mathbf{10}) + (\mathbf{10}, \mathbf{5}) + (\overline{\mathbf{10}}, \overline{\mathbf{5}})$$

of the adjoint representation of E_8, we get a prediction for the characters:

$$\chi_{E_8}(\mathbf{1}, q) = \chi_{SU(5)}(\mathbf{1}, q)^2 + 4\chi_{SU(5)}(\mathbf{5}, q)\chi_{SU(5)}(\mathbf{10}, q).$$

The $SU(5)$ characters can be shown to be

$$\begin{aligned}
\chi_{SU(5)}(\mathbf{1}, q) &= \frac{1}{\eta(\tau)^4} \sum_{\vec{m} \in \mathbf{Z}^4} q^{\left(\sum m_i^2 + (\sum m_i)^2\right)/2} \\
\chi_{SU(5)}(\mathbf{5}, q) &= \frac{1}{\eta(\tau)^4} \sum_{\vec{m} \in \mathbf{Z}^4, \sum m_i \equiv 1 \bmod 5} q^{\left(\sum m_i^2 - \frac{1}{5}(\sum m_i)^2\right)/2} \\
\chi_{SU(5)}(\mathbf{10}, q) &= \frac{1}{\eta(\tau)^4} \sum_{\vec{m} \in \mathbf{Z}^4, \sum m_i \equiv 2 \bmod 5} q^{\left(\sum m_i^2 - \frac{1}{5}(\sum m_i)^2\right)/2}.
\end{aligned}$$

It can be shown [**35, 36**] that

$$\chi_{E_8}(\mathbf{1}, q) = \chi_{SU(5)}(\mathbf{1}, q)^2 + 4\chi_{SU(5)}(\mathbf{5}, q)\chi_{SU(5)}(\mathbf{10}, q),$$

which is exactly the desired character identity.

Let us go back and re-examine these statements more carefully. In the character decomposition for Spin$(16)/\mathbf{Z}_2$, namely

$$\chi_{E_8}(\mathbf{1}, q) = \chi_{Spin(16)}(\mathbf{1}, q) + \chi_{Spin(16)}(\mathbf{128}, q),$$

there is a $\mathbf{Z}_2$ orbifold implicit – the $\mathbf{1}$ is from the untwisted sector, the $\mathbf{128}$ from the twisted sector. Similarly, in the expression

$$\chi_{E_8}(\mathbf{1}, q) = \chi_{SU(5)}(\mathbf{1}, q)^2 + 4\chi_{SU(5)}(\mathbf{5}, q)\chi_{SU(5)}(\mathbf{10}, q),$$

there is a $\mathbf{Z}_5$ orbifold implicit – the $\mathbf{1}$ is from the untwisted sector, the other four pieces are from four twisted sectors. This is exactly what we should find – the correct subgroup of E_8 is $SU(5)^2/\mathbf{Z}_5$, not $SU(5)^2$; there is a $\mathbf{Z}_5$ that should replace the left-moving $\mathbf{Z}_2$ GSO-analogue of the ordinary heterotic string construction.

It is straightforward to repeat this analysis for other subgroups of E_8. For example, let us next consider $SU(9)/\mathbf{Z}_3$. Again, the central charge of the current algebra at level 1 is 8, which matches that of the E_8 current algebra. From the decomposition of the E_8 conformal family

$$[\mathbf{1}] = [\mathbf{1}] + [\mathbf{84}] + [\overline{\mathbf{84}}]$$

one predicts a character identity

$$\chi_{E_8}(\mathbf{1}, q) = \chi_{SU(9)}(\mathbf{1}, q) + 2\chi_{SU(9)}(\mathbf{84}, q)$$

and it can be shown that this is correct. Furthermore, note that in the expression above, there is a $\mathbf{Z}_3$ orbifold implicit, exactly as desired since the correct subgroup is $SU(9)/\mathbf{Z}_3$. So, we can also describe the E_8 degrees of freedom with $SU(9)/\mathbf{Z}_3$, with a $\mathbf{Z}_3$ orbifold replacing the left-moving $\mathbf{Z}_2$ GSO-analogue of the ordinary heterotic string construction.

For many other maximal-rank subgroups, there is an analogous story. For at least one non-maximal-rank, non-simply-laced subgroup, however, matters are more complicated. E_8 has a $G_2 \times F_4$ subgroup, and one can check that the central charge of the $G_2 \times F_4$ current algebra at level 1 is 8, exactly right to match that of E_8, and furthermore there is a character identity relating $\chi_{E_8}(\mathbf{1}, q)$ to a sum over $G_2 \times F_4$ characters, as needed. However, that character sum includes more terms that just characters of identity representations, terms which ordinarily would arise in twisted sectors of orbifolds, but there is no group-theoretic orbifold to perform: $G_2 \times F_4$ has no finite center or indeed a finite normal subgroup that one could orbifold, and correspondingly the subgroup of E_8 is $G_2 \times F_4$, not any quotient thereof. Therefore, although there exists formally a character decomposition, we do not understand how to realize it physically.

Next, in order to make this useful, we must describe how one fibers more general current algebras over spaces, in order to describe a compactified theory. To do this, we will outline fibered WZW models, first introduced by [**37–41**] under the name "lefton, righton Thirring models," which we specialize to (0,2) theories.

Before describing fibered WZW models, let us first recall ordinary WZW models. An ordinary WZW model looks like a sigma model on the group manifold of a group G together with H flux:

$$S = -\frac{k}{2\pi}\int_{\Sigma} \mathrm{Tr}\left[g^{-1}\partial g g^{-1}\overline{\partial} g\right] - \frac{ik}{2\pi}\int_B d^3y \epsilon^{ijk} \mathrm{Tr}\left[g^{-1}\partial_i g g^{-1}\partial_j g g^{-1}\partial_k g\right].$$

The first term is the usual sigma model, the second or 'Wess-Zumino' term describes H flux. This theory has a $G_L \times G_R$ symmetry, with currents

$$J(z) = g^{-1}\partial g, \quad \overline{J}(\overline{z}) = \overline{\partial} g g^{-1}$$

which (thanks to the Wess-Zumino term) obey

$$\overline{\partial} J(z) = \partial \overline{J}(\overline{z}) = 0.$$

These currents each realize a chiral G-Kac-Moody or current algebra at level k.

Next, let us describe how to fiber these WZW models. Let P be a principal G bundle over X, with connection A. Given an ordinary heterotic string worldsheet theory, the idea is that we shall replace the left-moving fermions λ_- with a WZW model with left-multiplication gauged with the pullback of A. Schematically, the action has the form

$$\begin{aligned}
&\frac{1}{\alpha'}\int_{\Sigma}\left(g_{i\overline{\jmath}}\partial_\alpha\phi^i\partial^\alpha\phi^{\overline{\jmath}} + \cdots\right) \\
&\qquad - \frac{k}{4\pi}\int_{\Sigma}\mathrm{Tr}\left(g^{-1}\partial g g^{-1}\overline{\partial} g\right) - \frac{ik}{12\pi}\int_B d^3y\epsilon^{ijk}\mathrm{Tr}\left(g^{-1}\partial_i g g^{-1}\partial_j g g^{-1}\partial_k g\right) \\
&\qquad - \frac{k}{2\pi}\int_{\Sigma}\mathrm{Tr}\left((\partial\phi^\mu)A_\mu\overline{\partial} g g^{-1} + \frac{1}{2}(\partial\phi^\mu\overline{\partial}\phi^\nu)A_\mu A_\nu\right).
\end{aligned}$$

The first line describes a nonlinear sigma model on X; the second line describes a WZW model; the third line describes the operation of gauging left-multiplication. The third line will look very familiar to readers acquainted with gauged WZW models; the differences here are that we only gauge a chiral multiplication, and that the gauge field is the pullback of a gauge field on the target space, rather than a worldsheet gauge field.

Now, a WZW model action is invariant under gauging symmetric group multiplications, but not under the chiral group multiplications that we have here. Under

$$\begin{aligned}
g &\mapsto hg \\
A_\mu &\mapsto hA_\mu h^{-1} + h\partial_\mu h^{-1}
\end{aligned}$$

the classical action is not invariant.

On the one hand, this lack of invariance is expected – this is the bosonization of the chiral anomaly. On the other hand, this lack of invariance creates a potential well-definedness issue in the fibered WZW construction.

The fix is a quantum correction which cancels the classical non-invariance. In particular, the classical action picks up a quantum correction across coordinate patches, due to the right-moving chiral fermion anomaly.

To make the action gauge-invariant, we proceed in the usual form, by assigning a transformation law to the B field. To be able to do this globally implies that

$$k\mathrm{ch}_2(\mathcal{E}) = \mathrm{ch}_2(TX)$$

which is the analogue of the usual anomaly-cancellation condition for left-movers at level k. If this condition is obeyed, the action is well-defined globally.

Proceeding in the usual fashion for heterotic strings, the right-moving fermion kinetic terms on the worldsheet couple to H flux:

$$\frac{i}{2} g_{\mu\nu} \psi_+^\mu D_{\overline{z}} \psi_+^\nu$$

where

$$D_{\overline{z}} \psi_+^\mu = \overline{\partial} \psi_+^\mu + \overline{\partial} \phi^\mu \left(\Gamma^\nu{}_{\sigma\mu} \; - \; H^\nu{}_{\sigma\mu} \right) \psi_+^\sigma$$

(and we have switched from holomorphic indices i, $\overline{\jmath}$, to real indices μ, ν). To make the fermion kinetic terms gauge-invariant under the transformation of the B field, we must redefine H to be

$$H = dB + (\alpha') \left(kCS(A) - CS(\omega) \right)$$

so that H is a well-defined 3-form globally. This implies that

$$k\mathrm{ch}_2(\mathcal{E}) = \mathrm{ch}_2(TX)$$

which is, again, the anomaly cancellation condition.

Next, let us demand (0,2) supersymmetry. One discovers an old faux-supersymmetry-anomaly in subleading terms in α' in the heterotic string, one originally discussed by [**42, 43**]. The supersymmetry transformations in the ordinary heterotic string worldsheet are:

$$\delta \lambda_- = -i\epsilon \psi_+^\mu A_\mu \lambda_- .$$

Note that this is the same as a chiral gauge transformation with parameter $-i\epsilon\psi_+^\mu A_\mu$. Because of the chiral anomaly in the ordinary heterotic worldsheet, this means there is a quantum contribution to the supersymmetry transformations at order α'. In our bosonized description, this quantum effect appears at leading order.

The one-fermi terms in the supersymmetry transformations of the various components of the action are as follows. The nonlinear sigma model on the base contributes

$$\frac{1}{\alpha'} \int_\Sigma (i\alpha \psi^{\overline{\imath}}) \overline{\partial} \phi^\mu \partial \phi^\nu \left(H - dB \right)_{\overline{\imath}\mu\nu} .$$

The WZW fibers contribute

$$-k \int_{\Sigma} (i\alpha\psi^{\bar{\imath}})\overline{\partial}\phi^{\mu}\partial\phi^{\nu} CS(A)_{\bar{\imath}\mu\nu}.$$

This is the bosonization of the faux-supersymmetry-anomaly discussed above. Finally, there is an additional quantum contribution, a faux-supersymmetry-anomaly arising from the right-moving fermions, which contributes

$$\int_{\Sigma} (i\alpha\psi^{\bar{\imath}})\overline{\partial}\phi^{\mu}\partial\phi^{\nu} CS(\omega)_{\bar{\imath}\mu\nu}.$$

In order for the one-fermi terms to cancel out, we see that

$$H = dB + \alpha' \left(kCS(A) - CS(\omega) \right)$$

which is, again, the anomaly cancellation condition, the third time now that we have derived it. The three-fermi terms behave similarly.

Next, let us turn to the massless spectrum, or more properly the spectrum of chiral primaries. In an ordinary WZW model, the WZW primaries are associated to integrable representations of the group G. Here, for each integrable representation R of the principal G bundle P, we get an associated vector bundle $\mathcal{E}_R$. The chiral primaries are then easily checked to be given by $H^*(X, \mathcal{E}_R)$ for each integrable representation R.

For an example, consider $G = SU(n)$ at level 1. Here, the integrable representations are the fundamental $\mathbf{n}$ and its exterior powers, hence the chiral primaries are counted by $H^*(X, \Lambda^*\mathcal{E})$, in agreement with old results [**44**].

These fibered WZW constructions realize the 'new' elliptic genera of Ando and Liu [**45, 46**]. Ordinary elliptic genera describe left-movers coupled to a level 1 current algebra; these elliptic genera, on the other hand, have left-moving level k current algebras.

It would be very interesting if these elliptic genera could be applied to understand, for example, black hole entropy computations.

4. String compactifications on stacks

In this section we will review results in [**47–52**], describing string compactifications on stacks and applications to gauged linear sigma models.

Stacks are a mild generalization of spaces, on which there exist functions, metrics, spinors, and all the other machinery one needs to make sense of strings.

One would like to understand string compactifications on stacks, not only to understand the most general possible string compactifications, but also because they often appear physically inside various constructions.

So, what is a stack? We can cover a stack with coordinate charts, just like a manifold. So, locally, a stack looks just like a space. Globally, on a manifold, across triple overlaps, the coordinate charts close to the identity. On a stack, across triple overlaps, the coordinate charts need only close up to an automorphism. In other words, locally a stack is just like a space – the difference is global.

How does one make sense of string compactifications on stacks concretely? Every[3] smooth Deligne-Mumford stack can be presented as a global quotient $[X/G]$ for X a space and G a group. To such a presentation, we associate a G-gauged sigma model on X.

Following that program, we quickly run into a potential problem. If to $[X/G]$ we associate a G-gauged sigma model, then

- $[\mathbf{C}^2/\mathbf{Z}_2]$ defines a two-dimensional conformal field theory,
- $[X/\mathbf{C}^\times]$ for $X = \left(\mathbf{C}^2 \times \mathbf{C}^\times\right)/\mathbf{Z}_2$ is an isomorphic stack, but $[X/\mathbf{C}^\times]$ defines a two-dimensional theory without conformal invariance.

Thus, there is a potential presentation-dependence problem: the same stack, presented in two different ways, is associated with distinct quantum field theories. Just like in the physical realization of derived categories, where an analogous problem arises, our proposal is that this presentation-dependence is fixed by renormalization-group flow. In other words, stacks classify endpoints of renormalization-group flow.

There are some potential problems with this program, reasons to believe that presentation-independence fails. One example involves deformation theory. The first check of a geometrical interpretation of any physical structure is to compare physical deformations to mathematical deformations, and in all previously known cases, they match. For stacks, on the other hand, there is a mismatch: deformations of physical theories do not match mathematical deformations of stacks. Another example of a potential problem with the program outlined above involves cluster decomposition, or rather lack thereof, in the theories one associates to special kinds of stacks known as gerbes.

These potential problems can be fixed [**47**–**51**]. The results include: mirror symmetry for stacks, new Landau-Ginzburg models, physical calculations of quantum cohomology for stacks, and an understanding of noneffective quotients in physics.

For example, let us consider quantum cohomology. As outlined earlier, the quantum cohomology ring of $\mathbf{CP}^N$ is $\mathbf{C}[x]/(x^{N+1} - q)$, a deformation of the classical cohomology ring $\mathbf{C}[x]/(x^{N+1})$. The quantum cohomology ring of a $\mathbf{Z}_k$ gerbe over $\mathbf{CP}^N$ with characteristic class $-n$ mod k is

$$\mathbf{C}[x,y]/\left(y^k - q_2, x^{N+1} - y^n q_1\right).$$

In particular, even though the gerbe is not a space, one can still make sense of notions familiar from ordinary spaces.

For another example, the Toda dual of $\mathbf{CP}^N$ is described by the holomorphic function

$$W = \exp(-Y_1) + \cdots + \exp(-Y_N) + \exp(Y_1 + \cdots + Y_N).$$

[3] With minor caveats.

The analogous duals to $\mathbf{Z}_k$ gerbes over $\mathbf{CP}^N$ are described by

$$W = \exp(-Y_1) + \cdots + \exp(-Y_N) + \Upsilon^n \exp(Y_1 + \cdots + Y_N)$$

where Υ is a character-valued field [**47**–**49**].

More generally, there exists a notion of toric stacks [**53**] which allows us to realize sigma models on many stacks in terms of simple two-dimensional gauge theories. Standard mirror constructions now produce discrete-valued fields [**47**–**49**], a new effect, which ties into the stacky fan description of [**53**].

We now believe that for banded, abelian G-gerbes on spaces X, (2,2) supersymmetric strings cannot distinguish between

- the gerbe itself,
- disjoint union of copies of X (one for each character of G), with flat B fields, determined by the image of the map

$$H^2(X, G) \xrightarrow{G \to U(1)} H^2(X, U(1)).$$

In other words, these are described by the same conformal field theory. This fact comes up in mirror constructions, and the meaning of discrete-valued fields, and has impliciations for quantum cohomology computations [**50**].

This result can also be applied to understand gauged linear sigma models. For example, let us consider the GLSM for the complete intersection $\mathbf{P}^7[2,2,2,2]$. At the Landau-Ginzburg point, one has a superpotential of the form

$$\sum_a p_a G_a(\phi) = \sum_{ij} \phi_i A^{ij}(p) \phi_j$$

where the G_a are quadratic polynomials, and the $A^{ij}(p)$ is an 8×8 matrix with entries linear in the p's, and the ϕ_i are the homogeneous coordinates on $\mathbf{P}^7$. In particular, this superpotential describes mass terms for the ϕ_i, away from the locus $\{\det A = 0\}$, leaving just the p fields. Since the p fields have charge -2, this describes a $\mathbf{Z}_2$ gerbe, which physics sees as a double cover. Altogether, the Landau-Ginzburg point describes a branched double cover of $\mathbf{P}^3$. This is a non-birational twisted derived equivalence, and a physical realization of Kuznetsov's homological projective duality [**52**].

Although (2,2) models on gerbes decompose into a disjoint union, (0,2) models do not in general. Heterotic strings on gerbes with twisted bundles do not factorize into a disjoint union of target spaces. The prototype for such bundles is the "$\mathcal{O}(1/k)$" line bundle over the $\mathbf{Z}_k$ gerbe $\mathbf{P}^N_{[k,k,\ldots,k]}$ on $\mathbf{P}^N$, defined by a Fermi superfield of (minimal) charge 1. More generally, heterotic strings on gerbes give an understanding of some of the two-dimensional (0,4) theories appearing in the geometric Langlands program. Implicit here is a lesson for the landscape program: many more string vacua may exist than previously enumerated.

5. Conclusions

In this talk we have outlined three recent developments pertinent to heterotic strings. First, we discussed recent progress in understanding nonperturbative corrections in heterotic strings. Second, we discussed a potential heterotic swampland – the inability of ordinary heterotic worldsheet constructions to describe most E_8 bundles with connection – and its resolution via the construction of new heterotic worldsheet CFT's. Finally, we outlined string compactifications on stacks. These compactifications give new insight into many gauged linear sigma models, ranging from exotic-seeming GLSM's with nonminimal charges to seemingly ordinary GLSM's which have novel geometries appearing at various limits of their Kähler moduli spaces, physically realizing Kuznetsov's homological projective duality and Kontsevich's noncommutative spaces. Heterotic string compactifications on special stacks known as gerbes appear to provide a new class of string compactifications.

References

[1] M. Becker, L-S. Tseng, S-T. Yau, "Heterotic Kähler/non-Kähler transitions," arXiv: 0706.4290.
[2] K. Becker, M. Becker, J-X. Fu, L-S. Tseng, S-T. Yau, "Anomaly cancellation and smooth non-Kähler solutions in heterotic string theory," Nucl. Phys. **B751** (2006) 108–128, hep-th/0604137.
[3] K. Becker, L-S. Tseng, "Heterotic flux compactifications and their moduli," Nucl. Phys. **B741** (2006) 162–179, hep-th/0509131.
[4] K. Becker, M. Becker, K. Dasgupta, R. Tatar, "Geometric transitions, non-Kähler geometries and string vacua," Int. J. Mod. Phys. **A20** (2005) 3442–3448, hep-th/0411039.
[5] N. Halmagyi, I. Melnikov, S. Sethi, "Instantons, hypermultiplets and the heterotic string," arXiv: 0704.3308.
[6] S. Sethi, "A note on heterotic dualities via M-theory," arXiv: 0707.0295.
[7] A. Adams, M. Ernebjerg, J. Lapan, "Linear models for flux vacua," hep-th/0611084.
[8] A. Adams, "Conformal field theory and the Reid conjecture," hep-th/0703048.
[9] V. Braun, Y-H. He, B. Ovrut, T. Pantev, "The exact MSSM spectrum from string theory," JHEP **0605** (2006) 043, hep-th/0512177.
[10] V. Bouchard, M. Cvetic, R. Donagi, "Tri-linear couplings in a heterotic minimal supersymmetric standard model," Nucl. Phys. **B745** (2006) 62–83, hep-th/0602096.
[11] S. Katz, E. Sharpe, "Notes on certain (0,2) correlation functions," Comm. Math. Phys. **262** (2006) 611–644, hep-th/0406226.
[12] E. Sharpe, "Notes on certain other (0,2) correlation functions," hep-th/0605005.
[13] E. Sharpe, "Notes on correlation functions in (0,2) theories," pp 93–104 in *Snowbird lectures on string geometry*, Comtemp. Math. 401, Amer. Math. Soc, Providence, RI, 2006, hep-th/0502064.
[14] A. Adams, J. Distler, M. Ernebjerg, "Topological heterotic rings," Adv. Theor. Math. Phys. **10** (2006) 657–682, hep-th/0506263.
[15] J. Guffin, S. Katz, "Deformed quantum cohomology and (0,2) mirror symmetry," arXiv: 0710.2354.
[16] I. Melnikov, S. Sethi, "Half-twisted (0,2) Landau-Ginzburg models," arXiv: 0712.1058.
[17] J. McOrist, I. Melnikov, "Half-twisted correlators from the Coulomb branch," arXiv: 0712.3272.

[18] J. Guffin, E. Sharpe, "A-twisted Landau-Ginzburg models," arXiv: 0801.3836.
[19] J. Guffin, E. Sharpe, "A-twisted heterotic Landau-Ginzburg models," arXiv: 0801.3955.
[20] E. Silverstein, E. Witten, "Criteria for conformal invariance of $(0,2)$ models," Nucl. Phys. **B444** (1995) 161–190, hep-th/9503212.
[21] P. Berglund, P. Candelas, X. de la Ossa, E. Derrick, J. Distler, T. Hubsch, "On the instanton contributions to the masses and couplings of E_6 singlets," Nucl. Phys. **B454** (1995) 127–163, hep-th/9505164.
[22] C. Beasley, E. Witten, "Residues and worldsheet instantons," JHEP **0310** (2003) 065, hep-th/0304115.
[23] R. Blumenhagen, R. Schimmrigk, A. Wisskirchen, "(0,2) mirror symmetry," Nucl. Phys. **B486** (1997) 598–628, hep-th/9609167.
[24] R. Blumenhagen, S. Sethi, "On orbifolds of (0,2) models," Nucl. Phys. **B491** (1997) 263–278, hep-th/9611172.
[25] R. Blumenhagen, "Target space duality for (0,2) compactifications," Nucl. Phys. **B513** (1998) 573–590, hep-th/9707198.
[26] R. Blumenhagen, "(0,2) target space duality, CICY's, and reflexive sheaves," Nucl. Phys. **B514** (1998) 688–704, hep-th/9710021.
[27] A. Adams. A. Basu, S. Sethi, "$(0,2)$ duality," Adv. Theor. Math. Phys. **7** (2004) 865–950, hep-th/0309226.
[28] M. Atiyah, "Complex analytic connections in fibre bundles," Trans. Amer. Math. Soc. **85** (1957) 181–207.
[29] D. Morrison, R. Plesser, "Summing the instantons: quantum cohomology and mirror symmetry in toric varieties," Nucl. Phys. **B440** (1995) 279–354, hep-th/9412236.
[30] F. Cachazo, M. Douglas, N. Seiberg, E. Witten, "Chiral rings and anomalies in supersymmetric gauge theory," JHEP **0212** (2002) 071, hep-th/0211170.
[31] J. Distler, E. Sharpe, "Heterotic compactifications with principal bundles for general groups and general levels," hep-th/0701244.
[32] T. Banks, talk given at "String vacuum workshop" at Max-Planck Institute, Munich, November 2004, and elsewhere.
[33] C. Vafa, "The string landscape and the swampland," hep-th/0509212.
[34] R. Friedman, J. Morgan, E. Witten, "Vector bundles and F theory," Comm. Math. Phys. **187** (1997) 679–743, hep-th/9701162.
[35] E. Scheidegger, private communication.
[36] V. Kac, M. N. Sanielevici, "Decompositions of representations of exceptional affine algebras with respect to conformal subalgebras," Phys. Rev. **D37** (1988) 2231–2237.
[37] S. J. Gates, Jr., W. Siegel, "Leftons, rightons, nonlinear sigma models, and superstrings," Phys. Lett. **B206** (1988) 631–638.
[38] D. Depireux, S. J. Gates, Jr., Q-H. Park, "Lefton-righton formulation of massless Thirring models," Phys. Lett. **B224** (1989) 364–372.
[39] S. Bellucci, D. Depireux, S. J. Gates, Jr., "(1,0) Thirring models and the coupling of spin-zero fields to the heterotic string," Phys. Lett. **B232** (1989) 67–74.
[40] S. J. Gates, Jr., S. Ketov, S. Kuzenko, O. Soloviev, "Lagrangian chiral coset constructions of heterotic string theories in (1,0) superspace," Nucl. Phys. **B362** (1991) 199–231.
[41] S. J. Gates, Jr., "Strings, superstrings, and two-dimensional Lagrangian field theory," pp. 140–184 in *Functional integration, geometry, and strings*, proceedings of the XXV Winter School of Theoretical Physics, Karpacz, Poland (Feb. 1989), ed. Z. Haba, J. Sobczyk, Birkhauser, 1989.
[42] A. Sen, "Local gauge and Lorentz invariance of the heterotic string theory," Phys. Lett. **B166** (1986) 300–304.
[43] A. Sen, "Superspace analysis of local Lorentz and gauge anomalies in the heterotic string theory," Phys. Lett. **B174** (1986) 277–279.

[44] J. Distler, B. Greene, "Aspects of (2,0) string compactifications," Nucl. Phys. **B304** (1988) 1–62.

[45] K. Liu, "On modular invariance and rigidity theorems," J. Diff. Geom. **41** (1995) 343–396.

[46] M. Ando, "The sigma orientation for analytic circle-equivariant elliptic cohomology," Geom. Topol. **7** (2003) 91–153, math.AT/0201092.

[47] T. Pantev, E. Sharpe, "Notes on gauging noneffective group actions," hep-th/0502027.

[48] T. Pantev, E. Sharpe, "String compactifications on Calabi-Yau stacks," Nucl. Phys. **B733** (2006) 233–296, hep-th/0502044.

[49] T. Pantev, E. Sharpe, "GLSM's for gerbes (and other toric stacks)," Adv. Theor. Math. Phys. **10** (2006) 77–121, hep-th/0502053.

[50] S. Hellerman, A. Henriques, T. Pantev, E. Sharpe, M. Ando, "Cluster decomposition, T-duality, and gerby CFT's," hep-th/0606034.

[51] E. Sharpe, "Derived categories and stacks in physics," contribution to proceedings of Vienna homological mirror symmetry conference, June 2006, hep-th/0608056.

[52] A. Căldăraru, J. Distler, S. Hellerman, T. Pantev, E. Sharpe, "Non-birational twisted derived equivalences in abelian GLSM's," arXiv: 0709.3855.

[53] L. Borisov, L. Chen, G. Smith, "The orbifold Chow ring of toric Deligne-Mumford stacks," J. Amer. Math. Soc. **18** (2005) 193–215, math.AG/0309229.

Department of Mathematics, 155 South 1400 East, Room 233, University of Utah, Salt Lake City, UT 84112

Current address: Physics Department, Robeson Hall (0435), Virginia Tech, Blacksburg, VA 24061

E-mail address: ersharpe@vt.edu

AMS/IP Studies in Advanced Mathematics
Volume 44, 2008

Worldsheet Instantons and Torsion Curves

Volker Braun, Maximilian Kreuzer, Burt A. Ovrut,
and Emanuel Scheidegger

ABSTRACT. We study aspects of worldsheet instantons relevant to a heterotic standard model. The non-simply connected Calabi-Yau threefold used admits $\mathbb{Z}_3 \times \mathbb{Z}_3$ Wilson lines, and a more detailed investigation shows that the homology classes of curves are $H_2(X, \mathbb{Z}) = \mathbb{Z}^3 \oplus (\mathbb{Z}_3 \oplus \mathbb{Z}_3)$. We compute the genus-0 prepotential; this is the first explicit calculation of the Gromov-Witten invariants of homology classes with torsion (finite subgroups)[**1, 2, 3**]. In particular, some curve classes contain only a single instanton. This ensures that the Beasley-Witten cancellation of instanton contributions cannot happen on this (non-toric) Calabi-Yau threefold.

1. Heterotic standard models

1.1. Heterotic M-theory. Probably the most promising corner of string theory to construct models with realistic particle spectra is heterotic M-theory, also known as the Hořava-Witten setup [**4, 5**]. In it, the spacetime is taken to be Minkowski space $\mathbb{R}^{3,1}$ times a Calabi-Yau threefold X times an interval I in the eleventh direction. The two 10-dimensional boundaries each support a E_8 gauge theory, one of which should be broken by instantons and/or Wilson lines to the standard model gauge group. The other

2000 *Mathematics Subject Classification.* Primary 81T30, 14N35; Secondary 14D21, 53D45.

Key words and phrases. String theory, Gromov-Witten invariants.

This research was supported in part by the Department of Physics and the Math/Physics Research Group at the University of Pennsylvania under cooperative research agreement DE-FG02-95ER40893 with the U. S. Department of Energy and an NSF Focused Research Grant DMS0139799 for "The Geometry of Superstrings", in part by the Austrian Research Funds FWF grant number P18679-N16, in part by the European Union RTN contract MRTN-CT-2004-005104, in part by the Italian Ministry of University (MIUR) under the contract PRIN 2005-023102 "Superstringhe, brane e interazioni fondamentali", and in part by the Marie Curie Grant MERG-2004-006374.

V. Braun would like to thank the organizers of the Sowers Workshop, and in particular Mark Sowers, for the opportunity to present this work.

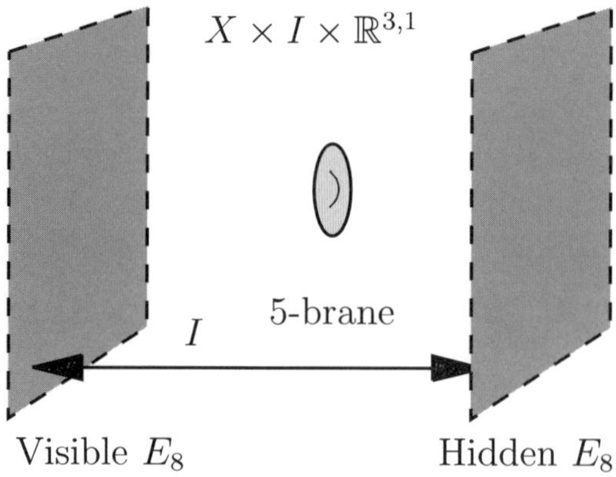

FIGURE 1. Hořava-Witten setup.

E_8^{hid} is then hidden and only couples gravitationally to the visible sector. In addition, there can be a number of 5-branes wrapping a curve of the Calabi-Yau threefold in the interior of the interval. In order to realize weak-scale supersymmetry breaking either the 5-brane or the hidden E_8^{hid} gauge sector should break supersymmetry and gravitationally mediate the effect to the ($N = 1$ supersymmetric) visible E_8^{vis} sector.

1.2. A Heterotic standard model. In order to find a compactification of heterotic M-theory one thus has to specify a Calabi-Yau threefold X and two E_8 gauge bundles on it. For our purposes it will be convenient to pick an $SU(n)$ subgroup in E_8, and use this group embedding to promote $SU(n)$ bundles (that is, vector bundles V_{vis} and V_{hid}) to E_8 bundles. The commutant of its holonomy $\text{Hol}(V_{\text{vis}})$ inside the visible E_8^{vis} is then the unbroken low-energy gauge group in 4 dimensions.

Here, we are going to be interested in the case where V_{vis} is a rank 4 holomorphic, slope stable (hence preserving $N = 1$ by itself) vector bundle together with a $\mathbb{Z}_3 \times \mathbb{Z}_3$ Wilson line. In other words,

$$\text{Hol}(V_{\text{vis}}) = SU(4) \times \mathbb{Z}_3 \times \mathbb{Z}_3 \quad \subset E_8^{\text{vis}}. \tag{1.1}$$

The unbroken gauge group is then

$$SU(3)_C \times SU(2)_L \times U(1)_Y \times U(1)_{B-L}, \tag{1.2}$$

the standard model gauge group times a $U(1)_{B-L}$ which constrains proton decay.

The (visible) matter fields of the effective 4-dimensional theory correspond to the zero modes of the Dirac operator $\not{D}_{E_8}$ on the Calabi-Yau threefold X. They can be computed as the H^1 cohomology groups of the vector bundle V_{vis} and associated bundles $V_{\text{vis}}^{\vee}$, $\wedge^2 V_{\text{vis}}$, ... In [**6, 7**], we constructed a suitable vector bundle yielding precisely the MSSM matter spectrum of

- three families of quarks and leptons,
- one (up, down) pair of Higgs, and
- no other charged fields.

One interesting possibility for the hidden sector (cancelling the heterotic anomaly) consists of an unbroken E_8^{hid} plus anti-five-branes wrapping a rigid curve in the bulk [**8**]. See also [**9, 10**] for other heterotic constructions.

1.3. Bundle superpotential. Mathematically, deformations of holomorphic vector bundles are unobstructed (at a smooth point in the bundle moduli space). Therefore, there cannot be any perturbative superpotential for the bundle moduli fields.

However, world-sheet instantons can, in principle, generate a superpotential. The action for a worldsheet instanton wrapped on a holomorphic curve C is

$$S[C \to X] = \int_C \omega, \tag{1.3}$$

where ω is the complexified Kähler class

$$\omega = iJ + B = it^a J_a + B. \tag{1.4}$$

The instanton-generated superpotential W_{bundle} is then the sum over all holomorphic curves and weighted with the Pfaffian (a function of those moduli which determine the bundle restricted to C):

$$W_{\text{bundle}}(\phi, c, t) = \sum_{C \subset X} \text{Pf}(\phi, c) e^{i \int_C \omega}, \tag{1.5}$$

where

- ϕ are the $\dim H^1(X, V \otimes V^\vee)$ vector bundle moduli,
- c are the $h^{2,1}(X)$ complex structure moduli, and
- $t = \{t^a\}$ are the $h^{1,1}(X)$ Kähler moduli.

Ignoring any torsion parts in homology and cohomology for the moment, we can generate $H^2(X, \mathbb{Z})$ by the basis two-forms J_a of eq. (1.4) and, therefore, write the homology class of the curve C as

$$[C] = (n_1, \dots, n_{h^{1,1}}) = \Big(\int_C J_1, \dots, \int_C J_{h^{1,1}} \Big) \in \mathbb{Z}^{h^{1,1}} \simeq H_2(X) \tag{1.6}$$

Usually, we use the Fourier-transformed Kähler moduli $q_a = e^{2\pi i t^a}$ and write the instanton action part as

$$e^{i \int_C t^a J_a} = \prod q_a^{n_a}. \tag{1.7}$$

Unlikely as this may seem, in common constructions, like gauged linear sigma models and monad constructions, the bundle moduli superpotential eq. (1.5) vanishes. One possible explanation was offered by Beasley and Witten, who have argued [**11**] that the sum over all instantons within a given homology class vanishes under certain circumstances. That is, rewrite eq. (1.5) as

$$W_{\text{bundle}} = \sum_{C\subset X} \mathrm{Pf}(\phi, c) \prod q_a^{n_a} = \sum_{\vec{n}\in H_2(X)} \underbrace{\left(\sum_{[C]=\vec{n}} \mathrm{Pf}(\phi, c) \right)}_{=0\ ?} \prod q_a^{n_a} \tag{1.8}$$

Then, for bundles coming from the toric ambient space, the inner sum actually does vanish. As we will see later in subsection 3.3, this argument does not apply to our vector bundle. Since our underlying Calabi-Yau threefold X is not a toric complete intersection, this should not be too surprising.

2. The $\mathbb{Z}_3 \times \mathbb{Z}_3$ Calabi-Yau threefold

2.1. The manifold. We now construct the Calabi-Yau threefold X as a free $\mathbb{Z}_3 \times \mathbb{Z}_3$-quotient of a simply connected Calabi-Yau threefold $\widetilde{X}$. Clearly, then, the fundamental group $\pi_1(X) = \mathbb{Z}_3 \times \mathbb{Z}_3$ allows for suitable Wilson lines; see subsection 1.2.

The covering space

$$\widetilde{X} \subset \mathbb{P}^2_{[x_0:x_1:x_2]} \times \mathbb{P}^1_{[t_0:t_1]} \times \mathbb{P}^2_{[y_0:y_1:y_2]} \tag{2.1}$$

is the complete intersection (see also [**12**])

$$\begin{aligned} t_0\Big(x_0^3 + x_1^3 + x_2^3\Big) + t_1\Big(x_0x_1x_2\Big) &= 0 \\ (\lambda_1 t_0 + t_1)\Big(y_0^3 + y_1^3 + y_2^3\Big) + (\lambda_2 t_0 + \lambda_3 t_1)\Big(y_0y_1y_2\Big) &= 0 \end{aligned} \tag{2.2}$$

where we labelled three particular complex structure moduli $\lambda_1, \lambda_2, \lambda_3 \in \mathbb{C}$. The significance of these three moduli is that they parametrize $\mathbb{Z}_3 \times \mathbb{Z}_3$ symmetric complete intersections. Explicitly, the group action is ($\zeta = e^{\frac{2\pi i}{3}}$)

$$\begin{aligned} g_1 &: \begin{cases} [x_0 : x_1 : x_2] \mapsto [x_0 : \zeta x_1 : \zeta^2 x_2] \\ [t_0 : t_1] \mapsto [t_0 : t_1] \text{ (no action)} \\ [y_0 : y_1 : y_2] \mapsto [y_0 : \zeta y_1 : \zeta^2 y_2] \end{cases} \\ g_2 &: \begin{cases} [x_0 : x_1 : x_2] \mapsto [x_1 : x_2 : x_0] \\ [t_0 : t_1] \mapsto [t_0 : t_1] \text{ (no action)} \\ [y_0 : y_1 : y_2] \mapsto [y_1 : y_2 : y_0]. \end{cases} \end{aligned} \tag{2.3}$$

Since the two equations are each cubic in one $\mathbb{P}^2$, one can easily see that $\widetilde{X}$ is a double elliptic fibration. Hence its Euler number must be $\chi(\widetilde{X}) = \chi(X) = 0$.

A straightforward geometric calculation yields the Hodge diamond, which turns out to be

$$h^{p,q}(\widetilde{X}) = \begin{matrix} & & & 1 & & & \\ & & 0 & & 0 & & \\ & 0 & & 19 & & 0 & \\ 1 & & 19 & & 19 & & 1 \\ & 0 & & 19 & & 0 & \\ & & 0 & & 0 & & \\ & & & 1 & & & \end{matrix}, \quad h^{p,q}(X) = \begin{matrix} & & & 1 & & & \\ & & 0 & & 0 & & \\ & 0 & & 3 & & 0 & \\ 1 & & 3 & & 3 & & 1 \\ & 0 & & 3 & & 0 & \\ & & 0 & & 0 & & \\ & & & 1 & & & \end{matrix} \tag{2.4}$$

Note that, as mentioned above, X has $h^{2,1}(X) = 3$ complex structure parameters.

2.2. Coinvariant homology. The Hurewicz isomorphism tells us that

$$H_1(X,\mathbb{Z}) = \pi_1(X)_{\mathrm{ab}} = \mathbb{Z}_3 \oplus \mathbb{Z}_3. \tag{2.5}$$

Together with the Hodge diamond eq. (2.4), the Universal Coefficient Theorem and Poincaré duality, this determines the integral homology up to a finite, Abelian group T ("torsion"). The resulting integral homology groups are given in Table 1. The hard part is to compute the torsion part $T = H_2(X,\mathbb{Z})_{\mathrm{tors}}$, that is, the torsion homology classes of curves in X. Clearly, such a torsion homology class cannot be represented by a holomorphic curve. However, as we will see, torsion homology classes can be represented by formal differences of holomorphic curves. It turns out that

$$H_2(X,\mathbb{Z}) = \mathbb{Z}^3 \oplus \underbrace{\mathbb{Z}_3 \oplus \mathbb{Z}_3}_{=T} \tag{2.6}$$

How did we obtain this? First, note that by definition we have a quotient map q which induces the push-forward in homology,

$$\begin{array}{ccc} \widetilde{X} & & H_2(\widetilde{X},\mathbb{Z}) = \mathbb{Z}^{19} \\ \Big\downarrow q & & \Big\downarrow q_* \\ X & & H_2(X,\mathbb{Z}) = \mathbb{Z}^3 \oplus \mathbb{Z}_3 \oplus \mathbb{Z}_3 \end{array} \tag{2.7}$$

TABLE 1. Integral homology and cohomology of X

i	0	1	2	3	4	5	6
$H_i(X,\mathbb{Z})$	$\mathbb{Z}$	$\mathbb{Z}_3 \oplus \mathbb{Z}_3$	$\mathbb{Z}^3 \oplus T$	$\mathbb{Z}^8 \oplus T$	$\mathbb{Z}^3 \oplus \mathbb{Z}_3 \oplus \mathbb{Z}_3$	0	$\mathbb{Z}$
$H^i(X,\mathbb{Z})$	$\mathbb{Z}$	0	$\mathbb{Z}^3 \oplus \mathbb{Z}_3 \oplus \mathbb{Z}_3$	$\mathbb{Z}^8 \oplus T$	$\mathbb{Z}^3 \oplus T$	$\mathbb{Z}_3 \oplus \mathbb{Z}_3$	$\mathbb{Z}$

The $\mathbb{Z}_3 \times \mathbb{Z}_3$ group action on $\widetilde{X}$ identifies curves

$$\widetilde{C} = g\widetilde{C} \qquad \forall g \in \mathbb{Z}_3 \times \mathbb{Z}_3, \ \widetilde{C} \in H_2(\widetilde{X}, \mathbb{Z}). \tag{2.8}$$

The quotient under this equivalence relation is called "coinvariant homology"[1], and denoted as $H_2\big(\widetilde{X}, \mathbb{Z}\big)_{\mathbb{Z}_3 \times \mathbb{Z}_3}$. By explicitly constructing a basis of curves on X and identifying the group action on them, one can explicitly determine the $\mathbb{Z}_3 \times \mathbb{Z}_3$-group action on $H_2\big(\widetilde{X}, \mathbb{Z}\big) \simeq \mathbb{Z}^{19}$ in terms of two commuting 19×19 matrices. We found that [**2**]

$$\begin{aligned} H_2\big(\widetilde{X}, \mathbb{Z}\big)_{\mathbb{Z}_3 \times \mathbb{Z}_3} &= H_2\big(\widetilde{X}, \mathbb{Z}\big) \Big/ \operatorname{Span}\big\{\widetilde{C} - g\widetilde{C}\big\} \\ &= \mathbb{Z}^3 \oplus \mathbb{Z}_3 \oplus \mathbb{Z}_3. \end{aligned} \tag{2.9}$$

In general, the relationship between coinvariant homology and the homology of the quotient $X = \widetilde{X}/(\mathbb{Z}_3 \times \mathbb{Z}_3)$ is determined by the Cartan-Leray spectral sequence

$$E^2_{p,q} = H_p\Big(\mathbb{Z}_3 \times \mathbb{Z}_3, H_q\big(\widetilde{X}, \mathbb{Z}\big)\Big) \quad \Rightarrow \quad H_{p+q}\big(X, \mathbb{Z}\big). \tag{2.10}$$

Without going into details, we point out that the $p = 0$ column of the E^2 tableau is the coinvariant homology,

$$E^2_{0,q} = H_0\Big(\mathbb{Z}_3 \times \mathbb{Z}_3, H_q\big(\widetilde{X}, \mathbb{Z}\big)\Big) = H_q\big(\widetilde{X}, \mathbb{Z}\big)_{\mathbb{Z}_3 \times \mathbb{Z}_3}. \tag{2.11}$$

But, as usual, there are more entries in the spectral sequence, as well as higher differentials and extension ambiguities. However, one can show [**2**] that these ultimately do not contribute to $H_2(X, \mathbb{Z})$.

To summarize, the homology classes of curves are given in eq. (2.6). The torsion curve classes can be understood as coinvariant curve classes on the covering space $\widetilde{X}$.

3. Counting curves

3.1. Prepotential. Having constructed the Calabi-Yau threefold X, we would like to identify the holomorphic curves on it. The well-known trick to count the curves is via the instanton corrections to the Yukawa couplings in the standard embedding,

$$W^{\text{np}}_{\text{Yuk}}\big(t, \underline{\mathbf{27}}\big) = \sum_{\vec{n} \in H_2(X)} \underline{\mathbf{27}}_i \underline{\mathbf{27}}_j \underline{\mathbf{27}}_k \underbrace{C_{ijk}(\vec{n})}_{\in \mathbb{Q}} \prod q_a^{n_a} \tag{3.1}$$

The instanton numbers are encoded in the power series $\sum_{\vec{n}} C_{ijk}(\vec{n}) \prod q_a^{n_a}$, multi-indexed by $i, j, k = 1, \ldots, h^{1,1}$. Fortunately, the $(h^{1,1})^3$ power series

---

[1] Since this is the dual of invariant cohomology.

come from a single prepotential

$$\mathcal{F}^{\mathrm{np}}_{X,0}(t) = \sum_{\vec{n}\in H_2(X)} \underbrace{n_{\vec{n}}}_{\in\mathbb{Z}} \operatorname{Li}_3\Big(\prod q_a^{n_a}\Big). \tag{3.2}$$

The trilogarithm in the prepotential takes care of multi-covers of a single instanton. Written as in eq. (3.2), the expansion coefficients $n_{\vec{n}}$ are the instanton numbers, and they count[2] the holomorphic curves in the homology class $\vec{n} \in H_2(X,\mathbb{Z})$.

3.2. Torsion curves. In our case, the curve homology classes are

$$H_2\big(X,\mathbb{Z}\big) = H_2\big(\widetilde{X},\mathbb{Z}\big)_{\mathbb{Z}_3\times\mathbb{Z}_3} = \mathbb{Z}^3 \oplus \mathbb{Z}_3 \oplus \mathbb{Z}_3, \tag{3.3}$$

and we have to take into account the finite subgroup. Therefore, the prepotential $\mathcal{F}^{\mathrm{np}}_{X,0}$ must be

- a power series in the $h^{1,1}(X)$ variables p, q, r, and
- a polynomial in b_1, b_2 with $b_i^3 = 1$.

After properly distinguishing the torsion curves, the prepotential must be of the form

$$\mathcal{F}^{\mathrm{np}}_{X,0} = \sum_{\substack{(n_1,n_2,n_3,m_1,m_2)\\ \in H_2(X,\mathbb{Z})}} n_{(n_1,n_2,n_3,m_1,m_2)} \operatorname{Li}_3\big(p^{n_1} q^{n_2} r^{n_3} b_1^{m_1} b_2^{m_2}\big). \tag{3.4}$$

Knowing the general form of the series expansion, how do we go about and compute it? We compared three different approaches [**2, 3**]

- Start on $\widetilde{X}$ and mod out $\mathbb{Z}_3 \times \mathbb{Z}_3$.
  The prepotential on $\widetilde{X}$ is partially known [**13**]. The quotienting by the $\mathbb{Z}_3 \times \mathbb{Z}_3$ action essentially amounts to a variable substitution [**2**], and one obtains

$$\begin{aligned} \mathcal{F}^{\mathrm{np}}_{X,0}(p,q,r,b_1,b_2) = \frac{1}{9}p & \\ &\times\Theta_{E_8}\big(q^3;\, q^2b_1^2b_2^2,\, qb_2,\, 1,\, b_1^2b_2,\, b_2^2,\, 1,\, b_1^2,\, q^{-1}\big)\\ &\times\Theta_{E_8}\big(r^3;\, r^2b_1b_2,\, rb_2^2,\, 1,\, b_1b_2^2,\, b_2,\, 1,\, b_1,\, r^{-1}\big)\\ &\times P\big(q^3\big)^{12} P\big(r^3\big)^{12} + O(p^2). \end{aligned} \tag{3.5}$$

- Identify and count curves directly on X.

[2]If all curves are rigid, then the instanton number is the naive number of curves in the given homology class. Otherwise, the expansion coefficients $n_{\vec{n}}$ are still integers, but may be negative.

By counting sections and singular Kodaira fibers in the double torus fibration X, one can directly count the curves up to linear order in p. The result is

$$\mathcal{F}^{\mathrm{np}}_{X,0}(p,q,r,b_1,b_2) = p\left(\sum_{i,j=0}^{2} b_1^i b_2^j\right) P(q)^4 P(r)^4 + O(p^2). \tag{3.6}$$

Setting $b_1 = b_2 = 1$, the equality of eq. (3.5) and eq. (3.6) is a known identity of E_8 theta functions [**14**], but we do not know a mathematical proof in general.

- Toric mirror symmetry.
  This approach yields the full power series numerically for the whole prepotential, with degree bounded only by computer resources. In practice, it is useful to look at (toric) Z_3-quotients of the toric complete intersection $\widetilde{X}$ [**3**].

3.3. Instanton Numbers. The instanton numbers $n_{(1,n_2,n_3,m_1,m_2)}$ turn out to be independent of m_1 and m_2, that is, the same for all $3 \times 3 = 9$ choices of the torsion part. We list them in Table 2. In particular, observe that

$$n_{(1,0,0,*,*)} = 1, \tag{3.7}$$

that is, the 9 smallest volume curves are isolated in their homology class. Hence, their contribution to the bundle moduli superpotential cannot cancel within their curve class and Beasley-Witten's cancellation cannot happen on the non-toric Calabi-Yau threefold X.

Although it is a priori surprising that the instanton numbers in Table 2 do not depend on the torsion part of the homology class, this can be argued from a remaining discrete symmetry. However, this phenomenon is special

TABLE 2. The instanton numbers $n_{(1,n_2,n_3,*,*)}$

n_2 \ n_3	0	1	2	3	4	5	6	7
0	1	4	14	40	105	252	574	1240
1	4	16	56	160	420	1008	2296	4960
2	14	56	196	560	1470	3528	8036	17360
3	40	160	560	1600	4200	10080	22960	49600
4	105	420	1470	4200	11025	26460	60270	130200
5	252	1008	3528	10080	26460	63504	144648	312480
6	574	2296	8036	22960	60270	144648	329476	711760
7	1240	4960	17360	49600	130200	312480	711760	1537600

TABLE 3. The instanton numbers $n_{(3,n_2,n_3,*,*)}$

$n_{(3,n_2,n_3,0,0)}$

$n_2 \backslash n_3$	0	1	2
0	0	**3**	**36**
1	**3**	**108**	
2	**36**		

$n_{(3,n_2,n_3,m_1,m_2)}$, $(m_1, m_2) \neq (0,0)$

$n_2 \backslash n_3$	0	1	2
0	0	**0**	**27**
1	**0**	**81**	
2	**27**		

to $n_1 = 1$ and does not hold in general. For example, the instanton numbers $n_{(3,n_2,n_3,m_1,m_2)}$ listed in Table 3 do explicitly depend on m_1 and m_2.

References

[1] V. Braun, M. Kreuzer, B. A. Ovrut and E. Scheidegger, *Worldsheet Instantons, Torsion Curves, and Non-Perturbative Superpotentials*, Phys. Lett. B **649**, 334 (2007) [arXiv:hep-th/0703134].

[2] V. Braun, M. Kreuzer, B. A. Ovrut and E. Scheidegger, *Worldsheet instantons and torsion curves. Part A: Direct computation*, JHEP **0710**, 022 (2007) [arXiv:hep-th/0703182].

[3] V. Braun, M. Kreuzer, B. A. Ovrut and E. Scheidegger, *Worldsheet Instantons and Torsion Curves, Part B: Mirror Symmetry*, JHEP **0710**, 023 (2007) [arXiv:0704.0449 [hep-th]].

[4] P. Hořava and E. Witten, *Eleven-Dimensional Supergravity on a Manifold with Boundary*, Nucl. Phys. B **475** (1996) 94 [arXiv:hep-th/9603142].

[5] A. Lukas, B. A. Ovrut and D. Waldram, *Non-standard embedding and five-branes in heterotic M-theory*, Phys. Rev. D **59**, 106005 (1999) [arXiv:hep-th/9808101].

[6] V. Braun, Y. H. He, B. A. Ovrut and T. Pantev, *The exact MSSM spectrum from string theory*, JHEP **0605**, 043 (2006) [arXiv:hep-th/0512177].

[7] V. Braun, Y. H. He and B. A. Ovrut, *Stability of the minimal heterotic standard model bundle*, JHEP **0606**, 032 (2006) [arXiv:hep-th/0602073].

[8] V. Braun and B. A. Ovrut, *Stabilizing moduli with a positive cosmological constant in heterotic M-theory*, JHEP **0607**, 035 (2006) [arXiv:hep-th/0603088].

[9] W. Buchmuller, K. Hamaguchi, O. Lebedev and M. Ratz, *Supersymmetric standard model from the heterotic string*, Phys. Rev. Lett. **96**, 121602 (2006) [arXiv:hep-ph/0511035].

[10] V. Bouchard and R. Donagi, *An SU(5) heterotic standard model*, Phys. Lett. B **633**, 783 (2006) [arXiv:hep-th/0512149].

[11] C. Beasley and E. Witten, *Residues and world-sheet instantons*, JHEP **0310**, 065 (2003) [arXiv:hep-th/0304115].

[12] P. Candelas, X. de la Ossa, Y. H. He and B. Szendroi, *Triadophilia: A Special Corner in the Landscape*, arXiv:0706.3134 [hep-th].

[13] S. Hosono, M. Saito, and J. Stienstra, *On the mirror symmetry conjecture for Schoen's Calabi-Yau 3-folds*, Integrable systems and algebraic geometry (Kobe/Kyoto, 1997), pp. 194–235. World Sci. Publ., River Edge, NJ, 1998. arXiv:alg-geom/9709027.

[14] D. Zagier, *A modular identity arising from mirror symmetry*, Integrable systems and algebraic geometry (Kobe/Kyoto, 1997), pp. 477–480. World Sci. Publ., River Edge, NJ, 1998.

Department of Physics, University of Pennsylvania, 209 S. 33rd Street, Philadelphia, PA 19104–6395, USA
E-mail address: vbraun@physics.upenn.edu

Institute for Theoretical Physics, Vienna University of Technology, Wiedner Hauptstr. 8-10, 1040 Vienna, Austria
E-mail address: Maximilian.Kreuzer@tuwien.ac.at

Department of Physics, University of Pennsylvania, 209 S. 33rd Street, Philadelphia, PA 19104–6395, USA
E-mail address: ovrut@physics.upenn.edu

Institut für Mathematik, Universität Augsburg, 86135 Augsburg, Germany
E-mail address: emanuel.scheidegger@math.uni-augsburg.de

Schedule of Talks

Monday May 14, 2007

9:00 – 9:30	Welcome and opening remarks	
9:30 – 10:30	V. Balasubramanian	Emergence of spacetime and gravitational entropy, I
11:00 – 12:00	Ori Ganor	Puff field theory
2:00 – 3:00	Rob Leigh	Mottness and strong coupling
3:30 – 4:30	Anastasios Petkou	Towards the holography of higher spins: the paradigm of conformal holography in AdS_4
4:30 – 5:30	Sumit Das	Null singularities and their gauge theory duals

Tuesday May 15, 2007

9:30 – 10:30	Keith Dienes	Statistical results from the heterotic landscape
11:00 – 12:00	Scott Thomas	Landscape naturalness
2:00 – 3:00	Nemanja Kaloper	Charting the landscape of modified gravity
3:30 – 4:30	Dan Kabat	Local bulk physics in AdS/CFT
4:30 – 5:30	Li-Sheng Tseng	Heterotic geometry and fluxes

Wednesday May 16, 2007

9:30 – 10:30	Djordje Minic	On the spectrum of pure QCD in 3 and 4d
11:00 – 12:00	Jan de Boer	Emergence of spacetime and gravitational entropy, II
Afternoon	Free	
Evening	Public lecture, presented by Vijay Balasubramanian	

Thursday May 17, 2007

9:30 – 10:30	A. Tomasiello	Progress on the geometry of supersymmetric string vacua
11:00 – 12:00	Finn Larsen	Precision entropy of black holes
2:00 – 3:00	Simeon Hellerman	Cosmological unification of string theories
3:30 – 4:30	Eric Sharpe	Recent developments in heterotic compactifications
4:30 – 5:30	Volker Braun	Worldsheet instantons and torsion curves

Friday May 18, 2007

9:30 – 10:30	Savdeep Sethi	Aspects of hypermultiplets
11:00 – 12:00	Kentaro Hori	Phases of N=2 theories in 1+1 dimensions with boundary
2:00 – 3:00	E. Diaconescu	Supersymmetry breaking in IIA vacua
3:30 – 4:30	Petr Horava	Quantum gravity at a Lifshitz point

List of Participants

Name	Institution
Zach Alsgaard	Virginia Tech
Vijay Balasubramanian*	University of Pennsylvania
Ed Barnes	University of Virginia
Aaron Bergman	Texas A&M
Jan de Boer*	University of Amsterdam
Volker Braun*	University of Pennsylvania
Zachary Burell	Auburn University
Jonathan Cates	Virginia Tech
Sumit Das*	University of Kentucky
Joshua Davis	UCLA
Emanuel Diaconescu*	Rutgers University
Keith Dienes*	University of Arizona
Josh Erlich	College of William and Mary
John Ficenec	Virginia Tech
Ori Ganor*	University of California, Berkeley
Elizabeth Gichana	Radford University
Eric Gimon	University of California, Berkeley
Peter Haskell	Virginia Tech
Simeon Hellerman*	Institute for Advanced Study, Princeton
Petr Horava*	University of California, Berkeley
Kentaro Hori*	University of Toronto
Vishnu Jejjala	University of Durham, UK
Dan Kabat*	Columbia University
Nemanja Kaloper*	University of California, Davis
Michael Kavic	Virginia Tech
Cynthia Keeler	University of California, Berkeley
Finn Larsen*	University of Michigan
Tommy Levi	New York University (NYU)
Rob Leigh*	University of Illinois, Urbana-Champaign
Bob McNeese	Brown University
Ilarion Melnikov	University of Chicago
Djordje Minic*	Virginia Tech

Name	Institution
Tetsuro Mizutani	Virginia Tech
Uma Murthy	Virginia Tech
Phil Nelson	Virginia Tech
Anastasios Peppas	Blinn College
Anastasios Petkou*	University of Crete
Michel Pleimling	Virginia Tech
Alexey Pronin	Virginia Tech
Matt Raum	Virginia Tech
Savdeep Sethi*	University of Chicago
Eric Sharpe*	University of Utah, Virginia Tech
Christopher Snell	Virginia Tech
Tatsu Takeuchi	Virginia Tech
Uwe Tauber	Virginia Tech
Scott Thomas*	Rutgers University
Alessandro Tomasiello*	Stanford University, SLAC
Li-Sheng Tseng*	Harvard University
Oleksandr Yelnykov	Virginia Tech

(* = speaker)

Titles in This Series

44 **Eric Sharpe and Arthur Greenspoon, Editors,** Advances in String Theory, 2008

43 **Lizhen Ji, Editor,** Arithmetic Groups and Their Generalizations, 2008

42.1 **Ka-Sing Lau, Zhou-Ping Xin, and Shing-Tung Yau, Editors,** Third International Congress of Chinese Mathematicians, 2008

41 **Wen-Ching Winnie Li, Editor,** Recent Trends in Coding Theory and its Applications, 2007

40 **Ovidiu Caliin, Der-Chen Chang, and Peter Greiner, Editors,** Geometric Analysis on the Heisenberg Group and Its Generalizations, 2007

39 **Zhijie Chen, Sheng-Li Tan, Jianpan Wang, and Stephen S.-T. Yau, Editors,** Proceedings of the International Conference on Complex Geometry and Related Fields, 2007

38 **Noriko Yui, Shing-Tung Yau, and James D. Lewis, Editors,** Mirror Symmetry V, 2006

37 **Lizhen Ji, Jian-Shu Li, H. W. Xu, and Shing-Tung Yau, Editors,** Lie Groups and Automorphic Forms, 2006

36 **Chuu-Lian Terng, Editor,** Integrable Systems, Geometry, and Topology, 2006

35 **Felix Finster,** The Principle of the Fermionic Projector, 2006

34 **Ren-Hong Wong, Editor,** Computational Geometry, 2003

33 **Eric D'Hoker, Duong Phong, and Shing-Tung Yau,** Mirror Symmetry IV, 2002

32 **Xi-Ping Zhu,** Lectures on Mean Curvature Flows, 2002

31 **Kiyoshi Igusa,** Higher Franz-Reidemeister Torsion, 2002

30 **Weiman Han and Mircea Sofonea,** Quasistatic Contact Problems in Viscoelasticity and Viscoplasticity, 2002

29 **S. T. Yau and Shuxing Chen, Editors,** Geometry and Nonlinear Partial Differential Equations, 2002

28 **Valentin Afraimovich and Sze-Bi Hsu,** Lectures on Chaotic dynamical Systems, 2002

27 **M. Ram Murty,** Introduction to p-adic Analytic Number Theory, 2002

26 **Raymond Chan, Yue-Kuen Kwok, David Yao, and Qiang Zhang, Editors,** Applied Probability, 2002

25 **Donggao Deng, Daren Huang, Rong-Qing Jia, Wei Lin, and Jian Zhong Wong, Editors,** Wavelet Analysis and Applications, 2002

24 **Jane Gilman, William W. Menasco, and Xiao-Song Lin, Editors,** Knots, Braids, and Mapping Class Groups—Papers Dedicated to Joan S. Birman, 2001

23 **Cumrun Vafa and S.-T. Yau, Editors,** Winter School on Mirror Symmetry, Vector Bundles and Lagrangian Submanifolds, 2001

22 **Carlos Berenstein, Der-Chen Chang, and Jingzhi Tie,** Laguerre Calculus and Its Applications on the Heisenberg Group, 2001

21 **Jürgen Jost,** Bosonic Strings: A Mathematical Treatment, 2001

20 **Lo Yang and S.-T. Yau, Editors,** First International Congress of Chinese Mathematicians, 2001

19 **So-Chin Chen and Mei-Chi Shaw,** Partial Differential Equations in Several Complex Variables, 2001

18 **Fangyang Zheng,** Complex Differential Geometry, 2000

17 **Lei Guo and Stephen S.-T. Yau, Editors,** Lectures on Systems, Control, and Information, 2000

16 **Rudi Weikard and Gilbert Weinstein, Editors,** Differential Equations and Mathematical Physics, 2000

15 **Ling Hsiao and Zhouping Xin, Editors,** Some Current Topics on Nonlinear Conservation Laws, 2000

14 **Jun-ichi Igusa,** An Introduction to the Theory of Local Zeta Functions, 2000

TITLES IN THIS SERIES

13 **Vasilios Alexiades and George Siopsis, Editors,** Trends in Mathematical Physics, 1999

12 **Sheng Gong,** The Bieberbach Conjecture, 1999

11 **Shinichi Mochizuki,** Foundations of p-adic Teichmüller Theory, 1999

10 **Duong H. Phong, Luc Vinet, and Shing-Tung Yau, Editors,** Mirror Symmetry III, 1999

9 **Shing-Tung Yau, Editor,** Mirror Symmetry I, 1998

8 **Jürgen Jost, Wilfrid Kendall, Umberto Mosco, Michael Röckner, and Karl-Theodor Sturm,** New Directions in Dirichlet Forms, 1998

7 **D. A. Buell and J. T. Teitelbaum, Editors,** Computational Perspectives on Number Theory, 1998

6 **Harold Levine,** Partial Differential Equations, 1997

5 **Qi-keng Lu, Stephen S.-T. Yau, and Anatoly Libgober, Editors,** Singularities and Complex Geometry, 1997

4 **Vyjayanthi Chari and Ivan B. Penkov, Editors,** Modular Interfaces: Modular Lie Algebras, Quantum Groups, and Lie Superalgebras, 1997

3 **Xia-Xi Ding and Tai-Ping Liu, Editors,** Nonlinear Evolutionary Partial Differential Equations, 1997

2.2 **William H. Kazez, Editor,** Geometric Topology, 1997

2.1 **William H. Kazez, Editor,** Geometric Topology, 1997

1 **B. Greene and S.-T. Yau, Editors,** Mirror Symmetry II, 1997